For Gwyneth,
with love and admiration
Christmas 1994,
Guido

Titles in This Series

Volume

6 Guido Mislin, Editor
The Hilton Symposium 1993
Topics in topology and group theory
1994

5 D. A. Dawson, Editor
Measure-valued processes, stochastic partial differential equations, and interacting systems
1994

4 Hershy Kisilevsky and M. Ram Murty, Editors
Elliptic curves and related topics
1994

3 Rémi Vaillancourt and Andrei L. Smirnov, Editors
Asymptotic methods in mechanics
1993

2 Philip D. Loewen
Optimal control via nonsmooth analysis
1993

1 M. Ram Murty, Editor
Theta functions
1993

Volume 6

CRM PROCEEDINGS & LECTURE NOTES

Centre de Recherches Mathématiques
Université de Montréal

The Hilton Symposium 1993

Topics in Topology and Group Theory

Guido Mislin
Editor

The Centre de Recherches Mathématiques (CRM) of the Université de Montréal was created in 1968 to promote research in pure and applied mathematics and related disciplines. Among its activities are special theme years, summer schools, workshops, postdoctoral programs, and publishing. The CRM is supported by the Université de Montréal, the Province of Québec (FCAR), and the Natural Sciences and Engineering Research Council of Canada. It is affiliated with the Institut des Sciences Mathématiques (ISM) of Montréal, whose constituent members are Concordia University, McGill University, the Université de Montréal, the Université du Québec à Montréal, and the Ecole Polytechnique.

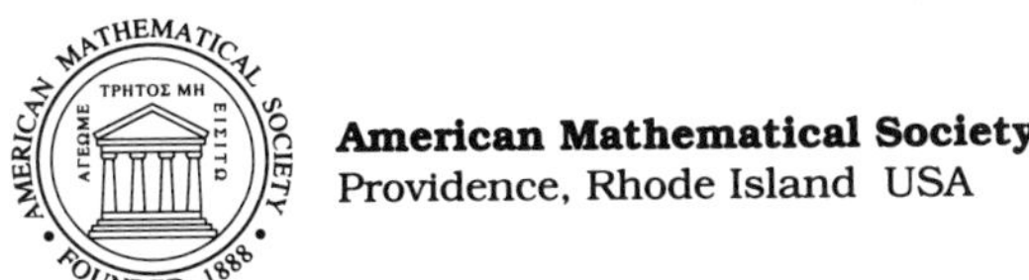

American Mathematical Society
Providence, Rhode Island USA

The production of this volume was supported in part by the Fonds pour la Formation de Chercheurs et l'Aide à la Recherche (Fonds FCAR) and the Natural Sciences and Engineering Research Council of Canada (CRSNG).

1991 *Mathematics Subject Classification.* Primary 55-02; Secondary 20J05.

Library of Congress Cataloging-in-Publication Data

The Hilton symposium 1993/Guido Mislin, editor.
p. cm. – (CRM proceedings & lecture notes; v. 6) ISBN 0-8218-0273-9 (acid-free)
1. Topology–Congresses. 2. Group theory–Congresses. I. Mislin, Guido. II. Series.
QA611.A1H55 1994 94-16852
514′.2–dc20 CIP

♾ The paper used in this book is acid-free and falls within the guidelines
established to ensure permanence and durability.
♻ Printed on recycled paper.

This volume was typeset using $\mathcal{AMS}$-TeX,
the American Mathematical Society's TeX macro system,
and submitted to the American Mathematical Society in camera-ready
form by the Centre de Recherches Mathématiques.

10 9 8 7 6 5 4 3 2 1 99 98 97 96 95 94

Contents

Foreword

In May of 1993 an international meeting on Algebra and Topology was held at the University of Montreal, in celebration of Peter Hilton's seventieth birthday. The contributions in this volume cover a great variety of topics and reflect in this way the wide spectrum of Hilton's own research. Several survey articles shed light on recent developments and lead the way for new research, much of which was initiated, influenced and clarified by Peter Hilton's own distinguished work.

Thanks are due to all participants and friends, who helped make the Conference a great success. The speakers and their collaborators are happy to dedicate this volume of essays and articles to you Peter, our great friend, colleague and teacher.

Guido Mislin

Centre de Recherches Mathématiques
CRM Proceedings and Lecture Notes
Volume 6, 1994

Recent Advances in Unstable Localization

Carles Casacuberta

1. Introduction

The essentials of localization of 1-connected spaces—or, more generally, nilpotent spaces—at a set of primes P were solidly established between 1970 and 1975. Since then, the monograph by Hilton, Mislin and Roitberg [**HMR**] has been a fundamental piece of reference which has made the theory available to a broad public of potential users. Indeed, many of these became successful users and have been enriching the theory during two decades.

The insight of Bousfield and Kan [**BK, Bo1, Bo3**] broadened largely the domain of application of localization techniques. However, it became clear that dealing with nonnilpotent spaces carried considerable difficulties, and often would lead to finding mysterious effects of certain functors on the homotopy groups of spaces.

When, shortly afterwards, Bousfield transferred that machinery to stable homotopy theory [**Bo4**], the scope of localization increased again. Nowadays, most attempts to determine the structure of the homotopy groups of spheres require localization as a basic tool. One of its deepest forms is localization with respect to Morava K-theories, which plays a key role in the study of periodicity in homotopy theory [**Ra1, DHS, HS**].

I cannot attempt to present the status of all what is being currently investigated in connection with localization. My aim in this article is to collect instead a certain number of new ideas which have arisen more or less since 1988. Although the basic concepts behind these ideas are not new to specialists, several observations and applications have been recognized as truly original. Furthermore, they seem to be far from exhausted as sources of new results.

The first half of the paper contains a summary of our own contribution. The results in Section 4 are going to appear in more detailed form as a joint paper with George Peschke [**CP**]. Indeed, the influence of Peschke's philosophy is obvious in many parts of the present survey, as well as the deep input from Markus Pfenniger.

Before presenting that material, we explain in Section 3 why the naïve attempt of constructing a reflection onto the class of spaces with P-local homotopy groups cannot possibly work in the based homotopy category of CW-complexes $\mathcal{H}$ (not necessarily nilpotent). The key point in overcoming this difficulty is to take full advantage of the topology of (based) mapping spaces $\mathrm{map}_*(X,Y)$, instead of restricting attention to the "purely categorical" device $[X,Y]$, the set of morphisms in $\mathcal{H}$ from X to Y. Thus, in order to make invertible a prime q in the homotopy

1991 *Mathematics Subject Classification.* Primary: 55P60; Secondary: 18A40.
This is the final form of the paper.

groups of spaces, one should not take as targets for the functor those spaces for which

$$q^* : \pi_n(X) \to \pi_n(X)$$

is bijective for all n, where q^* is induced by the degree q map of S^n; instead, one should consider spaces X for which

$$q^* : \Omega^n X \to \Omega^n X$$

is a homotopy equivalence for all n (in fact, there is no restriction in imposing this condition only for $n = 1$).

This idea was contained in embryonal form in [**Pe2**] and was pursued further in [**CP, CPP**]; but it only took its full force when Dror Farjoun revived and developed an old observation of Bousfield [**Bo3**], by showing that it was possible to "invert" *any* map $f\colon A \to B$ in $\mathcal{H}$ in that broader sense, not only $q\colon S^n \to S^n$. More precisely, the class of spaces X for which

$$f^* \colon \operatorname{map}_*(B, X) \to \operatorname{map}_*(A, X)$$

is a weak homotopy equivalence turns out to be reflective in $\mathcal{H}$ for every map f, while the class of spaces X for which

$$f^* \colon [B, X] \to [A, X]$$

is a bijection often fails to be reflective. This is good news for people having once tried to find optimal conditions on certain families of maps in $\mathcal{H}$ enabling these maps to be rendered invertible in a universal way (not by passing to the category of fractions, but by means of a suitable functor within $\mathcal{H}$). This is very tricky, if attacked with classical category-theoretical tools; cf. [**Ad2, DH**].

In the second half of the present article, a few aspects of that new approach to localization have been selected, by attempting to sketch the main ideas and including findings of Dror Farjoun [**Fa1, Fa2, Fa3, Fa4, Fa5**], Dror Farjoun-Smith [**FS**], Bousfield [**Bo5**], and Neisendorfer (unpublished). I am indebted to them for keeping me informed and for many tutorials. I also acknowledge several ideas from members of our topology seminar in Barcelona, which arose while reading the above preprints in early 1993, and have been included in my own presentation of the subject here.

Finally, I take the liberty to repeat under these acknowledgements certain feelings that I already expressed on the occasion of the meeting in Montréal. I belong to the youngest generation having entered into mathematics by reading the work of Peter Hilton, which is, as you know, a very pleasant experience. In addition, I am in big debt with him for teaching me further himself, not only localization, nor only mathematics. *This* is an even more pleasant experience. In the exposition which follows, the successful parts reflect what I could learn from him. For the unsuccessful, I apologize to him and to the reader.

2. Reflective subcategories

Homotopy theorists did not invent localization, but adapted it so as to be useful in homotopy theory. Therefore, most of the current terminology about localization originated in commutative algebra or category theory, and many results familiar to homotopy theorists admit in fact a more abstract formulation, which is often more enlightening.

Thus, we adopt in this paper the earlier point of view of Adams [**Ad2**], by emphasizing that localizations may be viewed as *idempotent monads*, about which there is an extensive literature available; see e.g. [**DFH**] or [**BW**] and their references. A *monad* or *triple* $\mathbf{E} = (E, \eta, \mu)$ in a category $\mathcal{C}$ consists of a functor $E\colon \mathcal{C} \to \mathcal{C}$ together with natural transformations $\eta\colon \mathrm{Id} \to E$ (called *unit*) and $\mu\colon E^2 \to E$ (called *multiplication*) such that

$$\mu \cdot \mu E = \mu \cdot E\mu \quad \text{and} \quad \mu \cdot \eta E = \mu \cdot E\eta = \mathrm{Id}\,.$$

A monad (E, η, μ) is *idempotent* if $E\eta = \eta E$; this condition is equivalent to μ being a natural equivalence of functors; see [**De**].

An idempotent monad $\mathbf{E} = (E, \eta, \mu)$ is characterized by either its class $\mathcal{S}(E)$ of *E-equivalences* (morphisms f such that Ef is invertible) or its class $\mathcal{D}(E)$ of *E-local objects* (objects in the image of E); see [**Ad2**] for further details. For every object X of $\mathcal{C}$, the morphism $\eta_X\colon X \to EX$ will be called the *E-localization* of X. It is an E-equivalence, which is in fact terminal among all E-equivalences going out of X, and at the same time initial among all morphisms from X to E-local objects. Thus, if we view E as a functor from $\mathcal{C}$ to $\mathcal{D}(E)$, then it is left adjoint to the inclusion of $\mathcal{D}(E)$ into $\mathcal{C}$ (we will not distinguish between a class of objects and the full subcategory with those objects). The term "localization" seems to be appropriate in this general setting, since the effect of E is precisely "inverting" a certain class of arrows, namely all E-equivalences.

As observed in [**CPP**], it turns out to be very convenient to distill one further abstract notion—the notion of *orthogonal pair*—from the properties of localizations. If $\mathcal{C}$ is any category, an object X and a morphism $f\colon A \to B$ are called *orthogonal* (this term has been borrowed from Freyd and Kelly [**FK**]) if

$$f^*\colon \mathcal{C}(B, X) \cong \mathcal{C}(A, X),$$

that is, if for any morphism $g\colon A \to X$ there is a unique $h\colon B \to X$ such that $hf = g$. For example, the Abelianization $\varphi\colon G \to G/[G, G]$ of a group G is orthogonal to all Abelian groups.

If $\mathcal{S}$ is a class of morphisms in $\mathcal{C}$, we denote by $\mathcal{S}^\perp$ the class of objects orthogonal to all morphisms in $\mathcal{S}$. If $\mathcal{S}$ consists of a single morphism f—which will often be the case in this article—we will abbreviate $\{f\}^\perp$ to $f^\perp$. For a class of objects $\mathcal{D}$, we use the notation $\mathcal{D}^\perp$ in the same way.

DEFINITION 2.1. An *orthogonal pair* $(\mathcal{S}, \mathcal{D})$ in $\mathcal{C}$ consists of a class of morphisms $\mathcal{S}$ and a class of objects $\mathcal{D}$ such that $\mathcal{S}^\perp = \mathcal{D}$ and $\mathcal{D}^\perp = \mathcal{S}$.

If $(\mathcal{S}, \mathcal{D})$ is an orthogonal pair, then $\mathcal{S}$ and $\mathcal{D}$ are *saturated*, meaning that $\mathcal{D}^{\perp\perp} = \mathcal{D}$ and $\mathcal{S}^{\perp\perp} = \mathcal{S}$. Also, taking three times the orthogonal class is exactly the same as doing it once. Hence, for example, $(\mathcal{S}^{\perp\perp}, \mathcal{S}^\perp)$ is an orthogonal pair for every class of morphisms $\mathcal{S}$, which we call the orthogonal pair *generated* by $\mathcal{S}$.

A major source of interesting orthogonal pairs is, of course, the theory of idempotent monads. Indeed, for every idempotent monad (E, η, μ), the classes $\mathcal{S}(E)$ and $\mathcal{D}(E)$ of E-equivalences and E-local objects form an orthogonal pair. At this point it might be asked if, given a monad $\mathbf{E} = (E, \eta, \mu)$, the fact that $\mathcal{S}(E)$, $\mathcal{D}(E)$ form an orthogonal pair implies that $\mathbf{E}$ is idempotent. This has been answered in the negative in [**CFT**].

It is not rare to find researchers looking for a proof of existence of certain localization functors (or, more generally, of left adjoints). Again, there is an extensive literature about that. We will be interested in those situations where a specific orthogonal pair $(\mathcal{S}, \mathcal{D})$ is given, and the question is to decide whether there is an idempotent monad (E, η, μ) such that $\mathcal{S} = \mathcal{S}(E)$ and $\mathcal{D} = \mathcal{D}(E)$. In fact, we will even be more restrictive, since all instances of this problem in the sequel will be of the following kind.

PROBLEM 2.2. Given a category $\mathcal{C}$ and a morphism f, is the orthogonal pair $(f^{\perp\perp}, f^{\perp})$ associated with an idempotent monad?

Note that, if coproducts exist in the category $\mathcal{C}$, then asking the above question for a single morphism f is equivalent to asking it for a set of morphisms $\{f_i \mid i \in I\}$, since the following conditions for an object X are equivalent:

1. X is orthogonal to each f_i.
2. X is orthogonal to the coproduct f of the set $\{f_i\}$.

In those cases when the answer to Problem 2.2 is affirmative, it is said that the subcategory $f^{\perp}$ is *reflective*. If (E, η, μ) is a solution, then it is of course unique up to an isomorphism of monads. For an object X, the E-localization $\eta_X : X \to EX$ is sometimes called a *reflection* of X onto the subcategory $f^{\perp}$. For example, Abelianization is a reflection in the category of groups onto the subcategory of Abelian groups, which is indeed of the form $f^{\perp}$; see the last paragraph of this article.

The most general form of Problem 2.2 (involving a possibly proper class of morphisms $\mathcal{S}$ instead of a single f) is called the *orthogonal subcategory problem*. It admits a solution in complete or cocomplete categories under rather mild assumptions; see [**Bo3, Ke, Pf**]. However, the situation is more complicated in categories such as the homotopy category of CW-complexes. In [**CPP**] we discuss what can be done in this category, and more generally in categories where coproducts and weak colimits exist.

3. Spaces with P-local homotopy groups

We next specialize to the realm of homotopy theory and give one fundamental example. Let us denote by $\mathcal{H}$ the based homotopy category of CW-complexes. Thus a space X and a (based) map $f : A \to B$ are orthogonal if $f^* : [B, X] \to [A, X]$ is a bijection of (based) homotopy classes of maps. We denote by $\mathcal{N}$ the full subcategory of $\mathcal{H}$ of nilpotent spaces.

From now on we fix a set of primes P (which may be empty) and denote, as usual, by P' its complement (which we normally assume not to be empty!). For every prime $q \in P'$, let

$$\rho_{q,n} : S^n \to S^n$$

be the standard map of degree q. Let f be the coproduct of the maps $\rho_{q,n}$ for $q \in P'$ and $n \geq 1$. We will be concerned with the orthogonal subcategory problem (Problem 2.2) for this particular map f, firstly in the category $\mathcal{N}$, and secondly in the whole category $\mathcal{H}$.

In any case, a space X belongs to $f^{\perp}$ if and only if

$$(\rho_{q,n})^* : \pi_n(X) \cong \pi_n(X)$$

for all $q \in P'$ and $n \geq 1$. But $(\rho_{q,n})^*$ is just multiplication by q if $n \geq 2$, and the qth power map for $n = 1$ (we will denote the fundamental group multiplicatively at all times). Hence, the class $f^{\perp}$ consists of spaces X whose higher homotopy groups admit a $\mathbb{Z}_P$-module structure (where $\mathbb{Z}_P$ denotes the ring of integers localized at P), and whose fundamental group is *uniquely* P'*-radicable*, i.e., the qth power map

$x \mapsto x^q$ is bijective in $\pi_1(X)$ for all $q \in P'$. We call such groups *P-local*, as in [**HMR**] or [**Ri1**], both in the commutative and the noncommutative case. That is, spaces in $f^\perp$ are spaces with P-local homotopy groups.

It is well known that $f^\perp$ is reflective in $\mathcal{N}$. The associated idempotent monad is the P-localization described in the early seventies by Hilton-Mislin-Roitberg and Bousfield-Kan, after the first insight of Sullivan [**Su**].

However, it may be surprising to discover that

THEOREM 3.1. *$f^\perp$ fails to be reflective in $\mathcal{H}$.*

The argument is related to the following observation of Mislin; cf. [**Fa1**]: The class of 1-connected spaces is not reflective in $\mathcal{H}$. For suppose it were; then there would be a map $\eta\colon \mathbb{R}P^2 \to Y$ universal among all maps from the real projective plane to 1-connected spaces. In particular, η would be orthogonal to a $K(\mathbb{Z},2)$, so that

$$H^2(Y) \cong H^2(\mathbb{R}P^2) \cong \mathbb{Z}/2.$$

But, since $H_1(Y) = 0$,

$$H^2(Y) \cong \operatorname{Hom}\bigl(H_2(Y),\mathbb{Z}\bigr),$$

and this group can never be isomorphic to $\mathbb{Z}/2$. In summary, it is not possible to "1-connectify" an arbitrary space X as if we were Abelianizing a group (the universal covering map $\widetilde{X} \to X$ goes in the opposite direction!).

Using a similar line of argument, in order to prove Theorem 3.1 we may use the fact that $H^2\bigl(G; A[G]\bigr) \neq 0$ for every noncyclic subgroup G of $\mathbb{Q}$ and every nonzero Abelian group A (see [**Ca**] and the references therein). Here $A[G]$ denotes the Abelian group of formal sums of elements of G with coefficients in A, which is a $\mathbb{Z}[G]$-module under the multiplication of G.

Suppose that there is a map $\eta\colon S^1 \to Y$ universal among all maps from the circle to spaces with P-local homotopy groups. Then η is orthogonal to the classifying space of any P-local discrete group G. But the bijection

$$\eta^*\colon \bigl[Y, K(G,1)\bigr] \cong \bigl[S^1, K(G,1)\bigr]$$

is equivalent to the bijection

$$\eta^*\colon \operatorname{Hom}\bigl(\pi_1(Y), G\bigr) \cong \operatorname{Hom}\bigl(\pi_1(S^1), G\bigr),$$

and this tells us that $\eta_*\colon \pi_1(S^1) \to \pi_1(Y)$ is orthogonal to all P-local groups G. Hence, $\pi_1(Y) \cong \mathbb{Z}_P$ (see the remarks about P-localization of arbitrary groups in Section 4).

Now let L be a "twisted Eilenberg-Mac Lane space" with

$$\pi_1(L) \cong \mathbb{Z}_P, \quad \pi_2(L) \cong \mathbb{Z}_P[\mathbb{Z}_P], \quad \pi_k(L) = 0 \text{ if } k \geq 3,$$

where the action of π_1 on π_2 is multiplication (this is a special instance of the action of a group G on $A[G]$). In our case, the single twisted k-invariant [**Hill**] of L is necessarily zero. Spaces of this kind classify cohomology with twisted coefficients, in a certain precise sense; see Theorem 7.18 in [**Gi**] or §3 of [**DH**].

Since L has P-local homotopy groups, there is a bijection

$$\eta^*\colon [Y, L] \cong [S^1, L].$$

By restricting this bijection to maps inducing the identity (resp. an inclusion) of fundamental groups, we infer that

$$H^2(Y;\mathbb{Z}_P[\mathbb{Z}_P]) \cong H^2(S^1;\mathbb{Z}_P[\mathbb{Z}_P]),$$

which is zero, while the Cartan-Leray spectral sequence for the universal cover $\widetilde{Y}$ shows that $H^2(Y;\mathbb{Z}_P[\mathbb{Z}_P])$ must contain a subgroup isomorphic to $H^2(\mathbb{Z}_P;\mathbb{Z}_P[\mathbb{Z}_P])$. But this group, as we said at the beginning of this discussion, is nonzero. This yields the desired contradiction.

Theorem 3.1 might deceive (or relieve) anyone having tried to extend the core of Hilton-Mislin-Roitberg to arbitrary CW-complexes, not necessarily nilpotent. It becomes clear that the most obvious approach cannot work.

But it is known, since Bousfield [**Bo1**], that there *are* idempotent monads in $\mathcal{H}$ extending P-localization of nilpotent spaces. In fact, as we shall discuss in Section 5, there are *many* such monads. Yet, the class of local spaces associated with any one of these monads must be more restricted than the class of spaces with P-local homotopy groups. For example, Bousfield gave in Theorem 5.5 of [**Bo1**] a purely algebraic description of $H_*(\ ;\mathbb{Z}_P)$-local spaces, which involves a complicated condition on the fundamental group and an analogous condition on its action on the higher homotopy groups. These conditions force the homotopy groups of $H_*(\ ;\mathbb{Z}_P)$-local spaces to be P-local, but much more than that. In the next section we describe another interesting class of spaces which includes all P-local nilpotent spaces and is reflective in $\mathcal{H}$.

Before closing this section, we ask a simple question which is related to our previous discussion.

QUESTION 3.2. Is the class $\mathcal{D}$ of 1-connected rational spaces reflective in $\mathcal{H}$?

The plausible answer is, of course, no; but we have been unable to prove it so far. The less ambitious reader may undertake the easier exercise of finding a map f such that $\mathcal{D} = f^{\perp}$.

4. Localizing with respect to self maps of the circle

It has been long known that a connected space X is nilpotent if and only if the groups $[W,\Omega X]$ are nilpotent for every finite CW-complex W; see [**Ro**] and Corollary X.3.8 in [**Wh**]. If we choose as a special case $W = S^{k-1}_+$ (where the subindex denotes a disjoint basepoint) we obtain

$$[S^{k-1}_+,\Omega X] \cong \pi_k(X) \rtimes \pi_1(X), \quad k \geq 1 \tag{4.1}$$

(see [**Pe1**]), where the semidirect product is referred to the ordinary action of the fundamental group on the higher homotopy groups. Of course, the assertion that $\pi_k(X) \rtimes \pi_1(X)$ is nilpotent is equivalent to the assertion that $\pi_1(X)$ is nilpotent and acts nilpotently on $\pi_k(X)$; cf. §2 in [**Hi2**]. Hence, the spaces S^0 and S^{k-1}_+ for $k \geq 2$ suffice to "recognize" nilpotent spaces by means of the above criterion. In fact, we might have equally well written S^{k-1}_+ for $k \geq 1$, since a group is nilpotent if and only if the conjugation action on itself is nilpotent.

Furthermore, as explained in [**Ro**], the natural map

$$[W,\Omega X] \to [W,\Omega(X_P)]$$

is a P-localization of groups when X is nilpotent, for every finite CW-complex W. Hence, $[W,\Omega Y]$ is a P-local nilpotent group for every P-local nilpotent space Y.

This suggests another possible approach—less naïve than the one sketched in Section 3—to extend the main existence results of Hilton-Mislin-Roitberg. Recall that we are calling *P-local* those groups G (nilpotent or not) in which the qth power map $x \mapsto x^q$ is bijective for all $q \in P'$.

DEFINITION 4.1. A CW-complex X will be called *P-local* if the group $[W, \Omega X]$ is P-local for every CW-complex W.

In fact, this definition turns out to be equivalent to imposing that $[W, \Omega X]$ be P-local for every *finite* CW-complex W. To understand this, suppose that the latter holds for a space X. Then, in particular, because of (4.1), the groups $\pi_k(X) \rtimes \pi_1(X)$ are P-local for $k \geq 1$. But these groups are the basepoint-free homotopy groups of ΩX. Thus, the qth power map will be bijective on all these groups if and only if the qth power map

$$\rho_q \colon \Omega X \to \Omega X,$$

sending every loop ω to ω^q, is a homotopy equivalence—here we are using, of course, the Whitehead theorem, together with the fact that the connected components of ΩX are simple. But this condition (for every $q \in P'$) implies that $[W, \Omega X]$ is indeed P-local for every CW-complex W, not necessarily finite.

In the course of this argument, we have proved the following, which we label for later reference.

THEOREM 4.2. *A CW-complex X is P-local if and only if the qth power map on the loop space ΩX is a homotopy equivalence for every $q \in P'$.*

In particular, if G is a discrete group, then the classifying space BG is P-local if and only if G is a P-local group. The next facts also follow from our previous discussion.

PROPOSITION 4.3. *A CW-complex X is P-local if and only if it is orthogonal to the degree q maps*

$$\rho_{q,k} \colon \Sigma\big(S_+^{k-1}\big) \to \Sigma\big(S_+^{k-1}\big), \quad k \geq 1,\ q \in P'.$$

Here we view $\Sigma\big(S_+^{k-1}\big)$ as $S^1 \wedge \big(S_+^{k-1}\big)$, and let $\rho_{q,k}$ be the product of the degree q map on the first factor and the identity on the second factor. Observe that

$$\Sigma\big(S_+^{k-1}\big) \simeq S^k \vee S^1,$$

although the obvious co-H-structures on these two homotopy types are different in general, for $\big[\Sigma\big(S_+^{k-1}\big), X\big]$ is isomorphic to the semidirect product $\pi_k(X) \rtimes \pi_1(X)$, while $[S^k \vee S^1, X]$ is isomorphic to the direct product $\pi_k(X) \times \pi_1(X)$.

PROPOSITION 4.4. *A CW-complex X is P-local if and only if $\pi_1(X)$ is a P-local group and, for every $k \geq 2$, the action*

$$\omega \colon \mathbb{Z}\big[\pi_1(X)\big] \to \operatorname{End}\big(\pi_k(X)\big)$$

has the property that

$$\omega(1 + x + x^2 + \cdots + x^{q-1})$$

is invertible for all $x \in \pi_1(X)$ and $q \in P'$.

To check this, develop $(a, x)^q$ for an arbitrary element (a, x) of $\pi_k(X) \rtimes \pi_1(X)$. This is an old trick which goes back at least to [**Ba2**].

It is now perfectly natural to call *P-local* a $\mathbb{Z}[G]$-module A (where G is any group) if $1 + x + x^2 + \cdots + x^{q-1}$ is an automorphism of A for every $x \in G$ and every $q \in P'$. Thus $A \rtimes G$ is P-local if and only if G is a P-local group and A is a P-local $\mathbb{Z}[G]$-module.

The concept of "P-local module" or "P-local action" has appeared—implicitly or explicitly—several times in the literature, always in connection with the study of roots in semidirect products of groups [**Ga, Pe2, Re**]. For an extensive account of algebraic properties of P-local modules, see [**CP**].

Let us denote by $\mathcal{D}_P$ the class of P-local spaces in the above sense. Proposition 4.3 tells us that $\mathcal{D}_P = f^{\perp}$ for a certain map f in $\mathcal{H}$; namely, the wedge of the maps $\rho_{q,k}$ for $k \geq 1$ and $q \in P'$, where we can replace, if we wish, $\rho_{q,1}$ by the degree q map $S^1 \to S^1$. Hence, $\mathcal{D}_P$ is saturated and we may consider the orthogonal pair $(\mathcal{S}_P, \mathcal{D}_P)$, where $\mathcal{S}_P = (\mathcal{D}_P)^{\perp}$. We call maps in $\mathcal{S}_P$ *P-equivalences* of spaces.

THEOREM 4.5. *The subcategory $\mathcal{D}_P$ is reflective in $\mathcal{H}$.*

In other words, for every space X there is a map $l\colon X \to X_P$ which is initial among all maps from X to spaces in $\mathcal{D}_P$. Thus, we have thrown away from our class of P-local spaces a bunch of conflictive spaces with "bad" actions of the fundamental group on the higher homotopy groups, such as the twisted Eilenberg-Mac Lane space L used in Section 3 to prove the non-reflectivity of the class of all spaces with P-local homotopy groups. Note that $\mathcal{D}_P$ is in fact a subclass of that class (just take $x = 1$ in Proposition 4.4).

The proof of Theorem 4.5 can be found in [**CP**]. For every space X, the P-localization map $l\colon X \to X_P$ is constructed as the homotopy direct limit of a system of P-equivalences

$$X = X_0 \to X_1 \to X_2 \to \cdots$$

where X_{i+1} is constructed by attaching cells to X_i so as to create P'-roots in the semidirect products $\pi_k(X_i) \rtimes \pi_1(X_i)$ and make these roots unique. It should be emphasized that the process stops at the first infinite ordinal, so that there is no need of resorting to transfinite direct limits.

The possibility of such a construction was suggested to us by Dror Farjoun; in fact, as we explain in Section 6, this is a special case of his own general construction. It is inspired by Bousfield's work [**Bo1, Bo3**] and, ultimately, by the work of Sullivan [**Su**], Mimura-Nishida-Toda [**MNT**], and Adams [**Ad1**].

It is also clear that the above construction is closely related to the procedure of P-localizing an arbitrary group, not necessarily nilpotent, by successively adjoining P'-roots and making them unique. This is far from being a new idea. It was first carried out by Baumslag in [**Ba1**] for free groups, and later generalized by Ribenboim to arbitrary groups in [**Ri1**]. It is fascinating that Ribenboim's construction of the P-localization $l\colon G \to G_P$ of an arbitrary group turns out to be completely analogous to Bousfield's construction of the localization with respect to a generalized homology theory. We tried to make clear the common pattern in [**CPP**], where we pointed out that the construction used to prove Theorem 4.5 is simply another instance of that general abstract procedure.

P-localization of groups may be viewed as the idempotent monad associated with the orthogonal pair $(f^{\perp\perp}, f^{\perp})$ where f is the free product of the maps

$$\rho_q\colon \mathbb{Z} \to \mathbb{Z},$$

where $\rho_q(1) = q$, for all $q \in P'$ (so that a group G belongs to $f^{\perp}$ if and only if G is P-local). Homomorphisms in $f^{\perp\perp}$ will be called *P-equivalences* of groups.

Since, for every space X, the P-localization map $l\colon X \to X_P$ is orthogonal to all classifying spaces of P-local discrete groups, we immediately obtain

THEOREM 4.6. *For every CW-complex X, the homomorphism*

$$l_*\colon \pi_1(X) \to \pi_1(X_P)$$

is a P-localization of groups.

However, we warn the reader that, in general, the map

$$[W, \Omega X] \to [W, \Omega(X_P)]$$

is far from being a P-localization of groups, contrary to what happens in the nilpotent case. The reason should be clear after the following discussion.

It is nowadays well known that Ribenboim's P-localization, which is defined on all groups, coincides with the classical P-localization when restricted to nilpotent groups [**Ga, Ri2**]. However, this was not obvious in its origins [**Ri1**]. Similarly, the functor provided by Theorem 4.5 does extend the classical P-localization of nilpotent spaces, i.e.,

$$\tilde{H}_k(X_P) \cong \mathbb{Z}_P \otimes \tilde{H}_k(X)$$

for all k and

$$\pi_k(X_P) \cong \mathbb{Z}_P \otimes \pi_k(X)$$

for $k \geq 2$ whenever X is nilpotent. Again, this is not obvious from the construction; for a detailed proof, see [**CP**].

Thus, it may be asked what $\tilde{H}_*(X_P)$ and $\pi_*(X_P)$ look like when X is non-nilpotent. Unfortunately, very little can be said in general. The right hint is given by the following result of [**CP**], which was obtained using arguments of obstruction theory with twisted coefficients.

THEOREM 4.7. *A map $f\colon X \to Y$ between connected CW-complexes is a P-equivalence if and only if $f_*\colon \pi_1(X) \to \pi_1(Y)$ is a P-equivalence of groups and*

1. *$f_*\colon H_*(X; A) \to H_*(Y; A)$ is an isomorphism for all P-local $\mathbb{Z}[\pi_1(Y)_P]$-modules A, or, equivalently,*
2. *$f^*\colon H^*(Y; A) \to H^*(X; A)$ is an isomorphism for all P-local $\mathbb{Z}[\pi_1(Y)_P]$-modules A.*

In particular, for every space X, the homomorphism

$$l_*\colon H_*(X; \mathbb{Z}_P) \to H_*(X_P; \mathbb{Z}_P)$$

is an isomorphism. However, the groups $\tilde{H}_*(X_P)$ (with integer coefficients) may fail to be P-local in general. An example is obtained by choosing X to be a wedge of at least two circles, for which $H_1(X_P)$ is the Abelianized of the P-localization of a free group. The reason why this contains a huge P'-torsion summand is explained in Theorem 37.3 of Baumslag's thesis [**Ba1**].

We are indebted to Shen Wenhuai for pointing out the following interesting duality:

1. $l_*\colon \tilde{H}_*(X) \to \tilde{H}_*(X_P)$ is a P-equivalence for all spaces X, but the groups $\tilde{H}_k(X_P)$ need not be P-local.
2. $l_*\colon \pi_*(X) \to \pi_*(X_P)$ need not be a P-equivalence, but the groups $\pi_k(X_P)$ are P-local for all spaces X.

Examples in which X is a $K(G,1)$ but X_P has a lot of higher homotopy are common. As a matter of fact, according to Theorem 4.7, $l\colon X \to X_P$ may be seen as "homology localization with twisted coefficients". The precise formulation is given in [**CP**], where it is pointed out that the construction of $(\ )_P$ can be modified so as to obtain homology localizations with "different degrees of twisting" on the coefficients. Hence, it is not surprising that $l\colon X \to X_P$ is closely related to the $H_*(\ ;\mathbb{Z}_P)$-localization $\eta\colon X \to E_P X$, which is the "totally untwisted" case. Indeed, Theorem 4.7 also tells us that there is a natural transformation of functors $(\ )_P \to E_P$, which is a homotopy equivalence in some cases. For example, as shown in [**CP**],

THEOREM 4.8. *Let G be a finite group and p a single prime. Then*

$$(BG)_p \simeq E_p(BG) \simeq (BG)\hat{}_p.$$

Hence, if G is finite and perfect, then $(BG)_p \simeq (BG^+)_p$, where the superscript denotes Quillen's plus-construction. It must be mentioned that this is false for infinite groups in general. For example, if G is locally free and perfect (such groups exist; see Lemma 3.1 in [**BDH**]), then G_p, the fundamental group of $(BG)_p$, contains a copy of G, while BG^+ is 1-connected. Still, P-localization of spaces is in many cases another instance in homotopy theory of a generic procedure, namely "doing something to the fundamental group by preserving homology to a certain extent", which tends, of course, to change quite drastically the higher homotopy groups in general.

5. Extending localization functors

The functor $(\ )_P$ described in the previous section is idempotent, it extends the classical P-localization of nilpotent spaces to all spaces, and it is distinct from Bousfield's $H_*(\ ;\mathbb{Z}_P)$-localization functor, which we keep denoting by E_P. As another example illustrating this last assertion, consider $X = S^1 \vee S^1$; then $\pi_1(X_P)$ is countable (since the P-localization of any countable group is countable), while $\pi_1(E_P X)$ is uncountable, by Proposition 4.4 in [**Bo2**]. The functor $(\ )_P$ has been called the "careful P-localization" by Dror Farjoun, in the sense that its effect on homotopy types is somehow less drastic than the effect of $\mathbb{Z}_P$-completion or $H_*(\ ;\mathbb{Z}_P)$-localization, while it still creates spaces with P-local homotopy groups.

This terminology suggests the possibility of partially ordering idempotent monads in $\mathcal{H}$ according to "how drastically" they change homotopy types. There is a simple way to develop this idea, which is again borrowed from [**FK**].

DEFINITION 5.1. Given two orthogonal pairs in any category $\mathcal{C}$, we write

$$(\mathcal{S}_1, \mathcal{D}_1) \le (\mathcal{S}_2, \mathcal{D}_2)$$

if $\mathcal{D}_1 \subseteq \mathcal{D}_2$.

Thus, the bigger orthogonal pair is the one with the bigger class of objects. Of course, this is equivalent to the condition $\mathcal{S}_2 \subseteq \mathcal{S}_1$.

It is readily checked that, given two idempotent monads $\mathbf{E}_1 = (E_1, \eta_1, \mu_1)$ and $\mathbf{E}_2 = (E_2, \eta_2, \mu_2)$ in the same category $\mathcal{C}$, the following facts are equivalent [**CFT**].

1. $\big(\mathcal{S}(E_2), \mathcal{D}(E_2)\big) \le \big(\mathcal{S}(E_1), \mathcal{D}(E_1)\big)$.
2. There is a (unique) morphism of monads $\mathbf{E}_1 \to \mathbf{E}_2$; i.e., a natural transformation of functors $E_1 \to E_2$ rendering commutative the obvious diagrams.

Now let $\mathcal{C}'$ be a full subcategory of $\mathcal{C}$, $(\mathcal{S}', \mathcal{D}')$ an orthogonal pair in $\mathcal{C}'$ and $(\mathcal{S}, \mathcal{D})$ an orthogonal pair in $\mathcal{C}$. We say that $(\mathcal{S}, \mathcal{D})$ *extends* $(\mathcal{S}', \mathcal{D}')$ if both $\mathcal{S}' \subseteq \mathcal{S}$ and $\mathcal{D}' \subseteq \mathcal{D}$. Then the following holds; cf. [**CPP**].

PROPOSITION 5.2. *In the above hypotheses,*

$$\big((\mathcal{D}')^{\perp}, (\mathcal{D}')^{\perp\perp}\big) \leq (\mathcal{S}, \mathcal{D}) \leq \big((\mathcal{S}')^{\perp\perp}, (\mathcal{S}')^{\perp}\big),$$

where orthogonality is meant in $\mathcal{C}$.

Accordingly, we call $\big((\mathcal{D}')^{\perp}, (\mathcal{D}')^{\perp\perp}\big)$ the *minimal extension* of $(\mathcal{S}', \mathcal{D}')$, and $\big((\mathcal{S}')^{\perp\perp}, (\mathcal{S}')^{\perp}\big)$ the *maximal extension.* If each of these three orthogonal pairs is associated with an idempotent monad, then the monad (E', η', μ') associated with $(\mathcal{S}', \mathcal{D}')$ admits an *initial* and a *terminal* extension over $\mathcal{C}$, namely the idempotent monads associated, respectively, with the maximal orthogonal pair and the minimal orthogonal pair. Of course, any of these orthogonal pairs might fail to be associated with an idempotent monad. In that case, the existence of a terminal or an initial extension of (E', η', μ') would not be guaranteed.

We will discuss two special cases of this situation, one in homotopy theory and the other one in group theory. Thus let us consider firstly the case $\mathcal{C} = \mathcal{H}$, the based homotopy category of CW-complexes, and $\mathcal{C}' = \mathcal{N}$, the full subcategory of nilpotent CW-complexes. Let $\mathbf{E}' = (E', \eta', \mu')$ correspond to P-localization in $\mathcal{N}$, and $(\mathcal{S}', \mathcal{D}')$ be the associated orthogonal pair. Then we have

THEOREM 5.3. $\big((\mathcal{D}')^{\perp}, (\mathcal{D}')^{\perp\perp}\big)$ *is the orthogonal pair associated with homology localization with* $\mathbb{Z}_P$ *coefficients. That is,* E_P *is terminal among all extensions of* E' *over* $\mathcal{H}$.

CONJECTURE 5.4. $\big((\mathcal{S}')^{\perp\perp}, (\mathcal{S}')^{\perp}\big)$ is the orthogonal pair associated with P-localization. That is, $(\ )_P$ is initial among all extensions of E' over $\mathcal{H}$.

The proof of Theorem 5.3 is elementary. A map $f\colon X \to Y$ in $(\mathcal{D}')^{\perp}$ is orthogonal to all P-local nilpotent spaces. Since these include all spaces $K(\mathbb{Z}_P, n)$ for $n \geq 1$, f is an $H^*(\ ;\mathbb{Z}_P)$-equivalence, and hence also an $H_*(\ ;\mathbb{Z}_P)$-equivalence. That is, $(\mathcal{D}')^{\perp}$ is contained in $\mathcal{S}(E_P)$. But all P-local nilpotent spaces are $H_*(\ ;\mathbb{Z}_P)$-local, and this implies that all $H_*(\ ;\mathbb{Z}_P)$-equivalences are in $(\mathcal{D}')^{\perp}$, so that $\mathcal{S}(E_P)$ is contained in $(\mathcal{D}')^{\perp}$, which completes the argument. (In fact, the second part of this argument is redundant, in view of Proposition 5.2.)

We have not been able to prove Conjecture 5.4 so far, although there is some strong evidence in its favour. If it turned out to be true, then we would have "captured" all instances of idempotent monads in $\mathcal{H}$ extending P-localization of nilpotent spaces. For any such monad (T, η, μ) and every space X, there would be natural maps

$$X_P \to TX \to E_P X$$

commuting with the units of the respective monads. This leads to results as the following one (which can be proved directly, without resorting to Conjecture 5.4).

THEOREM 5.5. *Let* G *be a finite group and* (T, η, μ) *any idempotent monad in* $\mathcal{H}$ *extending* p*-localization of nilpotent spaces, for a fixed single prime* p*. Then* $T(BG) \simeq (BG)^{\wedge}_p$.

In other words, there is only one (idempotent) way to "p-localize" the classifying space of a finite group; this improves Theorem 4.8.

Let us consider now the analogs in group theory of Theorem 5.3 and Conjecture 5.4. While the abstract setting is completely analogous, the conclusion turns out to be different. Recall from [**Bo2**] that for every Abelian group R there is an idempotent functor in the category of groups, called *HR-localization* and denoted by E_R, with the property that

$$E_R\big(\pi_1(X)\big) \cong \pi_1(E_R X)$$

for all spaces X, where $E_R X$ denotes the $H_*(\ ;R)$-localization of X. In the case $R = \mathbb{Z}_P$, we consistently abbreviate $E_{\mathbb{Z}_P}$ to E_P. The orthogonal pair associated with E_R in the category of groups is generated by the *HR-maps*, i.e., homomorphisms $\varphi\colon G \to K$ such that $H_1(\varphi)$ is iso and $H_2(\varphi)$ is epi. That is, HR-maps are precisely homomorphisms induced at the fundamental group by $H_*(\ ;R)$-equivalences of spaces; cf. [**Bo1**].

Take $\mathcal{C} = \mathcal{G}r$, the category of groups, and $\mathcal{C}'$ the full subcategory of nilpotent groups. Let $\mathbf{E}' = (E', \eta', \mu')$ correspond to P-localization of nilpotent groups, and let $(\mathcal{S}', \mathcal{D}')$ be the associated orthogonal pair. We have

THEOREM 5.6. *$\big((\mathcal{D}')^{\perp}, (\mathcal{D}')^{\perp\perp}\big)$ is associated with a certain idempotent monad $\mathbf{L}_P = (L_P, \eta, \mu)$, which is not isomorphic to $H\mathbb{Z}_P$-localization.*

THEOREM 5.7. *$\big((\mathcal{S}')^{\perp\perp}, (\mathcal{S}')^{\perp}\big)$ is the orthogonal pair associated with P-localization. That is, $(\ )_P$ is initial among all extensions of E' over $\mathcal{G}r$.*

Now it is Theorem 5.7 the one admitting an elementary proof. A group G in $(\mathcal{S}')^{\perp}$ is orthogonal to all P-equivalences of nilpotent groups. Since these include the maps $\rho_q\colon \mathbb{Z} \to \mathbb{Z}$, $q \in P'$, we infer that G is a P-local group. That is, $(\mathcal{S}')^{\perp}$ is contained in the class of P-local groups. The converse inclusion follows from Proposition 5.2.

Theorem 5.6 was proven, with different methods, in [**BT**] and [**CFT**]. The monad $\mathbf{L}_P$ turns out to be, in a certain precise sense, "the best idempotent approximation" to nilpotent $\mathbb{Z}_P$-completion. The nilpotent $\mathbb{Z}_P$-completion of a group G is defined as

$$\widehat{G}_P = \varprojlim (G/\Gamma^i G)_P,$$

where $\Gamma^i G$ denotes the lower central series of G. The group $\widehat{G}_P$ is not nilpotent in general, but it is "the best approximation of G by means of P-local nilpotent groups".

This functor $(\ )\widehat{_P}$ is part of a monad, which fails to be idempotent. For example, if F is a free group on a countably infinite set of free generators, then $(\widehat{F}_P)\widehat{_P}$ is not isomorphic to $\widehat{F}_P$; see §13 of [**Bo2**] and Proposition IV.5.4 in [**BK**]. However, in the category of groups (as in any category which is complete and well-powered), for every given monad $\mathbf{T} = (T, \eta, \mu)$ there is an idempotent monad $\mathbf{T}' = (T', \eta', \mu')$ with the same class of equivalences as $\mathbf{T}$, i.e., such that for a map φ, the map $T'\varphi$ is invertible if and only if $T\varphi$ is invertible; see [**Fak**] and [**CFT**]. In this situation, we call $\mathbf{T}'$ the *idempotentification* of $\mathbf{T}$. With this terminology, the idempotent monad $\mathbf{L}_P$ in Theorem 5.6 is the idempotentification of nilpotent $\mathbb{Z}_P$-completion.

The orthogonal pair associated with $\mathbf{L}_P$ is strictly smaller than the one associated with $H\mathbb{Z}_P$-localization, since it follows from the construction of $\mathbf{L}_P$ that $L_P G = \widehat{G}_P$ if G is finitely generated, while for a free group F on two free generators the natural map $E_P F \to \widehat{F}_P$ is not iso; see Proposition 4.4 in [**Bo2**]. Thus there are, for every group G, natural maps

$$G_P \to E_P G \to L_P G \to \widehat{G}_P,$$

which need not be isomorphisms. (Yet, if G is nilpotent, they are.)

Of course, one could investigate the same idea in homotopy theory, where the analog of nilpotent $\mathbb{Z}_P$-completion is Bousfield-Kan $\mathbb{Z}_P$-completion $(\mathbb{Z}_P)_\infty$, which fails to be idempotent on arbitrary spaces. The surprise is that, contrary to what happens in the category of groups, the idempotentification of $(\mathbb{Z}_P)_\infty$ is nothing else than $H_*(\ ;\mathbb{Z}_P)$-localization—which we know it is indeed the terminal extension of P-localization of nilpotent spaces to all spaces (Theorem 5.3). In other words, there is no gap left between E_P and $(\mathbb{Z}_P)_\infty$ which could be occupied by some other idempotent functor.

In fact, it is true in general that, using the same notation as above, if the right Kan extension of the inclusion of $\mathcal{D}'$ in $\mathcal{C}$ along itself exists, then it is part of a monad in $\mathcal{C}$, and if the idempotentification of this monad exists, then it provides the terminal extension of $\mathbf{E}'$ over $\mathcal{C}$; see [**CFT**]. We are not aware of any analogous procedure to obtain the initial extension in general.

Note that the efforts to prove Theorem 5.6 would have been superfluous if the following question had a negative answer:

QUESTION 5.8. Do there exist orthogonal pairs $(\mathcal{S}, \mathcal{D})$ in the category of groups which are not associated with any idempotent monad?

If $(\mathcal{S}, \mathcal{D})$ is generated by either a set of groups or a set of homomorphisms—where distinction is made between a "set" and a "proper class"—then $(\mathcal{S}, \mathcal{D})$ is associated with an idempotent monad; cf. [**Bo3, Pf, CPP, CFT**]. Thus the following question is equally relevant.

QUESTION 5.9. Do there exist orthogonal pairs $(\mathcal{S}, \mathcal{D})$ in the category of groups which are neither generated by a set of groups nor by a set of homomorphisms?

From Section 3 we know counterexamples to Question 5.8 in the based homotopy category of CW-complexes $\mathcal{H}$. However, as we shall point out later, Question 5.9 is still significant in $\mathcal{H}$.

6. Localizing with respect to any map

We have seen in Section 3 that certain orthogonal pairs of the form $(f^{\perp\perp}, f^{\perp})$ in the based homotopy category of CW-complexes are not associated with any idempotent monad. However, this is only due to the fact that standard orthogonality is not the "best" concept to look at in homotopy theory. Indeed, given two spaces X, Y, the set $[X, Y]$ is only part of the richer structure of the set $\mathrm{map}_*(X, Y)$ of based maps from X to Y, endowed with the compact-open topology. Therefore, it is a good idea to consider the following notion.

DEFINITION 6.1. Let $f\colon A \to B$ be any map. A CW-complex X is *f-local* or *f-periodic* if the induced map

$$f^*\colon \mathrm{map}_*(B, X) \to \mathrm{map}_*(A, X)$$

is a weak homotopy equivalence.

Of course, if X is f-local, then X is orthogonal to f in $\mathcal{H}$, since the condition imposed in the definition implies in particular that $f^*\colon [B, X] \to [A, X]$ is bijective. However, being f-local is much more restrictive in general than being orthogonal to f. For example, if $f\colon S^1 \to S^1$ is the degree q map for some prime q, then X is orthogonal to f if and only if $\pi_1(X)$ is uniquely q-radicable, while X is f-local if and only if the qth power map $\Omega X \to \Omega X$ is a weak homotopy equivalence (in this special case, we could delete "weak"), and, as we have explained in Section 4, this implies certain additional conditions on the higher homotopy groups of X.

THEOREM 6.2. *For every map $f\colon A \to B$, the class of f-local spaces is reflective in $\mathcal{H}$.*

This powerful result has been proved by Dror Farjoun in [**Fa1**]. An earlier version was sketched in §7 of [**Bo3**]; see also [**Bo5**]. We denote by $l\colon X \to L_f X$ the localization given by Theorem 6.2, and refer to it as *f-localization*. The corresponding equivalences will be called *f-equivalences*. The following result is implicit in the proof of Theorem 6.2.

THEOREM 6.3. *Let $f\colon A \to B$ be a map between n-connected spaces. Then the f-localization map $l\colon X \to L_f X$ induces isomorphisms $l_*\colon \pi_k(X) \to \pi_k(L_f X)$ for $k \le n$.*

Note that Theorem 4.5 is just a special case of Theorem 6.2. However, an important feature of Dror Farjoun's explicit construction is that it is functorial in the topological category. Thus, if we are given a commutative diagram of spaces and maps, the diagram obtained by applying L_f is again strictly commutative, not just up to homotopy. Other constructions, such as Sullivan's [**Su**], the one in Hilton-Mislin-Roitberg [**HMR**], or the one sketched in Section 4, are only functorial in the homotopy category. Since the class of local objects only determines an idempotent monad up to isomorphism, for every space X there is a wide variety of choices for the space X_P, all of which, however, are homotopy equivalent. In this context, Dror Farjoun's construction provides, as a special case, a functorial model for X_P.

It is interesting that the lack of functoriality of the earlier versions of P-localization motivated Anderson's paper [**An**], where a certain construction is described which turns out to be functorial in the topological category and coincides, up to homotopy, with Sullivan's P-localization of 1-connected spaces. Now we know that Anderson's construction is precisely g-localization, with respect to the following map g. Take the homotopy cofibre M of the map f giving rise to P-localization (i.e., the wedge of the degree q maps of S^1 for $q \in P'$) and consider the trivial map $g\colon M \to \mathrm{pt}$. Hence, Anderson's work is also a major precedent of Dror Farjoun's construction. In the same line of reasoning we find the work of Bendersky [**Be**], who constructed a functorial *semilocalization*, that is, a functor which preserves the fundamental group of an arbitrary space X and P-localizes its higher homotopy groups $\pi_k(X)$, $k \ge 2$. Again, Bendersky's functor is a special instance of localization with respect to a map, which turns out to be in this case Σf, the suspension of the map f inducing P-localization; see Example 3.6 in [**CPP**].

Dror Farjoun has asked the following question:

QUESTION 6.4. Is it true that every idempotent monad in $\mathcal{H}$ (or in some subcategory) is f-localization for a certain map f?

All examples which have been checked until now support a positive answer. However, a solution to this problem might require a deep input from set theory, for it is related to Question 5.9. First of all, it does not seem easy to decide under which hypotheses one can ensure that the class of equivalences associated with a given idempotent monad in $\mathcal{H}$ is generated by some set of maps, not even if "generated" is understood in the sense of Definition 6.1.

In order to construct the E_*-localization functor for a generalized homology theory E_*, Bousfield had to solve a difficulty of a similar kind. In that case, it is possible to consider the following map f.

THEOREM 6.5. *Given a generalized homology theory E_* (satisfying the limit axiom), let f be the wedge of a set of representatives of all isomorphism classes of E_*-equivalences $A \to B$ where the cardinality of the set of cells in A and B is not bigger than the cardinality of $E_*(\mathrm{pt})$. Then the following assertions are equivalent for a space X:*

1. *X is E_*-local.*
2. *X is f-local.*
3. *X is orthogonal to f.*

The key ingredient in the proof of this result is Lemma 11.3 in [**Bo1**]. Therefore, homology localizations are also special cases of localization with respect to a map.

If E_* is ordinary homology with integer coefficients, and we consider the map $g\colon C \to \mathrm{pt}$, where C is the homotopy cofibre of the map f defined in Theorem 6.5, then the g-localization $l\colon X \to L_g X$ turns out to be precisely Quillen's plus-construction relative to the maximal perfect subgroup of $\pi_1(X)$. Observe that this fact suggests a way to define a "plus-construction" associated with any generalized homology theory E_*, by considering the homotopy cofibre C of the map f inducing E_*-localization, and localizing with respect to $g\colon C \to \mathrm{pt}$. The properties of this functor are being currently studied by José Luis Rodríguez.

Localizations with respect to maps of the form $f\colon W \to \mathrm{pt}$ occur so often and have such pleasant properties that deserve some comments. If $f\colon W \to \mathrm{pt}$, then a space X is f-local if and only if $\mathrm{map}_*(W, X)$ is weakly contractible, which is equivalent to the condition

$$\pi_k(\mathrm{map}_*(W, X)) = 0 \quad \text{for all } k \geq 0.$$

But $[S^k, \mathrm{map}_*(W, X)] \cong [S^k \wedge W, X]$. Therefore, we have

PROPOSITION 6.6. *If $f\colon W \to \mathrm{pt}$, then a space X is f-local if and only if $[\Sigma^k W, X] = 0$ for $k \geq 0$.*

That is, X is f-local if and only if X is orthogonal to the maps $\Sigma^k W \to \mathrm{pt}$ for $k \geq 0$. It might be asked if it is possible to write down a similar description of f-local spaces in the case of a general map $f\colon A \to B$, i.e., in terms of orthogonality in the usual sense. Note that, if A and B are suspensions, then X is f-local if and only if X is orthogonal to $f\colon A \to B$ and to

$$f \wedge \mathrm{id}\colon A \wedge S^k_+ \to B \wedge S^k_+ \quad \text{for } k \geq 1;$$

(of course, Proposition 4.3 is a special case). Indeed, under the assumption that A and B are suspensions, the spaces $\mathrm{map}_*(A, X)$ and $\mathrm{map}_*(B, X)$ are H-spaces, and hence basepoint-free homotopy groups suffice to recognize a weak homotopy equivalence $\mathrm{map}_*(B, X) \to \mathrm{map}_*(A, X)$. The space $A \wedge S^k_+$ is sometimes denoted by $A \rtimes S^k$; cf. [**Fa1**].

If f is of the form $W \to \mathrm{pt}$, then it is appropriate to write P_W instead of L_f, where the letter P stands for "Postnikov" (!). This was suggested by Bousfield and Dror Farjoun after observing that a space X is local with respect to $f\colon S^n \to \mathrm{pt}$ if and only if $\pi_k(X) = 0$ for $k \geq n$; combining this fact with Theorem 6.3, one obtains

THEOREM 6.7. *For any connected space X and $n \geq 1$, $P_{S^n} X$ has the homotopy type of the $(n-1)$th term in the Postnikov tower of X. Thus*

$$\pi_k(P_{S^n} X) \cong \begin{cases} \pi_k(X) & \text{if } k \leq n-1, \\ 0 & \text{otherwise.} \end{cases}$$

Therefore, the homotopy fibre of $l\colon X \to P_{S^n}X$ is the $(n-1)$-connected cover $X\langle n-1\rangle$, which can be built out of copies of S^{n-1} via cofibrations, i.e., starting from one point and attaching cells of dimension $\geq n$.

In general, we can regard $P_W X$ as the result of reducing X "modulo all of its W-information"; moreover, the above example suggests that the homotopy fibre of the map $l\colon X \to P_{\Sigma W}X$ might be "the best approximation to X built out of copies of W via cofibrations"—possibly under additional hypotheses. This idea can indeed be made rigorous and turns out to be a rich source of interesting results [**Fa3, Fa4, Fa5**]. Among other things, if W is E_*-acyclic for a generalized homology theory E_*, so are all spaces obtained from W via (pointed) homotopy colimits of any kind. Hence, the study of P_W (and $P_{\Sigma^k W}$ for $k \geq 1$), together with the homotopy fibres of the maps $X \to P_{\Sigma^k W}X$, should lead to a better understanding of the class of E_*-acyclic spaces.

The use of the term *f-periodic*, which has been introduced in Definition 6.1 with the same meaning as f-local, is motivated by the following special situation. Let

$$v_n\colon \Sigma^d M \to M$$

be a v_n-self map of a finite p-local CW-complex M of type n (see Chapter 1 of [**Ra2**]); that is, v_n induces an isomorphism in $K(n)_*$ and is zero in $K(m)_*$ for $m \neq n$. The existence of such maps for every n is ensured by [**HS**]. Then for every space X and every $k \geq 0$ we have maps

$$(v_n)^*\colon [\Sigma^k M, X] \to [\Sigma^{k+d} M, X], \tag{6.1}$$

which are isomorphisms if X is v_n-local. But $[\Sigma^k M, X]$ is usually denoted by $\pi_k(X; M)$. Hence, v_n-local—or, so to say, v_n-periodic—spaces satisfy

$$\pi_k(X; M) \cong \pi_{k+d}(X; M)$$

for all k, that is, they have indeed periodic homotopy groups. However, a v_n-periodic space may have a lot more of algebraic structure in its homotopy groups than mere periodicity, unless v_n has been chosen to be a suspension. The situation is completely analogous to the phenomenon which we described in Section 4 in the case of P-local spaces (which can actually be viewed as the case $n = 0$). This has been studied in [**Fa1**] for $n = 1$.

In fact, the analogy goes further. We may think of localization with respect to $p\colon S^2 \to S^2$, which is a "v_0-map", as a geometric construction inducing a certain algebraic construction on homotopy, namely

$$\pi_k(L_p X) = \mathbb{Z}[1/p] \otimes \pi_k(X) \quad \text{for } k \geq 2$$

(while $\pi_1(X)$ remains unchanged), where p "acts" on $\pi_k(X)$ via the appropriate suspension $p\colon S^k \to S^k$. More generally, for any v_n-map

$$v_n\colon \Sigma^d M \to M,$$

we may view v_n as an operator of degree d on the homotopy groups $\pi_*(X; M)$, by (6.1). In this situation, it is common practice to invert algebraically the operator v_n by considering the groups

$$v_n^{-1}\pi_k(X; M) = \mathbb{Z}[v_n, v_n^{-1}] \otimes_{\mathbb{Z}[v_n]} \pi_k(X; M).$$

Now one could look for a geometric construction L_{v_n} inducing this algebraic construction on homotopy, that is, such that

$$\pi_k(L_{v_n}X; M) \cong v_n^{-1}\pi_k(X; M) \quad \text{for } k \geq 2.$$

This has been done by Bousfield in [**Bo5**]. The functor L_{v_n} is essentially localization with respect to a v_n-map, desuspended as many times as possible (but one!). However, for technical reasons, Bousfield considered instead localization with respect to a certain map of the form $f\colon \Sigma W \to \mathrm{pt}$, since these tend to preserve fibrations to a larger extent. Indeed, for any space W, if $E \to X$ is a fibration with fibre F, then the homotopy fibre of the natural map from $P_{\Sigma W}F$ to the homotopy fibre of

$$P_{\Sigma W}E \to P_{\Sigma W}X$$

is a product of Eilenberg-Mac Lane spaces [**FS**]. Furthermore, under suitable assumptions on W, this "error term" can be reduced to a single Eilenberg-Mac Lane space [**Bo5**].

If suitably defined, the functor L_{v_n} does not depend on the choice of a v_n-map. Hence, this construction allows to define for every space X a natural tower [**Bo5**]

$$L_{v_0}X \leftarrow L_{v_1}X \leftarrow L_{v_2}X \leftarrow \cdots$$

which deserves to be called the *unstable chromatic tower* of X, in view of the fact that it provides successive approximations to X by spaces showing higher and higher sorts of periodicity.

We conclude this section with a few additional remarks on the effect of L_f on fibrations. There are plenty of examples showing how far is L_f from preserving fibrations in general. However, the following holds.

THEOREM 6.8. *Let $f\colon A \to B$ be any map, and $p\colon E \to X$ a fibration with fibre F. If L_fF is contractible, then $L_f(p)$ is a homotopy equivalence.*

The idea here is to use fibrewise localization [**Fa3, Br**]. This is a map $E \to \overline{E}$ over X which is an f-equivalence and such that $\overline{E} \to X$ is a fibration with fibre L_fF. After ensuring the existence of such a functor, Theorem 6.8 follows immediately.

In the special case when f is of the form $W \to \mathrm{pt}$, more can be said. In this case, we have, among other results [**Fa3, Bo5**]:

THEOREM 6.9. *Let $f\colon W \to \mathrm{pt}$ be given, and $F \to E \to X$ be a fibration with X connected.*

1. *If F is f-local, then $L_{\Sigma f}$ preserves the fibration.*
2. *If X is f-local (or, more generally, if $L_{\Sigma f}X \simeq L_fX$), then L_f preserves the fibration.*
3. *There is a fibration of the form*

$$L_fF \to \overline{E} \to L_{\Sigma f}X,$$

together with an f-equivalence $E \to \overline{E}$. Moreover, the space $\overline{E}$ is Σf-local. If E is f-local, F is a group and the fibration $F \to E \to X$ is principal, then $\overline{E} \simeq E$.

The machinery presented in this section has been the starting point of a fruitful research by Dror Farjoun, Bousfield, Nofech [**No**], Blanc-Thompson [**BTh**], and others. We will be content to close this article with one illuminating application. Although the main result in the next section—Neisendorfer's theorem—dates from

1991 and has already become folklore, it has not been published by its author so far. The only written references are contained in [**Fa3**] and [**ABN**].

7. Neisendorfer localization

One of the most appealing possibilities offered by Theorem 6.2 is to play the following game. Choose a nicely looking map $f\colon A \to B$ and study the properties of the functor L_f, in the hope that they are relevant.

An interesting choice is $f\colon B(\mathbb{Z}/p) \to \mathrm{pt}$ for a fixed prime p. In this section we discuss the properties of the corresponding functor L_f, which we will abbreviate to L for simplicity.

Since f is an f-equivalence, $LB(\mathbb{Z}/p)$ is contractible. By recalling that finite p-groups are solvable and repeatedly using Theorem 6.8, we find that LBG is contractible for every finite p-group G.

Also, if $X = \operatorname{hocolim} X_i$, then

$$\operatorname{map}_*(X, Y) \simeq \operatorname{holim} \operatorname{map}_*(X_i, Y)$$

for every space Y. It follows that LBG is actually contractible for every locally finite p-torsion group G, not necessarily finite. But much more is true. Let G be any (discrete) Abelian group such that $LBG \simeq \mathrm{pt}$. Then $LK(G,n) \simeq \mathrm{pt}$ for $n \geq 1$. To prove this, use the universal fibrations

$$K(G, n-1) \to \mathrm{pt} \to K(G, n)$$

and an obvious induction, again relying on Theorem 6.8.

Hence, the functor L is quite destructive, for it annihilates all Eilenberg-Mac Lane spaces whose single homotopy group is p-torsion. However, as we next show, L is not harmful at all to p'-torsion. Note that BG is f-local if and only if $\operatorname{map}_*\big(B(\mathbb{Z}/p), BG\big)$ is weakly contractible, which is equivalent to $\big[B(\mathbb{Z}/p), BG\big] = 0$, or $\operatorname{Hom}(\mathbb{Z}/p, G) = 0$. Hence, BG is f-local if and only if G is p-torsionfree. Similarly, observe that the map $f\colon B(\mathbb{Z}/p) \to \mathrm{pt}$ is a p'-equivalence, and therefore

PROPOSITION 7.1. *Every p'-local space is f-local.*

Next, the group extension

$$0 \to \mathbb{Z} \to \mathbb{Z}[1/p] \to \mathbb{Z}/p^\infty \to 0,$$

which gives rise to a fibration

$$K(\mathbb{Z}/p^\infty, n-1) \to K(\mathbb{Z}, n) \to K\big(\mathbb{Z}[1/p], n\big),$$

tells us that

$$LK(\mathbb{Z}, n) \simeq K\big(\mathbb{Z}[1/p], n\big) \quad \text{if } n \geq 2.$$

Hence, L does not easily "digest" free Abelian groups. But p-completion helps digestion, for if a simple space X is p'-local, then $\widetilde{H}_*(X; \mathbb{Z}/p) = 0$, and hence $\widehat{X}_p \simeq \mathrm{pt}$. Therefore, if G is *any* Abelian group, then

$$\big(LK(G,n)\big)^{\widehat{}}_p \simeq \mathrm{pt} \quad \text{for } n \geq 2.$$

This might seem implausible, but note that we have omitted so far the study of the case $n = 1$, where the deepest input actually occurs. Indeed, $K(\mathbb{Z}, 1) = S^1$ is a finite CW-complex, and Miller's theorem [**Mill**], stating that $\operatorname{map}_*(BG, X)$

is weakly contractible for every locally finite group G and every finite-dimensional CW-complex X, tells us that all finite-dimensional CW-complexes are f-local.

In particular, $LK(\mathbb{Z},1) = K(\mathbb{Z},1)$. Hence, there *is* something escaping from the devastating effect of the composition of L followed by p-completion. This leads to the following result of Neisendorfer. It is interesting to observe the analogy with work of Mislin [**Mis**].

THEOREM 7.2. *Let X be a finite-dimensional 1-connected CW-complex. Suppose further that $\pi_2(X)$ is torsion. Then, for $n \geq 2$,*

$$\big(LX\langle n\rangle\big)^{\widehat{}}_p \simeq \widehat{X}_p,$$

where $X\langle n\rangle$ denotes the n-connected cover of X, and L is localization with respect to $f\colon B(\mathbb{Z}/p) \to \mathrm{pt}$.

The proof uses our previous remarks, together with part (2) of Theorem 6.9 applied to the fibration

$$F \to X\langle n\rangle \to X.$$

Since X is f-local by assumption, we have a fibration

$$LF \to LX\langle n\rangle \to X$$

of simple spaces. Thus

$$(LF)^{\widehat{}}_p \to \big(LX\langle n\rangle\big)^{\widehat{}}_p \to \widehat{X}_p$$

is again a fibration. Now it suffices to check that $(LF)^{\widehat{}}_p$ is contractible. Let Y be the homotopy fibre of the p'-localization map $F \to F_{p'}$. The assumption that $\pi_2(X)$ is torsion ensures that Y will be connected, for $\pi_1(F) \cong \pi_2(X)$. Since Y has finitely many homotopy groups, and these are p-torsion groups, LY is contractible. Hence, $LF \simeq L(F_{p'}) \simeq F_{p'}$, by Proposition 7.1, and the result follows.

As a consequence of Theorem 7.2, one obtains an easy proof of the following well known fact (cf. [**Se**], [**MN**]). All finite-dimensional 1-connected CW-complexes which are not contractible must have infinitely many nontrivial homotopy groups.

Theorem 7.2 has found another significant application in [**ABN**]. It also provides a counterexample to the following question. Given a map $f\colon A \to B$, one may consider the induced homomorphism $f_*\colon \pi_1(A) \to \pi_1(B)$. The orthogonal pair generated by f_* is reflective in the category of groups. In fact, the corresponding localization functor, which we denote by L_{f_*}, can be constructed exactly in the same way as Theorem 6.2 was proved. Then one could ask

QUESTION 7.3. For which maps f is it true that the groups $\pi_1(L_fX)$ and $L_{f_*}\pi_1(X)$ are isomorphic for all spaces X?

This turns out to be the case for most examples of f-localization which have been mentioned in this article, but not for $f\colon B(\mathbb{Z}/p) \to \mathrm{pt}$. Indeed, there exist finite CW-complexes X whose fundamental group contains nontrivial p-torsion, so that $\pi_1(L_fX)$ is not f_*-local in general.

Many results about f-localization can be transferred to the category of groups, where they have different descriptions and sometimes much more elementary proofs. For example, the group-theoretical analog of Theorem 6.8 is a triviality. Fibrewise P-localization of group extensions has been developed in [**CC**], after the first approach by Hilton [**Hi1**] in the nilpotent case. Its generalization to fibrewise φ-localization of group extensions for an arbitrary homomorphism $\varphi\colon \pi \to \nu$ does not offer major difficulties.

If φ is of the form $\varphi\colon \pi \to \{1\}$, then the effect of the functor L_φ can be easily described: For a given group G, the localization $L_\varphi G$ is the largest quotient of G on which all homomorphic images of π are trivial. For example, in the case when $\pi = \mathbb{Z}/p$, this amounts to killing the p-torsion of G. In general, if $\varphi\colon \pi \to \{1\}$, the kernel of $l\colon G \to L_\varphi G$ can be constructed as a direct limit, in a manner analogous to the construction of the *p-isolator* subgroup [**Ri1**] of an arbitrary group G (also called *p-radical*). However, for a more general homomorphism $\varphi\colon \pi \to \nu$, the kernel of $l\colon G \to L_\varphi G$ seems to be extremely difficult to characterize; see [**Ga, BC**].

It might also be asked, as in Question 6.4, if every idempotent monad in the category of groups is φ-localization for some homomorphism φ. We do not know the answer, which is again related to the questions posed at the end of Section 5. One of the first idempotent monads which comes to mind in the category of groups is Abelianization. We leave it to the reader to find a homomorphism φ for which the φ-local groups are precisely the Abelian groups (the answer should make it clear how φ is to be chosen in the case of the reflection onto any variety of groups).

References

[Ad1] J. F. Adams, *The sphere considered as an H-space mod p*, Quart. J. Math. Oxford **12** (1961), 52–60.

[Ad2] ——, *Localisation and completion*, lecture notes by Z. Fiedorowicz, University of Chicago, 1975.

[ABN] J. Aguadé, C. Broto, and D. Notbohm, *Homotopy classification of some spaces with interesting cohomology and a conjecture of Cooke* I, Topology (to appear).

[An] D. W. Anderson, *Localizing CW-complexes*, Illinois J. Math. **16** (1972), 519–525.

[BW] M. Barr and C. Wells, *Toposes, triples and theories*, Grund. math. Wiss., vol. 278, Springer-Verlag, Berlin, Heidelberg, New York, 1985.

[Ba1] G. Baumslag, *Some aspects of groups with unique roots*, Acta Math. **194** (1960), 217–303.

[Ba2] ——, *Roots and wreath products*, Proc. Cambridge Philos. Soc. **56** (1960), 109–117.

[BDH] G. Baumslag, E. Dyer, and A. Heller, *The topology of discrete groups*, J. Pure Appl. Algebra **16** (1980), 1–47.

[Be] M. Bendersky, *A functor which localizes the higher homotopy groups of an arbitrary CW-complex*, Lecture Notes in Math., vol. 418, Springer-Verlag, Berlin, Heidelberg, New York, 1974, pp. 13–21.

[BC] A. J. Berrick and C. Casacuberta, *Groups and spaces with all localizations trivial*, Lecture Notes in Math., vol. 1509, Springer-Verlag, Berlin, Heidelberg, New York, 1992, pp. 20–29.

[BT] A. J. Berrick and G. C. Tan, *The minimal extension of P-localization on groups* (to appear).

[BTh] D. Blanc and R. D. Thompson, *A suspension spectral sequence for v_n-periodic homotopy groups* (to appear).

[Bo1] A. K. Bousfield, *The localization of spaces with respect to homology*, Topology **14** (1975), 133–150.

[Bo2] ——, *Homological localization towers for groups and π-modules*, Mem. Amer. Math. Soc., vol. 10, no. 186, 1977.

[Bo3] ——, *Constructions of factorization systems in categories*, J. Pure Appl. Algebra **9** (1977), 207–220.

[Bo4] ——, *The localization of spectra with respect to homology*, Topology **18** (1979), 257–281.

[Bo5] ——, *Localization and periodicity in unstable homotopy theory*, J. Amer. Math. Soc. (to appear).

[BK] A. K. Bousfield and D. M. Kan, *Homotopy limits, localizations and completions*, Lecture Notes in Math., vol. 304, Springer-Verlag, Berlin, Heidelberg, New York, 1972.

[Br] C. Broto, *Fibrewise localization*, in preparation.

[Ca] C. Casacuberta, *On the rationalization of the circle*, Proc. Amer. Math. Soc. **118** (1993), 995–1000.

[CC] C. Casacuberta and M. Castellet, *Localization methods in the study of the homology of virtually nilpotent groups*, Math. Proc. Cambridge Philos. Soc. **112** (1992), 551–564.

[CFT] C. Casacuberta, A. Frei, and G. C. Tan, *Extending localization functors*, J. Pure Appl. Algebra (to appear).

[CP] C. Casacuberta and G. Peschke, *Localizing with respect to self maps of the circle*, Trans. Amer. Math. Soc. **339** (1993), 117–140.

[CPP] C. Casacuberta, G. Peschke, and M. Pfenniger, *On orthogonal pairs in categories and localisation*, London Math. Soc. Lecture Note Ser., vol. 175, Cambridge University Press, Cambridge, 1992, pp. 211–223.

[De] A. Deleanu, *Idempotent codensity monads and the profinite completion of topological groups*, London Math. Soc. Lecture Note Ser., vol. 86, Cambridge University Press, Cambridge, 1983, pp. 154–163.

[DFH] A. Deleanu, A. Frei, and P. Hilton, *Idempotent triples and completion*, Math. Z. **143** (1975), 91–104.

[DH] A. Deleanu and P. Hilton, *On Postnikov-true families of complexes and the Adams completion*, Fund. Math. **106** (1980), 53–65.

[DHS] E. S. Devinatz, M. J. Hopkins, and J. H. Smith, *Nilpotence and stable homotopy theory* I, Ann. of Math. **128** (1988), 207–241.

[Fa1] E. Dror Farjoun, *Homotopy localization and v_1-periodic spaces*, Lecture Notes in Math., vol. 1509, Springer-Verlag, Berlin, Heidelberg, New York, 1992, pp. 104–113.

[Fa2] ——, *Higher homotopies of natural constructions* (to appear).

[Fa3] ——, *Localizations, fibrations and conic structures* (to appear).

[Fa4] ——, *Cellular spaces* (to appear).

[Fa5] ——, *Cellular inequalities* (to appear).

[FS] E. Dror Farjoun and J. H. Smith, *Homotopy localization nearly preserves fibrations* (to appear).

[Fak] S. Fakir, *Monade idempotente associée à une monade*, C. R. Acad. Sci. Paris Sér. A **270** (1970), 99–101.

[FK] P. J. Freyd and G. M. Kelly, *Categories of continuous functors* I, J. Pure Appl. Algebra **2** (1972), 169–191.

[Ga] A. García Rodicio, *Métodos homológicos en grupos P-locales*, Ph.D. thesis, Universidad de Santiago de Compostela, 1986.

[Gi] S. Gitler, *Cohomology operations with local coefficients*, Amer. J. Math. **85** (1963), 156–188.

[Hill] R. O. Hill, *A geometric interpretation of a classical group cohomology obstruction*, Proc. Amer. Math. Soc. **54** (1976), 405–412.

[Hi1] P. Hilton, *Relative nilpotent groups*, Lecture Notes in Math., vol. 915, Springer-Verlag, Berlin, Heidelberg, New York, 1982, pp. 136–147.

[Hi2] ——, *Nilpotente Gruppen und nilpotente Räume*, lecture notes by M. Pfenniger, Lecture Notes in Math., vol. 1053, Springer-Verlag, Berlin, Heidelberg, New York, 1984.

[HMR] P. Hilton, G. Mislin, and J. Roitberg, *Localization of nilpotent groups and spaces*, North-Holland Math. Studies, vol. 15, North-Holland, Amsterdam, 1975.

[HS] M. J. Hopkins and J. H. Smith, *Nilpotence and stable homotopy* II (to appear).

[Ke] G. M. Kelly, *A unified treatment of transfinite constructions for free algebras, free monoids, colimits, associated sheaves, and so on*, Bull. Austral. Math. Soc. **22** (1980), 1–83.

[Mill] H. Miller, *The Sullivan conjecture on maps from classifying spaces*, Ann. of Math. **120** (1984), 39–87.

[MN] C. A. McGibbon and J. A. Neisendorfer, *On the homotopy groups of a finite dimensional space*, Comment. Math. Helv. **59** (1984), 253–257.

[MNT] M. Mimura, G. Nishida, and H. Toda, *Localization of CW-complexes and its applications*, J. Math. Soc. Japan **23** (1971), 593–624.

[Mis] G. Mislin, *Lannes' T-functor and the cohomology of BG*, Compositio Math. **86** (1993), 177–187.

[No] A. Nofech, *On localization of inverse limits* (to appear).

[Pe1] G. Peschke, *H-semidirect products*, Canad. Math. Bull. **30** (1987), 402–411.

[Pe2] ——, *Localizing groups with action*, Publ. Mat. **33** (1989), 227–234.

[Pf] M. Pfenniger, *On completions*, U.C.N.W. Maths Preprint 91.17, Bangor (1991).

[Ra1] D. Ravenel, *Localization with respect to certain periodic homology theories*, Amer. J. Math. **106** (1984), 351–414.

[Ra2] ——, *Nilpotence and periodicity in stable homotopy theory*, Ann. of Math. Studies, vol. 128, Princeton University Press, Princeton, 1992.

[Re] A. Reynol, *P-localization of some classes of groups*, Publ. Mat. **37** (1993), 19–44.

[Ri1] P. Ribenboim, *Torsion et localisation de groupes arbitraires*, Lecture Notes in Math., vol. 740, Springer-Verlag, Berlin, Heidelberg, New York, 1978, pp. 444–456.

[Ri2] ——, *Equations in groups, with special emphasis on localization and torsion* II, Portugal. Math. **44** (1987), 417–445.

[Ro] J. Roitberg, *Note on nilpotent spaces and localization*, Math. Z. **137** (1974), 67–74.

[Se] J.-P. Serre, *Cohomologie module* 2 *des complexes d'Eilenberg-MacLane*, Comment. Math. Helv. **27** (1953), 198–232.

[Su] D. Sullivan, *Geometric topology* I: *Localization, periodicity and Galois symmetry*, MIT press, 1970.

[Wh] G. W. Whitehead, *Elements of homotopy theory*, Graduate Texts in Math., vol. 61, Springer-Verlag, Berlin, Heidelberg, New York, 1978.

Department of Mathematics, Universitat Autònoma de Barcelona, E-08193 Bellaterra, Spain

E-mail address: icrm1@cc.uab.es

Centre de Recherches Mathématiques
CRM Proceedings and Lecture Notes
Volume **6**, 1994

Hurwitz-Radon Matrices Revisited: From Effective Solution of the Hurwitz Matrix Equations to Bott Periodicity

Beno Eckmann

The matrix equations $A_j^2 = -E$, $A_jA_k + A_kA_j = 0$, $j, k = 1, \dots, s$, $j \neq k$ were discussed independently by Hurwitz and Radon around 1920*. Solutions have played, during the years, an important rôle in many parts of mathematics and mathematical physics: as coefficient matrices of elliptic differential operators, in the topology of vector fields and fiber bundles, in combinatorial analysis, in the composition problem for quadratic forms—this was the original motivation and starting point, and in that problem the (complex) matrices have to be *orthogonal.* A closely related variant is to solve the equations by *unitary* matrices.

In this survey we describe a procedure for constructing effectively the solutions, first in the unitary and from there in the orthogonal case. The essential tool is a "product" which reduces everything to small values of s; this product exhibits at the same time a very close relation with the well-known Bott periodicity in homotopy and K-theory.

I have tried to keep large parts of the presentation (except for those concerning complex and real K-theory) accessible to non-specialists, in order to emphasize the elementary aspects of the problem and the solution. I hope that this is somewhat in the spirit of Peter Hilton's beautiful contributions to the teaching of mathematics at various levels.

1. Orthogonal and unitary H-R matrices

1.1. Complex $n \times n$ matrices $A_1, A_2, \dots, A_s$ are called *Hurwitz-Radon matrices* if they fulfill the conditions

$$\text{(I)}\qquad \begin{cases} (1) & A_j^2 = -E \\ (2) & A_jA_k + A_kA_j = 0 \\ (3) & A_jA_j^T = E \end{cases}$$

for $j, k = 1, 2, \dots, s$, $j \neq k$.

NOTATIONS. A^T is the transposed of the matrix A. We write E (or E_n if the *size* n is to be emphasized) for the $n \times n$ unit matrix. Hurwitz-Radon will be abbreviated by H-R.

1991 *Mathematics Subject Classification.* Primary: 15A24, 15A63, 55R45; Secondary: 20C15.
This is the final form of the paper.

*Hurwitz died in 1919. His paper [**H**] appeared in 1923. Radon's work was submitted in 1922 and published also in 1923 [**R**].

(3) means that the A_j are *orthogonal.* In view of (1) it is equivalent to $A_j^T = -A_j$, i.e., the A_j are skew-symmetric.

1.2. An equivalent formulation of (I) is obtained as follows. Consider $s+1$ matrices $A_0, A_1, \ldots, A_s$ and require that their linear combination

$$x_0 A_0 + x_1 A_1 + \cdots + x_0 A_0$$

be an *orthogonal* matrix for all (real or complex) values of the x_j "except for a factor $\sum_0^s x_j^2$"; i.e.,

$$\left(\sum_0^s x_j A_j\right)\left(\sum_0^s x_k A_k^T\right) = \sum_0^s x_j^2 E.$$

This is the case if and only if

$$\text{(II)} \qquad \begin{cases} A_j A_j^T = E \\ A_j A_k^T + A_k A_j^T = 0 \end{cases}$$

for $j, k = 0, 1, \ldots, s$, $j \neq k$. To find all solutions it is clearly enough to consider those with $A_0 = E$. Substituting this for A_k we get $A_j + A_j^T = 0$ for $j = 1, 2, \ldots, s$, and then (II) for $j, k = 1, 2, \ldots, s$ is equivalent to (I).

1.3. The problem considered (independently) by Hurwitz [**H**] and Radon [**R**] concerns the "composition of quadratic forms"

$$\text{(III)} \qquad \left(\sum_0^s x_j^2\right)\left(\sum_1^n y_\ell^2\right) = \sum_1^n z_\ell^2$$

where the z_ℓ are complex bilinear forms of the $x_0, \ldots, x_s$ and $y_1, \ldots, y_n$. They determined for given n the maximum number $s+1$ for which such bilinear forms exist. If $n = \text{odd} \cdot 16^\alpha \cdot 2^\beta$, $\beta = 0, 1, 2, 3$, the maximal $s+1$ is $\rho(n) = 8\alpha + 2^\beta$, the "Radon number". The bilinear forms can be written in matrix notation as

$$\begin{pmatrix} z_1 \\ \vdots \\ z_n \end{pmatrix} = \left(\sum_0^s x_j A_j\right)\begin{pmatrix} y_1 \\ \vdots \\ y_n \end{pmatrix}.$$

Then (III) simply tells that $\sum_0^s x_j A_j$ is orthogonal "except for the factor $\sum_0^s x_j^2$".

The H-R problem thus consist in solving (I), for given n, by a maximum number s of matrices $A_1, \ldots, A_s$. Or, in other words, given s to find the minimum n for which such $n \times n$ matrices exist. In this note we will present a method for not only determing the minimum n but also to construct effectively *all* solutions of (I). We will do this in detail for the *unitary* variant (I′) of (I), see below 1.4. The original *orthogonal* problem (I) will then only be sketched; it is more complicated but is dealt with by essentially the same procedure.

It turns out that this effective procedure leads to an interesting relation with the homotopy groups of the infinite unitary, or orthogonal respectively, group (Bott periodicity).

1.4. The *unitary* matrix problem (I′) is obtained by simply replacing (3) by

$$A_j \bar{A}_j^T = E, \tag{3'}$$

i.e., one asks for unitary matrices fulfilling (1) and (2). Condition (3′) is then equivalent to $\bar{A}_j^T = -A$.

Again, if we put $A_0 = E$, the problem (I′) is equivalent to

$$x_0 A_0 + x_1 A_1 + \cdots + x_s A_s$$

being a *unitary* matrix for all (real) values of the x_j "except for the factor $\sum_0^s x_j^2$"; i.e.,

$$\left(\sum_0^s x_j A_j\right)\left(\sum_0^s x_k \bar{A}_k^T\right) = \left(\sum_0^s x_j^2\right) E.$$

2. Irreducible unitary solutions

2.1. We deal in the following, until Section 4, with *unitary* H-R matrices exclusively; i.e., with unitary matrices fulfilling (1) and (2). Given two solutions for the same s of size n and m respectively

$$\sigma = \{A_1, \ldots, A_s\}, \quad \sigma' = \{A'_1, \ldots, A'_s\}$$

their *sum* $\sigma + \sigma'$ is defined by

$$\sigma + \sigma' = \left\{ \begin{pmatrix} A_1 & \\ & B_1 \end{pmatrix}, \ldots, \begin{pmatrix} A_s & \\ & B_s \end{pmatrix} \right\}$$

(empty entries always mean 0 or 0-matrices). $\sigma + \sigma'$ is easily checked to be again a solution, of size $n + m$. If M is an invertible matrix then

$$M\sigma M^{-1} = \{MA_1M^{-1}, \ldots, MA_sM^{-1}\}$$

is again a solution of (1) and (2) of size n, and two solutions σ and τ are called *equivalent* if there is an M such that $\tau = M\sigma M^{-1}$.

2.2. A unitary solution is called *irreducible* if it is not equivalent to a sum of two solutions. Minimal size n solutions, for given s, must of course be irreducible. It turns out that there are very few irreducible solutions, and that only *one* size, depending on s, occurs:

PROPOSITION 2.1. *If s is even there is only one irreducible solution, up to equivalence, of size $n = 2^{s/2}$. If s is odd there are two irreducible solutions, up to equivalence, of size $n = 2^{(s-1)/2}$.*

COROLLARY 2.2. *The maximum number s of unitary H-R matrices of size $n = \text{odd}2^t$ is $2t + 1$.*

The proof of Proposition 2.1 is contained in the author's paper **[E]** published 50 year ago. It is an application of some basic facts of the theory of *group representations*. A reader familiar with that theory will have noticed that the terminology above is taken from group representations (except that the "size" is called there "degree"). We recall in 2.3 some of the arguments of **[E]** although these are not needed explicitly in the sequel of the present note.

2.3. Let G_s be the group given by the presentation

$$\langle a_1, \ldots, a_s, \varepsilon \mid \varepsilon^2 = 1, a_j^2 = \varepsilon, a_j a_k = \varepsilon a_k a_j \rangle$$

where $j, k = 1, \ldots, s$, $j \neq k$. A unitary solution of (1), (2) of size n yields a representation of G_s of degree n by assigning the matrix A_j to the generator a_j and $-E$ to ε. And vice-versa: any unitary representation of G_s with $\varepsilon \mapsto -E$ yields a solution.

G_s is a finite group of order 2^{s+1}. The commutator subgroup consists, for $s > 1$, of 1 and ε only; G_1 is Abelian (cyclic of order 4). The center of G_s consists of 1 and ε if s is even; and of 1, ε and two further elements z, εz if s is odd ($z = a_1 a_2 \cdots a_s$).

G_1 has two irreducible representations, of degree 1, with $\varepsilon \mapsto (-1)$; namely $A_1 = (i)$ or $(-i)$. For $s > 1$ representations of degree 1 cannot be used since they assign the matrix (1) to ε; their number is 2^s. Irreducible representations of degree > 1 automatically assign $-E$ to ε; their number is 1 for even s and 2 for odd s, and one has for their degrees d_s, d_s' (= powers of 2)

$$\begin{aligned} 2^{s+1} &= 2^s + d_s^2 && \text{for even } s, \\ 2^{s+1} &= 2^s + d_s^2 + {d_s'}^2 && \text{for odd } s \end{aligned}$$

which immediately yields $d_s = 2^{s/2}$ or $d_s = d_s' = 2^{(s-1)/2}$ respectively.

By a general result any representation of a finite group is equivalent to a unitary one, and equivalent to a sum of irreducible representations. Thus the term "irreducible solution" in 2.2 is in agreement with the usual meaning of irreducible representation.

2.4. Since any unitary solution of (1), (2) is equivalent to a sum of irreducible solutions, which we will denote by ρ_s for even s and by ρ_s, ρ_s' for odd s, it is convenient to consider the free Abelian group D_s generated by (the equivalence class of) ρ_s or of ρ_s, ρ_s' respectively. Thus

$$\begin{aligned} D_s &\cong \mathbb{Z} && \text{if } s \text{ is even}, \\ D_s &\cong \mathbb{Z} \oplus \mathbb{Z} && \text{if } s \text{ is odd}. \end{aligned}$$

Positive integers k are to be interpreted, for even s, as sums of k solutions ρ_s; and positive pairs $(k, k') \in \mathbb{Z} \oplus \mathbb{Z}$ as sums of k solutions ρ_s and k' solutions ρ_s', for odd s. The zeros and the negative integers have a symbolic meaning only ("Grothendieck groups").

2.5. Solutions for small values of s.

$$s = 1 \quad \rho_1 : A_1 = (i) \quad \text{or} \quad \rho_1' : A_1 = (-i)$$

$$s = 2 \quad \rho_2 : A_1 = \begin{pmatrix} i & \\ & -i \end{pmatrix}, \quad A_2 = \begin{pmatrix} & 1 \\ -1 & \end{pmatrix}$$

$$s = 3 \quad \rho_3 : A_1 = \begin{pmatrix} i & \\ & -i \end{pmatrix}, \quad A_2 = \begin{pmatrix} & 1 \\ -1 & \end{pmatrix}, \quad A_3 = \begin{pmatrix} & i \\ i & \end{pmatrix}$$

or

$$\rho_3' : A_1 = \begin{pmatrix} -i & \\ & i \end{pmatrix}, \quad A_2 = \begin{pmatrix} & -1 \\ 1 & \end{pmatrix}, \quad A_3 = \begin{pmatrix} & -i \\ -i & \end{pmatrix}$$

$$s=4 \quad \rho_4 : A_1 = \begin{pmatrix} i & & & \\ & -i & & \\ & & -i & \\ & & & i \end{pmatrix}, \quad A_2 = \begin{pmatrix} & 1 & & \\ -1 & & & \\ & & & -1 \\ & & 1 & \end{pmatrix}$$

$$A_3 = \begin{pmatrix} & i & & \\ i & & & \\ & & & -i \\ & & -i & \end{pmatrix}, \quad A_4 = \begin{pmatrix} & & 1 & \\ & & & 1 \\ -1 & & & \\ & -1 & & \end{pmatrix}$$

It is convenient to include in the list of the irreducible solutions the case $s = 0$: The size is 1, and ρ_0 is the empty set of matrices.

3. How to construct unitary solutions

3.1. We describe in this section a procedure for constructing explicitly all unitary solutions of (1), (2). "All" means, of course, up to equivalence; our notations will often not distinguish between equivalence classes and their representatives. In Section 5 the procedure will be restated in a more formal framework.

As stated in the previous section we only have to produce the irreducible solutions ρ_s, s even, and ρ_s, ρ'_s, s odd.

An almost trivial observation shows that one can forget about the ρ_s for even s. Namely, let $\rho_{s+1} = \{A_1, A_2, \ldots, A_s, A_{s+1}\}$; if s is even the size of ρ_{s+1} is $2^{((s+1)-1)/2} = 2^{s/2}$. If A_{s+1} is omitted $\{A_1, A_2, \ldots, A_s\}$ is a solution of size $2^{s/2}$, whence irreducible. Since there is only *one* such it must be ρ_s. Of course one might omit, instead of A_{s+1}, any other A_j to yield the same ρ_s up to equivalence.

PROPOSITION 3.1. *If s is even then omitting a matrix from ρ_{s+1} or ρ'_{s+1} yields the irreducible solution ρ_s.*

3.2. What happens if the same procedure is applied to ρ_{s+1} for odd s? The size of ρ_{s+1} is $2^{(s+1)/2} = 2 \cdot 2^{(s-1)/2}$. Therefore omitting A_{s+1} from ρ_{s+1} yields the sum of *two* irreducible solutions. What are they? For reasons of symmetry one will guess that one obtains $\rho_s + \rho'_s$.

Here is a proof. We consider $\{A_1, \ldots, A_s\}$ from ρ_{s+1}. Since $s+1$ is even $\{A_1, A_2, \ldots, A_{s+1}\}$ is equivalent to $\{-A_1, A_2, \ldots, A_{s+1}\}$. The matrices $A_1 A_2 \ldots A_s$ and $-A_1 A_2 \ldots A_s$ are similar and therefore have the same *trace*:

$$\operatorname{trace}(A_1 A_2 \ldots A_s) = -\operatorname{trace}(A_1 A_2 \ldots A_s) = 0.$$

Let further ρ_s be $\{B_1, B_2, \ldots, B_s\}$. The matrix $B_1 B_2 \ldots B_s$ commutes with all B_j and thus is a non-zero multiple of E, cE (here again a result of representation theory is used: A matrix which commutes with all matrices of an irreducible representation is a multiple of E). This also shows that one can take $\rho'_s = \{-B_1, \ldots, -B_s\}$; for the product is $-cE$; and since $\operatorname{trace}(cE) \neq \operatorname{trace}(-cE)$ these two solutions cannot be equivalent.

It is clear now that $\{A_1, \ldots, A_s\}$ must be $\rho_s + \rho'_s$ in order to make the trace of the product equal to 0.

PROPOSITION 3.2. *If s is odd and $\rho_{s+1} = \{A_1, \ldots, A_s, A_{s+1}\}$ then omitting A_{s+1} yields $\rho_s + \rho'_s$.*

PROPOSITION 3.3. *If s is odd then ρ'_s is obtained from ρ_s by changing the signs of all matrices (or of an odd number of matrices).*

3.3. We now describe our crucial tool, a *product of solutions*. We first recall that $A_1, A_2, \dots, A_s$ are unitary H-R matrices of size n if and only if

$$f(x) = \sum_0^s x_j A_j$$

(with $A_0 = E_n$) is a unitary matrix for all $x = (x_0, \dots, x_s) \in \mathbb{R}^{s+1}$ with $\sum_0^s x_j^2 = 1$.

Given two solutions $\sigma_s = \{A_1, \dots, A_s\}$, $\sigma_t = \{B_1, \dots, B_t\}$ of size n and m respectively we put

$$g(y) = \sum_0^s y_k B_k,$$

$B_0 = E_m$, $y = (y_0, \dots, y_t) \in \mathbb{R}^{t+1}$ and set

$$F(x,y) = \begin{pmatrix} f(x) \otimes E_m & E_n \otimes g(y) \\ -E_n \otimes \overline{g(y)}^T & \overline{f(x)}^T \otimes E_m \end{pmatrix}. \tag{4}$$

This is a matrix of size $2nm$. The entry $f(x) \otimes E_m$, e.g., means $\sum_0^s x_j A_j \otimes E_m$ where $A_j \otimes E_m$ is the matrix of the tensor product of the linear transformations corresponding to A_j and E_m. One easily checks that

$$F(x,y)\overline{F(x,y)}^T = \left(\sum_0^s x_j^2 + \sum_0^t y_k^2\right) E_{2nm}.$$

Thus if $\sum_0^s x_j^2 + \sum_0^t y_k^2 = 1$ then $F(x,y)$ is a unitary matrix. The coefficient matrix of x_0 is E_{2nm}; therefore the coefficient matrices of $x_1, \dots, x_s$, $y_0, y_1, \dots, y_t$ are $s+t+1$ unitary H-R matrices of size $2nm$ and constitute a solution $\sigma_{s+t+1} = \{C_1, \dots, C_s, C_{s+1}, \dots, C_{s+t+1}\}$.

Explicitly:

$$\begin{cases} C_j = \begin{pmatrix} A_j \otimes E_m & \\ & -A_j \otimes E_m \end{pmatrix} & \text{for } j = 1, \dots, s, \\ C_j = \begin{pmatrix} & E_{nm} \\ -E_{nm} & \end{pmatrix} & \text{for } j = s+1, \\ C_j = \begin{pmatrix} & E_n \otimes B_k \\ E_n \otimes B_k & \end{pmatrix} & \text{for } j = s+2, \dots, s+t+1, \\ & \text{with } k = j - s - 1. \end{cases} \tag{5}$$

We call σ_{s+t+1} the *product* of σ_s and σ_t, written $\sigma_s \cdot \sigma_t$.

3.4. If we take, in particular, the product $\rho_s \cdot \rho_t$ with t odd then its size is as follows:

$$\begin{aligned} 2 \cdot 2^{(s-1)/2}\, 2^{(t-1)/2} &= 2^{(s+t)/2} && \text{if } s \text{ is odd} \\ 2 \cdot 2^{s/2}\, 2^{(t-1)/2} &= 2^{(s+t+1)/2} && \text{if } s \text{ is even;} \end{aligned}$$

i.e., in both cases $\rho_s \cdot \rho_t = \rho_{s+t+1}$.

As remarked in 3.1 above the case s even, $s+t+1$ even, is not important, while s and t odd explicitly produce new solutions. In particular, $\rho_s \cdot \rho_1 = \rho_{s+2}$ (or ρ'_{s+2}) so that all ρ_s, s odd, can be constructed inductively from ρ_1.

THEOREM 3.4. *All irreducible unitary solutions ρ_{s+2} for odd s are obtained inductively from $\rho_1 = \{(i)\}$ by the product $\rho_s \cdot \rho_1$ (and ρ'_{s+2} by changing the signs of all matrices).*

In explicit form, let $\rho_s = \{A_1, \ldots, A_s\}$ of size $n = 2^{(s-1)/2}$. Then ρ_{s+2} is given by the $s+2$ matrices

$$(6) \qquad \begin{pmatrix} A_j & \\ & -A_j \end{pmatrix}, j = 1, \ldots, s; \quad \begin{pmatrix} & E_n \\ -E_n & \end{pmatrix}; \quad \begin{pmatrix} & iE_n \\ iE_n & \end{pmatrix}.$$

Note that the examples in 2.4 are exactly of the above form.

COROLLARY 3.5. *All unitary H-R matrices can be written with entries* 0, ± 1, $\pm i$ *alone.*

REMARK 3.6.. The short expression for (6) corresponding to (4), with $A_0 = E$ and $f(x) = \sum_0^s x_j A_j$, $g(y) = y_0 + iy_1$ is

$$F(x, z) = \begin{pmatrix} f(x) & zE_n \\ -\bar{z}E_n & \overline{f(x)}^T \end{pmatrix}$$

where we have written z for $y_0 + iy_1$.

3.5. In 2.4 we have introduced the notation D_s for the free Abelian group generated by the (equivalence classes of) irreducible solutions ρ_s, or ρ_s, ρ'_s respectively. D_s is $\cong \mathbb{Z}$ for even s, $\cong \mathbb{Z} \oplus \mathbb{Z}$ for odd s; only positive integers have a "concrete" meaning.

Omitting in a solution $\sigma_{s+1} \in D_{s+1}$ the last matrix A_{s+1} is compatible with the sum of solutions. It thus defines a homomorphism $h\colon D_{s+1} \to D_s$ for all $s \geq 0$. (In defining h the rôle of A_{s+1} can also be played by another matrix A_j). We write $\mathcal{E}_s$ for the cokernel $D_s/h(D_{s+1})$ of h and call it the *reduced unitary* H-R *group.* Since by Proposition 3.1 $h(\rho_{s+1}) = \rho_s$ for even s the map h is surjective, whence $\mathcal{E}_s = 0$ for even s. For odd s, $h(\rho_{s+1}) = \rho_s + \rho'_s$, by Proposition 3.2; the elements of $\mathcal{E}_s$ are classes in D_s modulo the "diagonal" $\mathbb{Z}(\rho_s + \rho'_s)$. Thus $\mathcal{E}_s \cong \mathbb{Z}$ (choose ρ_s and $\rho_s + \rho'_s$ as a basis of D_s instead of ρ_s, ρ'_s).

For odd s the elements of $\mathcal{E}_s$ can be described as follows. For positive $k \in \mathbb{Z}$ they are represented by sums of solutions ρ_s, the element $0 \in \mathbb{Z}$ by $\rho_s + \rho'_s$; and for negative k by sums of ρ'_s. In this way all elements of $\mathcal{E}_s$ have a "concrete" meaning in terms of solutions.

THEOREM 3.7. *The reduced unitary H-R group $\mathcal{E}_s$, $s \geq 0$, is 0 for even s; and is $\mathbb{Z}$ for odd s, generated by ρ_s or by $\rho'_s = -\rho_s$.*

4. How to construct orthogonal H-R matrices

4.1. The discussion of orthogonal solutions of the H-R matrix problem (1), (2) is quite parallel to the unitary case, though a little more complicated.

Clearly, if a unitary solution σ_s is equivalent to a *real* one then we get an orthogonal solution. This is actually the only way to obtain orthogonal solutions: By a theorem of Frobenius-Schur in representation theory a unitary representation is equivalent to a complex orthogonal one (if and) only if it is equivalent to a real one. If σ_s is *not* equivalent to a real solution we have to combine it with its conjugate-complex $\bar{\sigma}_s$, as $\sigma_s + \bar{\sigma}_s$; it is then easy to find a transformation T such that $T(\sigma_s + \bar{\sigma}_s)T^{-1}$ is real.

The following classification of the real—i.e., orthogonal—irreducible representations τ_s is taken from [**E**]. It is based on determining for which s the unitary irreducible solution ρ_s is equivalent to a real solution; and if not, equivalent to $\bar{\rho}_s$ or not. A different procedure can be found in [**E2**]; it applies an elegant approach of T. Y. Lam and T. Smith [**LS**] deviced in a more general framework.

4.2. The statements concerning the τ_s depend on the value of s modulo 8.

If $s \equiv 0,6,7$ then ρ_s is (equivalent to a) *real* solution, thus $\tau_s = \rho_s$; and the size is $2^{s/2}$ for $s \equiv 0,6$, $2^{(s-1)/2}$ for $s \equiv 7$ (where one has another orthogonal solution $\tau'_s = \rho'_s$).

If $s \equiv 2,3,4$ then ρ_s is *not real*, and equivalent to $\bar{\rho}_s$, thus $\tau_s = \rho_s + \bar{\rho}_s$, and the size is $2^{(s+2)/2}$ for $s \equiv 2,4$, $2^{(s+1)/2}$ for $s \equiv 3$ (where one has another solution $\tau'_s = \rho'_s + \bar{\rho}'_s$).

If $s \equiv 1,5$ then ρ_s is *not real*, and *not* equivalent to $\bar{\rho}_s$ (whence $\bar{\rho}_s = \rho'_s$), thus $\tau_s = \rho_s + \bar{\rho}_s = \rho_s + \rho'_s$; and the size is $2^{(s+1)/2}$. There is only *one* irreducible orthogonal solution.

4.3. As in the unitary case, we consider the free Abelian groups generated by the irreducible orthogonal solutions τ_s, or τ_s, τ'_s and denote them by $D_s^{\mathcal{O}}$. They are $\cong \mathbb{Z}$ for $s \equiv 0,1,2,4,5,6$ modulo 8, and $\cong \mathbb{Z} \oplus \mathbb{Z}$ for $s \equiv 3,7$ modulo 8.

The *reduced orthogonal* H-R groups $D_s^{\mathcal{O}}/h(D_{s+1}^{\mathcal{O}})$ are denoted by $\mathcal{E}_s^{\mathcal{O}}$. They are easily computed from the list in 4.2 and the sizes of the τ_s, for $s \equiv 0,1,\ldots,7$ modulo 8:

$$
\begin{array}{ll}
s \equiv 0: & h(\tau_1) = h(\rho_1 + \bar{\rho}_1) = 2\rho_0 = 2\tau_0 \\
s \equiv 1: & h(\tau_2) = h(\rho_2 + \bar{\rho}_2) = \rho_1 + \rho'_1 + \bar{\rho}_1 + \bar{\rho}'_1 = 2\tau_1 \\
s \equiv 2: & h(\tau_3) = h(\rho_3 + \bar{\rho}_3) = \rho_2 + \bar{\rho}_2 = \tau_2 \\
s \equiv 3: & h(\tau_4) = h(\rho_4 + \bar{\rho}_4) = \rho_3 + \rho'_3 + \bar{\rho}_3 + \bar{\rho}'_3 = \tau_3 + \tau'_3 \\
s \equiv 4: & h(\tau_5) = h(\rho_5 + \bar{\rho}_5) = \rho_4 + \bar{\rho}_4 = \tau_4 \\
s \equiv 5: & h(\tau_6) = h(\rho_6) = \rho_5 + \rho'_5 = \tau_5 \\
s \equiv 6: & h(\tau_7) = h(\rho_7) = \rho_6 = \tau_6 \\
s \equiv 7: & h(\tau_8) = h(\rho_8) = \rho_7 + \rho'_7 = \tau_7 + \tau'_7.
\end{array}
$$

This yields

THEOREM 4.1.. *The reduced orthogonal H-R groups $\mathcal{E}_s^{\mathcal{O}}$ are $\mathbb{Z}/2$ for $s \equiv 0$ and 1 modulo 8; 0 for $s \equiv 2,4,5,6$; and $\mathbb{Z}$ for $s \equiv 3,7$.*

4.4. The explicit construction of "all" orthogonal H-R matrices is again based on the product defined in 3.3. It yields a product $D_s^{\mathcal{O}} \times D_t^{\mathcal{O}} \to D_{s+t+1}^{\mathcal{O}}$.

We first note that

$$\tau_s \cdot \tau_7 = \tau_{s+8}.$$

Indeed, as can be seen from the list in 4.2, the size of τ_{s+8} is $16n$ where n is the size of τ_s; and this is precisely the size of $\tau_s \cdot \tau_7$. We thus get all τ_s from those for $0 \leq s \leq 7$ by multiplying with powers of τ_7.

We can forget about $\tau_2, \tau_4, \tau_5, \tau_6$; omitting the last matrix in τ_3 yields τ_2, the last, the two last, and the three last matrices in τ_7 yields τ_6, τ_5, τ_4 respectively. τ_0 is the empty set of matrices of size 1, and $\tau_1 = \left\{\begin{pmatrix} & 1 \\ -1 & \end{pmatrix}\right\}$; note that this is $= \tau_0^2$. And

$\tau_8 = \tau_7 \cdot \tau_0 = \{ \begin{pmatrix} & A_j \\ -A_j & \end{pmatrix}, j = 1, \ldots, 7, \begin{pmatrix} & E_8 \\ -E_8 & \end{pmatrix} \}$ where the A_j are the matrices of τ_7.

As for τ_3 and τ_7, they are described by the well-known multiplication tables of the quaternions and the Cayley octaves respectively, as follows.

Let $\tau_3 = \{A_1, A_2, A_3\}$, $A_0 = E_4$, and $f(x) = \sum_0^3 x_j A_j$. Then

$$f(x) = \begin{pmatrix} x_0 & x_1 & x_2 & x_3 \\ -x_1 & x_0 & x_3 & -x_2 \\ -x_2 & -x_3 & x_0 & x_1 \\ -x_3 & x_2 & -x_1 & x_0 \end{pmatrix},$$

and $f(x)y$, where $y = (y_0, y_1, y_2, y_3)$, is the quaternion product of x and $y \in \mathbb{R}^4$.

Similarly, for $\tau_7 = \{A_1, \ldots, A_7\}$, $A_0 = E_8$,

$$f(x) = \begin{pmatrix} x_0 & x_1 & x_2 & x_3 & x_4 & x_5 & x_6 & x_7 \\ -x_1 & x_0 & x_3 & -x_2 & x_5 & -x_4 & -x_7 & x_6 \\ -x_2 & -x_3 & x_0 & x_1 & x_6 & x_7 & -x_4 & -x_5 \\ -x_3 & x_2 & -x_1 & x_0 & x_7 & -x_6 & x_5 & -x_4 \\ -x_4 & -x_5 & -x_6 & -x_7 & x_0 & x_1 & x_2 & x_3 \\ -x_5 & x_4 & -x_7 & x_6 & -x_1 & x_0 & -x_3 & x_2 \\ -x_6 & x_7 & x_4 & -x_5 & -x_2 & x_3 & x_0 & -x_1 \\ -x_7 & -x_6 & x_5 & x_4 & -x_3 & -x_2 & x_1 & x_0 \end{pmatrix}$$

The coefficient matrices of the x_j have, in both cases, entries $0, \pm 1$ only.

THEOREM 4.2. *All irreducible orthogonal solutions ρ_s are obtained from $\tau_0 = \varnothing$, $\tau_1 = \begin{pmatrix} & 1 \\ -1 & \end{pmatrix}$, and τ_3, τ_7 as above by multiplying with powers of τ_7 (and omitting matrices for $s \equiv 2, 4, 5, 6$ modulo 8).*

COROLLARY 4.3. *All orthogonal H-R matrices can be written with entries 0, ± 1 alone.*

REMARK 4.4. $\tau_0^2 = \tau_1$ and $\tau_3^2 = 4\tau_7$.

The second assertion follows from $\tau_3 = \rho_3 + \bar{\rho}_3 = 2\rho_3$ and $\tau_7 = \rho_7$.

REMARK 4.5. From the sizes of the minimal, i.e., irreducible orthogonal solutions τ_s for the values of s modulo 8 (see 4.2 above) one easily obtains the "Radon number" mentioned in 1.3, as follows.

We write $n = \text{odd} \cdot 2^{4\alpha+\beta}$, $\beta = 0, 1, 2, 3$, and look for the largest s such that the corresponding size is $2^{4\alpha+\beta}$. Since solutions for $s = 2, 4, 5, 6$ are obtained be *omitting* matrices we need only consider the other s.

$$\begin{array}{llll} \beta = 0: & s = 8\alpha & \text{yields} & 2^{s/2} = 2^{4\alpha} \\ \beta = 1: & s = 8\alpha + 1 & \text{yields} & 2^{(s+1)/2} = 2^{4\alpha+1} \\ \beta = 2: & s = 8\alpha + 3 & \text{yields} & 2^{(s+1)/2} = 2^{4\alpha+2} \\ \beta = 3: & s = 8\alpha + 7 & \text{yields} & 2^{(s-1)/2} = 2^{4\alpha+3}. \end{array}$$

In summary: The maximum number of orthogonal H-R matrices of size $n = \text{odd} \cdot 2^{4\alpha+\beta}$ is $8\alpha + 2^\beta - 1$.

5. The reduced H-R rings and Bott periodicity

5.1. In both the unitary and the orthogonal case properties of the *product* $\sigma_s \cdot \sigma_t = \sigma_{s+t+1}$ can be expressed in terms of the groups D_s and $\mathcal{E}_s$ (or $D_s^{\mathcal{O}}$ and $\mathcal{E}_s^{\mathcal{O}}$). It is easy to see that it is distributive relative to the sum $\sigma_s + \sigma_s'$ of solutions. It is a little harder to check that it is associative.

With regard to $h\colon D_{s+1} \to D_s$, it is immediate from the matrix list (5) of $\sigma_{s+1} \cdot \sigma_t$ or of $\sigma_s \cdot \sigma_{t+1}$ that

$$h(\sigma_{s+1}) \cdot \sigma_t = h(\sigma_{s+1} \cdot \sigma_t),$$
$$\sigma_s \cdot h(\sigma_{t+1}) = h(\sigma_s \cdot \sigma_{t+1}).$$

The product of representatives thus yields a product $\mathcal{E}_s \times \mathcal{E}_t \to \mathcal{E}_{s+t+1}$, or $\mathcal{E}_s^{\mathcal{O}} \times \mathcal{E}_t^{\mathcal{O}} \to \mathcal{E}_{s+t+1}^{\mathcal{O}}$ respectively. It is *graded* if $\mathcal{E}_s$ is given the degree $s+1$.

The direct sum $\mathcal{E}_* = \bigoplus_{s=-1}^{\infty} \mathcal{E}_s$, where we have added a term $\mathcal{E}_{-1} = \mathbb{Z}$ generated by the unit for the product, is a graded ring. We call it the *reduced unitary* H-R *ring*. Similarly we have a graded ring $\mathcal{E}_*^{\mathcal{O}} = \bigoplus_{s=-1}^{\infty} \mathcal{E}_s^{\mathcal{O}}$, the *reduced orthogonal* H-R *ring*. The discussion in Sections 3 and 4 gives the complete structure of these rings, as follows.

5.2. In the unitary case: $\mathcal{E}_s = 0$ for even s, and for odd s, $\mathcal{E}_s \cong \mathbb{Z}$ generated by ρ_s, where $\rho_{2k-1} = \rho_1^k$, $k = 1, 2, \ldots$.

THEOREM 5.1. *The reduced unitary H-R ring $\mathcal{E}_*$ is the polynomial ring $\mathbb{Z}[\rho_1]$ generated by $\rho_1 = \{(i)\}$.*

In the orthogonal case: $\mathcal{E}_s^{\mathcal{O}} = 0$ for $s \equiv 2, 4, 5, 6$ modulo 8. $\mathcal{E}_0^{\mathcal{O}} \cong \mathbb{Z}/2$ generated by τ_0, $\mathcal{E}_1^{\mathcal{O}} \cong \mathbb{Z}/2$ generated by $\tau_1 = \{\left(\begin{smallmatrix} & 1 \\ -1 & \end{smallmatrix}\right)\}$. $\mathcal{E}_3$ and $\mathcal{E}_7$ are $\cong \mathbb{Z}$ generated by τ_3 and τ_7 (see 4.4), and the product with τ_7 is an isomorphism $\mathcal{E}_s^{\mathcal{O}} \cong \mathcal{E}_{s+8}^{\mathcal{O}}$ for all s. In particular, τ_7 generates a polynomial subring $\mathbb{Z}[\tau_7]$ of $\mathcal{E}_*^{\mathcal{O}}$.

The ring $\mathcal{E}_*^{\mathcal{O}}$ is commutative. For products involving $\mathcal{E}_s^{\mathcal{O}}$, $s \equiv 2, 4, 5, 6$ modulo 8, there is nothing to prove. $\tau_0\tau_1 = \tau_1\tau_0 = \tau_0^3 = 0$ in $\mathcal{E}_2^{\mathcal{O}}$; $\tau_0\tau_3 = \tau_3\tau_0 = 0$; $\tau_0\tau_7 = \tau_7\tau_0 = \tau_8$ in $\mathcal{E}_8^{\mathcal{O}} = \mathbb{Z}/2$. Similarly $\tau_1\tau_3 = \tau_3\tau_1$ in $\mathcal{E}_4^{\mathcal{O}}$; $\tau_1\tau_7 = \tau_7\tau_1 = \tau_9$ in $\mathcal{E}_9^{\mathcal{O}} = \mathbb{Z}/2$. Finally $\tau_3\tau_7 = 4\tau_3^3 = \tau_7\tau_3$. We summarize:

THEOREM 5.2. *The reduced orthogonal H-R ring $\mathcal{E}_*^{\mathcal{O}}$ is the commutative ring, graded by $s+1$ for $\mathcal{E}_s^{\mathcal{O}}$, generated by three elements τ_0, τ_3, τ_7 with relations $2\tau_0 = 0$, $\tau_0^3 = 0$, $\tau_3^2 = 4\tau_7$.*

ADDENDA 5.3. $\tau_0^2 = \tau_1$; τ_7 generates the polynomial ring $\mathbb{Z}[\tau_7] \subset \mathcal{E}_*^{\mathcal{O}}$; the product with τ_7 is an isomorphism $\mathcal{E}_s^{\mathcal{O}} \cong \mathcal{E}_{s+8}^{\mathcal{O}}$; $\mathcal{E}_s = 0$ for $s \equiv 2, 4, 5, 6$ modulo 8.

5.3. All this is so reminiscent of the *Bott periodicity theorems* (see [**K**], for example) that there has to be an explicit relationship.

In its original form the Bott periodicity is about the homotopy groups of the infinite unitary group $U = \lim U(n)$ and the infinite orthogonal group $\mathcal{O} = \lim \mathcal{O}(n)$. The limit is understood with respect to the imbeddings $U(n) \to U(n+1)$, $\mathcal{O}(n) \to \mathcal{O}(n+1)$ by adding a 1 to the diagonal, $n = 1, 2, 3, \ldots$. The homotopy groups $\pi_s(U)$ are the "stable" homotopy groups of $U(n)$; if f is odd, $\pi_s\big(U(n)\big) \cong \pi_s\big(U((s+1)/2)\big)$ for $n \geq (s+1)/2$, and if s is even, $\cong \pi_s\big(U((s+2)/2)\big)$ for $n \geq (s+2)/2$. Similarly for $\pi_s(\mathcal{O})$. Bott proved in 1956 ff (see [**K**]) that the $\pi_s(U)$ are periodic with period 2, the $\pi_s(\mathcal{O})$ with period 8; and that $\pi_s(U) = \mathbb{Z}$ for odd s, $= 0$ for even s; and that $\pi_s(\mathcal{O}) \cong \mathbb{Z}/2$ for $s \equiv 0, 1$ modulo 8, $= 0$ for $s \equiv 2, 4, 5, 6$, $\cong \mathbb{Z}$ for $s \equiv 3, 7$.

Thus $\mathcal{E}_s \cong \pi_s(U)$ *and* $\mathcal{E}_s^{\mathcal{O}} \cong \pi_s(\mathcal{O})$ *for all* $s \geq 0$. Of course one wants to find a *map* which provides these isomorphisms. The natural choice is as follows.

Given a solution $\sigma_s = \{A_1, A_2, \ldots, A_s\}$, unitary or orthogonal, consider as earlier $f(x) = \sum_0^s x_j A_j$ with $A_0 = E$, $x = (x_0, \ldots, x_s) \in \mathbb{R}^{s+1}$, $\sum_0^s x_j^2 = 1$. Then f is a map of $S^s \subset \mathbb{R}^{s+1}$ into some $U(n)$ or $\mathcal{O}(n)$. Passing to U or $\mathcal{O}$ and to the homotopy class of f we get a map $\Phi\colon D_s \to \pi_s(U)$ or $\Psi\colon D_s \to \pi_s(\mathcal{O})$ respectively. If $\sigma_s = h(\sigma_{s+1})$ then f can be extended to $S^{s+1} \subset \mathbb{R}^{s+2}$ and is thus nullhomotopic. The maps $\mathcal{E}_s \to \pi_s(U)$ and $\mathcal{E}_s^{\mathcal{O}} \to \pi_s(\mathcal{O})$ thus obtained are again written Φ and Ψ.

Φ and Ψ are easily seen to be (additive) homomorphisms: A sum $\sigma_s + \sigma_s'$ becomes in U, e.g.,

$$\begin{pmatrix} f(x) & & & & & \\ & f'(x) & & & & \\ & & 1 & & & \\ & & & \ddots & & \\ & & & & 1 & \\ & & & & & \ddots \end{pmatrix}$$

which is homotopic to the matrix product of

$$\begin{pmatrix} f(x) & & \\ & 1 & \\ & & \ddots \end{pmatrix} \quad \text{and} \quad \begin{pmatrix} f'(x) & & \\ & 1 & \\ & & \ddots \end{pmatrix}$$

But the homotopy group addition, in the case of a group space, is by *multiplying* the values in the group.—That they are *isomorphisms* is easily checked for small values of s:

Case U. $s = 1 : \rho_1 = (i)$; $f(x) = (x_0 + ix_1) \in U(1)$ with $x_0^2 + x_1^2 = 1$. This is a generator a_1 of $\pi_1(U(1)) = \pi_1(U) = \mathbb{Z}$.

$$s = 3 : \rho_3 = \left\{ \begin{pmatrix} i & \\ & -i \end{pmatrix}, \begin{pmatrix} & 1 \\ -1 & \end{pmatrix}, \begin{pmatrix} & i \\ i & \end{pmatrix} \right\}; \; f(x) = \begin{pmatrix} x_0 + ix_1 & x_2 + ix_3 \\ -x_2 + ix_3 & x_0 - ix_1 \end{pmatrix}$$

with $x_0^2 + x_1^2 + x_2^2 + x_3^2 = 1$. $f\colon S^3 \to SU(2) = S^3$ is a generator of $\pi_3(SU(3)) = \pi_3(U(3)) = \pi_3(U) = \mathbb{Z}$.

Case $\mathcal{O}$. $s = 0 : \tau_0 = \varnothing$; $f(x_0) = (x_0)$ with $x_0^2 = 1$, $x_0 = \pm 1$. This is a generator b_0 of $\pi_0(\mathcal{O}(1)) = \pi_0(\mathcal{O}) = \mathbb{Z}/2$.

$$s = 1 : \tau_1 = \left\{ \begin{pmatrix} & 1 \\ -1 & \end{pmatrix} \right\}; \; f(x) = \begin{pmatrix} x_0 & x_1 \\ -x_1 & x_0 \end{pmatrix} \quad \text{with} \quad x_0^2 + x_1^2 = 1.$$

This is a generator of $\pi_1(S\mathcal{O}(2)) = \mathbb{Z}$; under $S\mathcal{O}(2) \to S\mathcal{O}(3) \to \mathcal{O}$ it becomes the generator b_1 of $\pi_1(S\mathcal{O}(3)) = \pi_1(\mathcal{O}) = \mathbb{Z}/2$.

$s = 3 : \tau_3$ has been described in 4.4 by a matrix $f(x)$, $x \in S^3 \subset \mathbb{R}^4$. The map $f\colon S^3 \to S\mathcal{O}(4)$ becomes, under $S\mathcal{O}(4) \to S\mathcal{O}(5)$ a generator b_3 of $\pi_3(S\mathcal{O}(5)) \cong \pi_3(\mathcal{O}) = \mathbb{Z}$.

For higher values of s it is not easy to establish directly the isomorphisms. This is due to the fact, in the case U for example, that ρ_s has the size $2^{(s-1)/2}$; one then obtains a map $f\colon S^s \to U(2^{(s-1)/2})$ while the usual generators of $\pi_s(U)$ are in $\pi_s\Big(U\big((s+1)/2\big)\Big)$, and $2^{(s-1)/2}$ is $> (s+1)/2$ for $s > 3$.

5.4. Using the ring structures the isomorphism proof becomes simple, as follows.

$\pi_*(U) = \bigoplus_{s=-1}^{\infty} \pi_s(U)$ (with a term $s = -1$ being $\mathbb{Z}$) is turned into a ring by exactly the same product formula (4) used for $\sigma_s \cdot \sigma_t$, which makes sense for continuous maps $f(x)$, $g(y)$; similarly for the case $\mathcal{O}$. Then trivially Φ and Ψ are ring homomorphisms

$$\Phi\colon \mathcal{E}_* \to \pi_*(U),$$

$$\Psi\colon \mathcal{E}_*^{\mathcal{O}} \to \pi_*(\mathcal{O}).$$

We have to determine the ring structure of $\pi_*(U)$ and $\pi_*(\mathcal{O})$. To this end we pass to (complex and real) *K-theory* and recall that $\pi_s(U) \cong \tilde{K}_{\mathbb{C}}(S^{s+1})$ and $\pi_s(\mathcal{O}) \cong \tilde{K}_{\mathbb{R}}(S^{s+1})$. As shown in [**E2**] the product (4) for elements of the homotopy groups $\pi_s(U)$, or $\pi_s(\mathcal{O})$ respectively, corresponds under these isomorphisms to the exterior tensor product of stable vector bundles, i.e., of elements of $\tilde{K}_{\mathbb{C}}(S^{s+1})$ or of $\tilde{K}_{\mathbb{R}}(S^{s+1})$. With regard to this product the complete K-theoretic formulation of Bott periodicity yields the ring structure of $\bigoplus_{-1}^{\infty} K_{\mathbb{C}}(S^{s+1}) = \pi_*(U)$, and similarly for $K_{\mathbb{R}}$ and $\pi_*(\mathcal{O})$. Namely (see [**K**]):

$\pi_*(U)$ is the polynomial ring $\mathbb{Z}[a_1]$ in the generator $a_1 \in \pi_1(U)$. And $\pi_*(\mathcal{O})$ is the (graded) commutative ring generated by the generators $b_0 \in \pi_0(\mathcal{O})$, $b_3 \in \pi_3(\mathcal{O})$, and $b_7 \in \pi_7(\mathcal{O})$ with only relations $2b_0 = 0$, $b_0^3 = 0$, $b_3^2 = 4b_7$.

Thus all we have to prove is that $\Phi(\rho_1) = a_1$ and $\Psi(\tau_0) = b_0$, $\Psi(\tau_3) = b_3$. But this has been done above.

THEOREM 5.7. *The maps $\Phi\colon \mathcal{E}_* \to \pi_*(U)$ and $\Psi\colon \mathcal{E}_*^{\mathcal{O}} \to \pi_*(\mathcal{O})$ are ring isomorphisms.*

COROLLARY 5.8. *For all s, $\Phi(\rho_s)$ is a generator of $\pi_s(U)$ and $\Psi(\tau_s)$ is a generator of $\pi_s(\mathcal{O})$. In other words, these generators of $\pi_s(U)$ and $\pi_s(\mathcal{O})$ are given by maps $S^s \to U$ or $S^s \to \mathcal{O}$ which are linear with respect to the coordinates of $S^s \subset \mathbb{R}^{s+1}$, the coefficient matrices being E and the minimal size H-R solutions.*

COROLLARY 5.9. *(a) Any continuous map $S^s \to U$ is homotopic to a linear map $f\colon S^s \to U(n) \to U$, for suitable n, of the form $f(x) = \sum_0^s x_j A_j$, $x = (x_0, x_1, \ldots, x_s) \in S^s \subset \mathbb{R}^{s+1}$ with $A_0 = E$ and where $A_1, \ldots, A_s$ are H-R matrices of size n.*

(b) If a linear map $f\colon S^s \to U(n)$ is nullhomotopic in U then it is linearly nullhomotopic, i.e., restricted from a linear map $S^{s+1} \to U(n)$.

Similarly for maps $S^s \to \mathcal{O}$.

PROBLEM. Prove Corollary 5.9 directly, i.e., without using Bott periodicity. This would reduce the proof of the Theorems of Bott to the algebra of H-R matrices as described in this survey.

References

[E] B. Eckmann, *Gruppentheoretischer Beweis des Satzes von Hurwitz-Radon über die Komposition quadratischer Formen*, Comment. Math. Helvetici **15** (1942/43), 358–366.

[E2] ——, *Hurwitz-Radon matrices and periodicity modulo* 8, L'Ens. Math. **35** (1989), 77–91.

[H] A. Hurwitz, *Über die Komposition der quadratischen Formen*, Math. Ann. **88** (1923), 1–25.

[LS] T. Y. Lam and T. Smith.
[K] M. Karoubi, *K-Theory*, Springer-Verlag, Berlin-Heidelberg-New York, 1978.
[R] J. Radon, *Lineare Scharen orthogonaler Matrizen*, Abh. Math. Sem. Hamburg (1922), 1–14.

Mathematik, ETH-Zentrum, 8092 Zurich, Switzerland
E-mail address: eckmann@math.ethz.ch

Centre de Recherches Mathématiques
CRM Proceedings and Lecture Notes
Volume 6, 1994

The First Order Euler Characteristic

Ross Geoghegan and Andrew Nicas

ABSTRACT. We introduce a new invariant of finite complexes, the "first order Euler characteristic". We compute it in a variety of situations and give some applications to geometric group theory.

From our point of view, the classical Euler characteristic is "zero-th order". We are going to describe a higher order analog which we call the "first order Euler characteristic". Indeed, for all $n \geq 0$ there is an n-th order version, but that will be treated elsewhere.

This paper is a preliminary report on our forthcoming paper [**GN4**]. That work relies heavily on earlier work in our paper [**GN1**]. Therefore, in order to give the reader a reasonably self-contained account, we include here (in §3 and §6) an exposition of parts of [**GN1**].

We particularly appreciate Peter Hilton's interest in and encouragement of our recent work, [**GN1, GN2, GN3, GN4**]. We also thank Boris Okun for pointing out the relevance of [**Kn**].

1. Definitions and main theorems

Recall three definitions of the classical Euler characteristic, $\chi(X)$, of a finite complex X.

DEFINITION A_0. $\chi(X) = \sum_{k\geq 0}(-1)^k$ number of k-cells in X.

DEFINITION B_0. $\chi(X;R) = \sum_{k\geq 0}(-1)^k \operatorname{rank}_R H_k(X;R)$ where R is a principal ideal domain. (This integer is independent of R.)

When X is an oriented manifold, M, we also have:

DEFINITION C_0. $\chi(M)$ = intersection number of the graph of the identity map of M with itself.

In this paper we introduce a higher order analog which we call the first order Euler characteristic of X. There will be three analogous definitions, labelled A_1, B_1, and C_1 corresponding to A_0, B_0 and C_0. We indicate in §6 and §7 why, under appropriate hypotheses, these new definitions are equivalent.

First, we establish some notation. Let X be a finite connected CW complex with base vertex v. Write $G \equiv \pi_1(X,v)$ and $\Gamma \equiv \pi_1(X^X, \mathrm{id})$ where X^X is the function space of all continuous maps $X \to X$. Each $\gamma \in \Gamma$ can be represented by a cellular homotopy $F^\gamma \colon X \times I \to X$ such that $F_0^\gamma = F_1^\gamma = \mathrm{id}_X$. Orient the cells

1991 *Mathematics Subject Classification.* Primary 55M20; Secondary 19D55, 20F28, 20F32.
The first author was supported in part by NSF Grant #DMS-9005508.
The second author was supported in part by NSERC Grant #OGP0038057.
The final version of this paper will be submitted for publication elsewhere.

of X, thus establishing a preferred basis for the integral cellular chains $(C_*(X), \partial)$. Choose a lift, $\tilde{e}$, in the universal cover, $\widetilde{X}$, for each cell e of X, and orient $\tilde{e}$ compatibly with e. Regard the cellular chain complex $(C_*(\widetilde{X}), \widetilde{\partial})$ as a free right $\mathbb{Z}G$-module chain complex with preferred basis $\{\tilde{e}\}$. Let $D_*^\gamma\colon C_*(X) \to C_{*+1}(X)$ be the chain homotopy induced by F^γ.

Sign Convention. If e is an oriented k-cell of X then $D_k^\gamma(e)$ is the $(k+1)$-chain $(-1)^{k+1}F_*^\gamma(e \times I) \in C_{k+1}(X)$, where $e \times I$ is given the product orientation.

Let R be a commutative ring. Regard ${}_R\widetilde{\partial}_k \equiv \widetilde{\partial}_k \otimes \mathrm{id}\colon C_k(\widetilde{X}) \otimes R \to C_{k-1}(\widetilde{X}) \otimes R$ and ${}_RD_k^\gamma \equiv D_k^\gamma \otimes \mathrm{id}\colon C_k(X) \otimes R \to C_{k+1}(X) \otimes R$ as matrices over RG and R respectively using the the preferred bases. Let G_{ab} be the Abelianization of G, i.e. the quotient of G by its commutator subgroup. The natural homomorphism ("Abelianization") $A\colon G \to G_{\mathrm{ab}} \cong H_1(X)$ extends to a homomorphism of R-modules $A\colon RG \to H_1(X; R) = H_1(X) \otimes R$.

We can now state the first definition of our *first order Euler characteristic* with coefficients in a commutative ring R. It is a homomorphism $\chi_1(X; R)\colon \Gamma \to H_1(X; R)$. When $R = \mathbb{Z}$ we write, in abbreviated form, $\chi_1(X)\colon \Gamma \to H_1(X)$. Note that Γ is Abelian, and when X is aspherical, $\Gamma \cong Z(G)$, the center of G; see Proposition 2.1.

Definition A_1. Let R be a commutative ring of coefficients. Then

$$\chi_1(X; R)(\gamma) = \sum_{k \geq 0} (-1)^{k+1} A\left(\mathrm{trace}({}_R\widetilde{\partial}_{k+1}\, {}_RD_k^\gamma)\right).$$

Here, we are multiplying RG-matrices by R-matrices to obtain RG-matrices. Note that $\chi_1(X; R)(\gamma) = \chi_1(X)(\gamma) \otimes 1$. This formula is independent of the various choices that have been made; see §5. Note that in order to know the right hand side, we must have information at the chain level, namely the matrices ${}_R\widetilde{\partial}_{k+1}$ and ${}_RD_k^\gamma$.

Our second definition requires the assumption that $H_*(X; R)$ be a free R-module where R is a principal ideal domain. This will be true, for example, if R is a field. For each $k \geq 0$, choose a basis $\{b_1^k, \ldots, b_{\beta_k}^k\}$ for $H_k(X; R)$. Let $\{\bar{b}_j^k\}$ be the corresponding dual basis for $H^k(X; R)$. Let $\Phi^\gamma\colon X \times S^1 \to X$ be the obvious quotient obtained from F^γ, above. By means of the Künneth formula, Φ^γ induces $\Phi_*^\gamma\colon H_k(X; R) \otimes H_1(S^1; R) \to H_{k+1}(X; R)$. Let $u \in H_1(S^1; R)$ be the generator which defines the usual orientation on S^1.

Definition B_1. Let R be a principal ideal domain. Suppose that $H_*(X; R)$ is a free R-module. Then

$$\chi_1(X; R)(\gamma) = \sum_{k \geq 0} (-1)^{k+1} \sum_j \bar{b}_j^k \cap \Phi_*^\gamma(b_j^k \otimes u)$$

where $\cap$ is the cap product in the sense of Dold's textbook.

It is straightforward to show that the formula in Definition B_1 is independent of the choice of basis for $H_*(X; R)$.

The above expression for $\chi_1(X; R)(\gamma)$ can also be written:

$$\chi_1(X; R)(\gamma) = \sum_{i=1}^{\beta_1} \sum_{k,j} (-1)^{k+1} \langle \bar{b}_i^1 \cup \bar{b}_j^k, \Phi_*^\gamma(b_j^k \otimes u) \rangle b_i^1$$

where $\langle \cdot, \cdot \rangle$ denotes the Kronecker pairing. A trace formula of this kind, for parametrized maps $X \times Y \to X$ was introduced by R. J. Knill in [**Kn**]. In order to know the right hand side in Definition B_1, we only need homological information about Φ^γ and cup product information about $H^*(X; R)$.

Our third definition, Definition C_1 below, is an analog of the geometric Definition C_0 of $\chi(X)$. Let M be a compact oriented smooth (or PL) manifold with boundary. The *fixed point set* of F^γ is $\mathrm{Fix}(F^\gamma) \equiv \{(x,t) \mid F^\gamma(x,t) = x\}$, i.e. the coincidence set of F^γ and the projection $p\colon M \times I \to M$. As before, we form $\Phi^\gamma\colon M \times S^1 \to M$. We may perturb Φ^γ to a smooth (or PL) map Ψ^γ whose graph meets the graph of the projection p transversely. Then $\mathrm{Fix}(\Psi^\gamma) \equiv \{(x,t) \mid \Psi^\gamma(x,t) = x\}$ is a closed 1-manifold which naturally carries the "intersection orientation", using the order (graph of p, graph of Ψ^γ), as explained in §6. This oriented 1-manifold defines an integral 1-cycle, $U(\gamma)$, in $X \times S^1$. For any commutative ring R of coefficients, we define the *intersection class*, $\theta_R(\gamma) \in H_1(M; R)$, to be the image of the homology class represented by the cycle $U(\gamma) \otimes 1$ under $p_*\colon H_1(M \times S^1; R) \to H_1(M; R)$. When $R = \mathbb{Z}$ we write $\theta_{\mathbb{Z}}(\gamma) = \theta(\gamma)$.

Definition C_1. Let R be a commutative ring of coefficients. Then

$$\chi_1(M; R)(\gamma) = -\theta_R(\gamma).$$

Definitions A_1, B_1 and C_1 define homomorphisms $\Gamma \to H_1(X; R)$ which are related as follows:

Theorem 1.1 (Equivalence).
(i) *When R is a principal ideal domain and $H_*(X; R)$ is a free R-module, Definitions A_1 and B_1 agree;*
(ii) *when X is an oriented manifold and R is any commutative coefficient ring, Definitions A_1 and C_1 agree.*

We will explain in Remark 5.9 why Definition A_1 is a "higher order" version of Definition A_0 (this requires a discussion of Hochschild homology). On the other hand, the reader can probably see that Definition B_1, involving cup products with 1-dimensional classes, is analogous to Definition B_0 which can be thought of as similarly involving a cup product with a 0-dimensional class, the collection of maps Φ^γ being replaced by the identity map of X. (The obvious analogy between Definitions C_0 and C_1 should allay any remaining doubts.) The agreement of Definitions A_1 and B_1 should be viewed as a "higher" analog of the classical Hopf trace formula which asserts that a chain level trace is equal to a trace on homology.

Let $h\colon X \to Y$ be a homotopy equivalence where Y is a finite CW complex. Let $h^{-1}\colon Y \to X$ be a homotopy inverse for h. Then the map $h_\#\colon X^X \to Y^Y$ given by $f \mapsto hfh^{-1}$ is a homotopy equivalence. In particular, $h_\#$ induces an isomorphism $(h_\#)_*\colon \Gamma \cong \Gamma' \equiv \pi_1(Y^Y, \mathrm{id})$. The assertion that $\chi_1(X; R)$ is a "homotopy invariant" means that the diagram:

$$\begin{array}{ccc} \Gamma & \xrightarrow{\chi_1(X;R)} & H_1(X;R) \\ {\scriptstyle (h_\#)_*}\downarrow & & \downarrow{\scriptstyle h_*} \\ \Gamma' & \xrightarrow{\chi_1(Y;R)} & H_1(Y;R) \end{array}$$

is commutative. Note that the vertical arrows are isomorphisms.

Theorem 1.2. *$\chi_1(X; R)$ is a homotopy invariant.*

Theorem 1.2 allows one to extend the definition of $\chi_1(X; R)$ to any topological space X which is homotopy equivalent to a finite complex.

Here is a summary of what follows. In the first part of the paper, §§3–7, we sketch our proofs of Theorems 1.1 and 1.2. Then we exhibit a variety of computations of $\chi_1(X)$ in §§8–10. We end with some applications to geometric group theory in §§11–12.

In more detail, the study of Definition A_1 leads us to a more sophisticated "lift" of $\chi_1(X)$ to the universal cover. As explained in §5, when we regard $\chi_1(X)\colon \Gamma \to H_1(G)$ as an element of $H^1(\Gamma, H_1(G))$, we are led to a cohomology class $\widetilde{\chi}_1(X) \in H^1(\Gamma, HH_1(\mathbb{Z}G))$, where $HH_1(\mathbb{Z}G)$ is the Hochschild homology of the groupring $\mathbb{Z}G$. Under a natural homomorphism, $\widetilde{\chi}_1(X)$ is mapped to $\chi_1(X)$, but $\widetilde{\chi}_1(X)$ carries more information (e.g. $\widetilde{\chi}_1(X)$ may be non-zero when $\chi_1(X) = 0$, see Example 8.2). By contrast, the "zero-th" order analog of this is rather uninteresting (see §5), for the classical Euler characteristic can be calculated equally from the the cellular chain complex of X or from the cellular $\mathbb{Z}G$ complex of the universal cover $\widetilde{X}$.

We should not expect to express $\widetilde{\chi}_1(X)$ in the language of Definition B_1, but we state a precise version of Definition C_1 for $\widetilde{\chi}_1(X)$ in §6. Whenever possible, we compute $\widetilde{\chi}_1(X)$ as well as $\chi_1(X)$.

Definitions A_1 and B_1 are both useful. Using Definition B_1, we compute $\chi_1(X)$ when X is the total space of an orientable Serre fibration with circle fiber. Using Definition A_1, we compute $\chi_1(X)$ when X is the mapping torus of a homotopy equivalence. As might be expected, the results are usually simpler and more elegant for $\chi_1(X; \mathbb{Q})$.

The Euler characteristic of suitable groups has been profitably studied in geometric group theory; see [**Bass, B, Eck, Got, Stall**]. We define the analogous $\chi_1(G)$ in §2. In a number of our computed examples, X is, in fact, a finite aspherical complex, yielding a variety of examples of $\chi_1(G)$. In addition, we obtain some general group theoretic theorems:

THEOREM 11.4. *Suppose G is a group of type $\mathcal{F}$* (i.e. *$K(G,1)$ is homotopy equivalent to a finite complex*) *and satisfies Bass's Weak Conjecture over $\mathbb{Q}$. If $\chi_1(G; \mathbb{Q}) \neq 0$, then the center of G is infinite cyclic.*

This should be viewed as a "first order" version of Gottlieb's theorem [**Got**] which asserts: $\chi(G) \neq 0$ implies the center of G is trivial.

Another generalization of Gottlieb's theorem is:

THEOREM 12.1. *Let G be as in the previous theorem and let $\theta\colon G \to G$ be a "special" automorphism whose order in the group of outer automorphisms of G is $r > 1$. If the sum of the Lefschetz numbers $\sum_{i=1}^{r-1} L(\theta^i)$ is non-zero, then θ moves every non-trivial element of the center of G* (i.e. $Z(G) \cap \operatorname{Fix}(\theta) = (1)$).

Here, $\operatorname{Fix}(\theta) = \{g \in G \mid \theta(g) = g\}$ is the subgroup of G fixed by θ. Such an automorphism θ is "special" whenever $\theta^r(\cdot) = g(\cdot)g^{-1}$ where some positive power of g lies in $Z(G)\operatorname{Fix}(\theta)$ (see Definition 10.5 and the remark which follows it).

In a future paper we hope to study the role of the first order Euler characteristic in differential geometry.

2. The first order Euler characteristic of a group

As in §1, let X be a connected CW complex with base vertex v. Let $\mathcal{E}(X) \subset X^X$ be the subset of self-homotopy equivalences of X and $\mathcal{E}(X, v) \subset \mathcal{E}(X)$ consist of those homotopy equivalences which fix v. There is an *evaluation fibration* $\mathcal{E}(X, v) \hookrightarrow \mathcal{E}(X) \xrightarrow{\eta} X$, where $\eta(f) = f(v)$. The homotopy exact sequence of this fibration yields the exact sequence:

$$\pi_1\big(\mathcal{E}(X, v), \mathrm{id}\big) \longrightarrow \Gamma \xrightarrow{\eta_\#} G \longrightarrow \pi_0\big(\mathcal{E}(X, v)\big) \longrightarrow \pi_0\big(\mathcal{E}(X)\big)$$

where $\Gamma \equiv \pi_1(X^X, \mathrm{id}) = \pi_1\big(\mathcal{E}(X), \mathrm{id}\big)$ and $G = \pi_1(X, v)$.

DEFINITION. The group $\mathcal{G}(X) \equiv \eta_\#(\Gamma)$ is called the *Gottlieb subgroup* of G.

Gottlieb showed ([**Got**, Theorem I.4]) that the image of $\eta_\#$ lies in the subgroup consisting of those elements of G which act trivially on $\pi_n(X, v)$, for all $n \geq 1$; in particular, $\mathcal{G}(X) \subset Z(G)$, the center of G.

PROPOSITION 2.1 ([**Got**, Theorem III.2]). *If X is aspherical then $\eta_\#\colon \Gamma \to Z(G)$ is an isomorphism.*

COROLLARY 2.2 ([**Got**, Corollary I.13]). *If X is aspherical then $\mathcal{G}(X) = Z(G)$.*

In particular, if X is aspherical then Γ, $\mathcal{G}(X)$ and $Z(G)$ coincide. The example of $X = S^2$ shows that this is not true for an arbitrary finite complex.

A group G *is of type* $\mathcal{F}$ if there exists a $K(G, 1)$ which is a finite complex. By Theorem 1.2, the first order Euler characteristic is a homotopy invariant. In particular, applying these definitions to any finite $K(G, 1)$ complex we obtain the *first order Euler characteristic of the group G of type $\mathcal{F}$*. For any commutative ring R of coefficients, it is a homomorphism $\chi_1(G; R)\colon Z(G) \to G_{\mathrm{ab}} \otimes R$.

PROPOSITION 2.3. *Let G be of type $\mathcal{F}$. If $\chi(G) \neq 0$ then $\chi_1(G; R)$ is trivial for any coefficient ring R.*

PROOF. The center, $Z(G)$, is trivial, by [**Got**, Theorem IV.1]. Indeed, a short proof of this fact is included below as Proposition 5.3. $\square$

Most of §§3–10 applies to all finite complexes. We take up the special case of groups of type $\mathcal{F}$ again in §§10–12.

3. An informal discussion of one-parameter fixed point theory

The formula appearing in Definition A_1 first occurs in our paper [**GN1**] in which we developed an analog of Nielsen fixed point theory for homotopies. Roughly, such a theory should have the following features:

1. It should apply to a homotopy $F\colon X \times I \to X$, where X is a finite complex, just as Nielsen fixed point theory applies to a map $f\colon X \to X$.
2. Defining $\mathrm{Fix}(F) = \big\{(x, t) \mid F(x, t) = x\big\}$ and $\mathrm{Fix}(f) = \big\{x \mid f(x) = x\big\}$, the new theory should detect $\mathrm{Fix}(F)$ by means of a "trace" computed at the level of chains and/or homology, just as the Lefschetz number detects $\mathrm{Fix}(f)$.
3. In the case of an oriented manifold, M, when the map F is in transverse general position, $\mathrm{Fix}(F)$ is an oriented 1-manifold, and defines a 1-cycle on $M \times I$, which the "trace" should detect. Hence this "trace" should take values in $H_1(M)$ and so, in the case of an arbitrary finite complex X, it should lie in $H_1(X)$. This analogy is valid because, geometrically, $L(f) \in H_0(M)$.
4. There should be a parallel, deeper theory making full use the fundamental group, G, and partitioning $\mathrm{Fix}(F)$ into fixed point classes. This "deeper trace", involving the groupring $\mathbb{Z}G$, should decompose into separate components each of which detects a fixed point class. In Nielsen fixed point

theory the analogous invariant, $R(f)$, variously known as the *Reidemeister trace* or the *generalized Lefschetz number* of f, is the principal invariant of that theory and has these properties.

5. This "deeper trace" should map to the "trace" in a natural manner, just as the sum of the coefficients in $R(f)$ is $L(f)$.

Our paper [**GN1**] presents a theory having all these features. A summary can be found in §2 of [**GN2**].

Just as the Euler characteristic is the Lefschetz number of the identity, it should not be surprising that a special case of our "trace" deserves to be called a first order Euler characteristic. And just as the classical Euler characteristic is algebraic, having features not obviously related to fixed point theory, the same is true of the first order invariant. Indeed, knowledge of [**GN1**] is only needed in §6 of this paper.

In preparation for §6, we include here a more precise version of (4) above.

Let $F\colon X \times I \to X$ where X is a finite complex. Assume F induces the identity on the fundamental group G. Two fixed points (x,t) and (y,t') of F are *in the same fixed point class* if and only if for some path ν from (x,t) to (y,t'), the loop $(p \circ \nu)(F \circ \nu)^{-1}$ is homotopically trivial. This defines an equivalence relation on the set of fixed points, $\operatorname{Fix}(F)$. Just as in the classical case, there is an injective function Φ from the set of fixed point classes of F to the set, G_1, of conjugacy classes in G: the class containing (x,t) is mapped to the conjugacy class containing $[(p\circ\mu)(F\circ\mu)^{-1}\tau^{-1}]$, where μ is any path from the basepoint $(v,0)$ to (x,t), and τ is a basepath from v to $F(v,0)$. It is straightforward to check that Φ is well-defined, that F has only finitely many fixed point classes, and that fixed points in the same path component of $\operatorname{Fix}(F)$ are in the same fixed point class.

Choose a lift $\tilde{e}$ for each cell e of X, as before, with compatible orientations; choose $\tilde{e} \times I$ as the lift of $e \times I$. If (x,t) is a fixed point of F, then $\widetilde{F}(\tilde{x},t) = \tilde{x}g$ for some $g \in G$. It is routine to show:

PROPOSITION 3.1. *Under the injection Φ described above, the fixed point class of F containing (x,t) is mapped to the conjugacy class containing g.*

4. Hochschild homology

To better explain where Definition A_1 comes from, we must review some basic facts about Hochschild homology.

Let R be a commutative ground ring and let S be an associative R-algebra with unit. If M is an S-S bimodule (i.e. a left and right S-module satisfying $(s_1m)s_2 = s_1(ms_2)$ for all $m \in M$, and $s_1, s_2 \in S$), the *Hochschild chain complex* $\{C_*(S,M), d\}$ consists of $C_n(S,M) = S^{\otimes n} \otimes M$ where $S^{\otimes n}$ is the tensor product of n copies of S and

$$\begin{aligned} d(s_1 \otimes \cdots \otimes s_n \otimes m) &= s_2 \otimes \cdots \otimes s_n \otimes ms_1 \\ &\quad + \sum_{i=1}^{n-1} (-1)^i s_1 \otimes \cdots \otimes s_i s_{i+1} \otimes \cdots \otimes s_n \otimes m \\ &\quad + (-1)^n s_1 \otimes \cdots \otimes s_{n-1} \otimes s_n m. \end{aligned}$$

The tensor products are taken over R. The n-th homology of this complex is the n-th *Hochschild homology of S with coefficient bimodule M*. It is denoted by $HH_n(S,M)$. If $M = S$ with the standard S-S bimodule structure then we write $HH_n(S)$ for $HH_n(S,M)$.

We will be concerned mainly with HH_1 and HH_0 which are computed from

$$\begin{array}{ccccc} \cdots \longrightarrow S\otimes S\otimes M & \xrightarrow{d} & S\otimes M & \xrightarrow{d} & M \\ s_1\otimes s_2\otimes m & \mapsto & s_2\otimes ms_1 - s_1s_2\otimes m + s_1\otimes s_2m & & \\ & & s\otimes m & \mapsto & ms - sm. \end{array}$$

Next, we consider traces in Hochschild homology. If A is a square matrix over M, we interpret its trace $\sum_i A_{ii}$ as an element of M (i.e. as a Hochschild 0-cycle). The corresponding homology class is denoted by $T_0(A) \in HH_0(S,M)$. When A is idempotent, $T_0(A)$ is called the *Hattori-Stallings trace* of A.

If A^i, $i = 1, \dots, n$, are $q_i \times q_{i+1}$ matrices over S and B is a $q_{n+1} \times q_1$ matrix over M, we define $A^1 \otimes \cdots \otimes A^n \otimes B$ to be the $q_1 \times q_1$ matrix with entries in the R-module $S^{\otimes n} \otimes M$ given by

$$(A^1 \otimes \cdots \otimes A^n \otimes B)_{ij} = \sum_{k_2,\dots,k_{n+1}} A^1_{i,k_2} \otimes A^2_{k_2,k_3} \otimes \cdots \otimes A^n_{k_n,k_{n+1}} \otimes B_{k_{n+1},j}.$$

The *trace* of $A^1 \otimes \cdots \otimes A^n \otimes B$, written $\mathrm{trace}(A^1 \otimes \cdots \otimes A^n \otimes B)$, is

$$\sum_{k_1,k_2,\dots,k_{n+1}} A^1_{k_1,k_2} \otimes A^2_{k_2,k_3} \otimes \cdots \otimes A^n_{k_n,k_{n+1}} \otimes B_{k_{n+1},k_1}$$

which we interpret as a Hochschild n-chain. Observe that the 1-chain $\mathrm{trace}(A\otimes B)$ is a cycle if and only if $\mathrm{trace}(AB) = \mathrm{trace}(BA)$, in which case we denote its homology class by $T_1(A \otimes B) \in HH_1(S,M)$. In the application below, S will be a groupring over the ground ring R and $M = S$.

We will use the notation G_1 for the set of conjugacy classes of a group G. The partition of G into the union of its conjugacy classes induces a direct sum decomposition of $HH_*(\mathbb{Z}G)$ as follows: each generating chain $c = g_1 \otimes \cdots \otimes g_n \otimes m$ can be written in *canonical form* as $g_1 \otimes \cdots \otimes g_n \otimes g_n^{-1} \cdots g_1^{-1} g$ where we think of $g = g_1 \cdots g_n m \in G$ as "marking" a conjugacy class. All the generating chains occurring in the boundary $d(c)$ are easily seen to have markers in $C(g)$ when put into canonical form. For $C \in G_1$ let $C_*(\mathbb{Z}G)_C$ be the subgroup of $C_*(\mathbb{Z}G)$ generated by those generating chains whose markers lie in C. The decomposition $\mathbb{Z}G \cong \bigoplus_{C\in G_1} \mathbb{Z}C$ as a direct sum of Abelian groups determines a decomposition of chain complexes $C_*(\mathbb{Z}G) \cong \bigoplus_{C\in G_1} C_*(\mathbb{Z}G)_C$. There results a natural isomorphism $HH_*(\mathbb{Z}G) \cong \bigoplus_{C\in G_1} HH_*(\mathbb{Z}G)_C$ where the summand $HH_*(\mathbb{Z}G)_C$ corresponds to the homology classes of Hochschild cycles marked by the elements of C. We call this summand the *C-component*. Given any $\mathbb{Z}G$-$\mathbb{Z}G$ bimodule N let $\overline{N}$ be the left $\mathbb{Z}G$ module whose underlying Abelian group is N and whose left module structure is given by $gm = g \cdot m \cdot g^{-1}$. There is a natural isomorphism $HH_*(\mathbb{Z}G, N) \cong H_*(G, \overline{N})$ which is induced from an isomorphism of the Hochschild complex to the bar complex for computing group homology; see [**I**, Theorem 1.d]. The decomposition $\overline{\mathbb{Z}G} \cong \bigoplus_{C\in G_1} \mathbb{Z}C$ is a direct sum of left $\mathbb{Z}G$ modules, inducing a direct sum decomposition $H_*(G, \overline{\mathbb{Z}G}) \cong \bigoplus_{C\in G_1} H_*(G, \mathbb{Z}C)$. Choosing representatives $g_C \in C$ we have an isomorphism of left $\mathbb{Z}G$ modules $\mathbb{Z}C \cong \mathbb{Z}\big(G/Z(g_C)\big)$ where $Z(h) = \{g \in G \mid h = ghg^{-1}\}$ denotes the centralizer of $h \in G$. Since $H_*\big(G, \mathbb{Z}(G/Z(g_C))\big)$ is naturally isomorphic to $H_*\big(Z(g_C)\big)$, we obtain a natural isomorphism $HH_*(\mathbb{Z}G) \cong \bigoplus_{C\in G_1} H_*\big(Z(g_C)\big)$; furthermore, $HH_*(\mathbb{Z}G)_C$ corresponds to the summand $H_*\big(Z(g_C)\big)$ under this identification. In particular

$HH_0(\mathbb{Z}G) \cong \mathbb{Z}G_1$, the free Abelian group generated by the conjugacy classes, and $HH_1(\mathbb{Z}G) \cong \bigoplus_{C\in G_1} H_1(Z(g_C))$, the direct sum of the Abelianizations of the centralizers. Indeed, if $g \otimes g^{-1}g_C$ is a cycle then its homology class in $HH_1(\mathbb{Z}G)$ corresponds to $\{g\} \in H_1\big(Z(g_C)\big)$.

The augmentation $\varepsilon\colon \mathbb{Z}G \to \mathbb{Z}$ can be viewed as a morphism of $\mathbb{Z}G$-$\mathbb{Z}G$ bimodules, where $\mathbb{Z}$ is given the trivial bimodule structure, or as a morphism $\varepsilon\colon \overline{\mathbb{Z}G} \to \overline{\mathbb{Z}}$ of left $\mathbb{Z}G$-modules. Then there is an induced chain map $C_*(\mathbb{Z}G,\mathbb{Z}G) \xrightarrow{\varepsilon} C_*(\mathbb{Z}G,\mathbb{Z})$ and a commutative diagram:

$$\begin{array}{ccc} HH_*(\mathbb{Z}G,\mathbb{Z}G) & \xrightarrow{\ \varepsilon\ } & HH_*(\mathbb{Z}G,\mathbb{Z}) \\ {\scriptstyle\mu}\downarrow & & {\scriptstyle\mu}\downarrow \\ H_*(G,\overline{\mathbb{Z}G}) & \xrightarrow{\ \varepsilon\ } & H_*(G,\overline{\mathbb{Z}}) \end{array}$$

where the vertical arrows are isomorphisms

Recall the Abelianization homomorphism $A\colon \mathbb{Z}G \to G_{\text{ab}} = H_1(X) = H_1(G)$ used in Definition A_1.

Proposition 4.1. *If $\sum_i c_i \otimes n_i \in C_1(\mathbb{Z}G,\mathbb{Z})$ is a Hochschild 1-cycle representing $z \in HH_1(\mathbb{Z}G,\mathbb{Z})$, where $c_i \in \mathbb{Z}G$ and $n_i \in \mathbb{Z}$, then $\mu(z) = \sum_i A(c_i n_i) \in H_1(G)$.*

Proof. This follows from the fact that $d\colon \mathbb{Z}G \otimes \mathbb{Z}G \otimes \mathbb{Z} \to \mathbb{Z}G \otimes \mathbb{Z}$ becomes $g_1 \otimes g_2 \otimes 1 \mapsto (g_2 - g_1g_2 + g_1) \otimes 1$. One easily shows that the map $g \otimes 1 \mapsto A(g)$ induces μ. □

Corollary 4.2. *If $g \otimes g^{-1}g_C$ is a Hochschild 1-cycle then the μ-image of its homology class in $H_1(G) \equiv G_{\text{ab}}$ is $\{g\}$.*

5. Discussion of Definition $\mathbf{A_1}$

With notation as in §1, let $\widetilde{D}^\gamma_k\colon C_k(\widetilde{X}) \to C_{k+1}(\widetilde{X})$ be the lift of D^γ_k. Write $\widetilde{\partial} = \bigoplus_k \widetilde{\partial}_k$, $\widetilde{D}^\gamma = \bigoplus(-1)^{k+1}\widetilde{D}^\gamma_k$ and $\widetilde{I} = \bigoplus_k(-1)^k \operatorname{id}_k$ (viewed as matrices). The chain homotopy relation becomes $\widetilde{D}^\gamma\widetilde{\partial} - \widetilde{\partial}\widetilde{D}^\gamma = \widetilde{I}\big(1 - \eta_\#(\gamma)^{-1}\big)$. [Explanation: the minus sign occurs on the left because of the sign convention built into the matrix $\widetilde{D}^\gamma$; the right hand side is thus because the 0-end of the homotopy F^γ is lifted to the identity, while the 1-end is lifted to the covering translation corresponding to $\eta_\#(\gamma)$; the inversion occurs because we have G acting on the right.]

Proposition 5.1. *$\chi_1(X;R)(\gamma)$, as given in Definition A_1, is independent of the choice of the cellular homotopy F^γ representing γ.*

Direct calculation yields:

$$d\left(\operatorname{trace}(\widetilde{\partial} \otimes \widetilde{D}^\gamma)\right) = \chi(X)\big(1 - \eta_\#(\gamma)^{-1}\big). \tag{5.2}$$

This leads to a quick proof (essentially due to Stallings [**Stall**]) of an important theorem of Gottlieb [**Got**, Theorem IV.1]:

Proposition 5.3. *If $\chi(X) \neq 0$ then $\mathcal{G}(X)$ is trivial.*

Proof. Since $\chi(X) \neq 0$, (5.2) shows that every $\big(1 - \eta_\#(\gamma)^{-1}\big)$ represents $0 \in HH_0(\mathbb{Z}G)$. This implies that $\eta_\#(\gamma) = 1$. □

COROLLARY 5.4. *In the Hochschild complex,* $C_*(\mathbb{Z}G, \mathbb{Z}G)$, $\operatorname{trace}(\widetilde{\partial} \otimes \widetilde{D}^\gamma)$ *is a cycle.*

The simplicity of (5.2) arises from the "lucky" fact that $\operatorname{trace}(\widetilde{I})$ ="trace(id)". Thus the distinction mentioned in §3 between the "trace" and the "deeper trace" does not occur in this case. Nevertheless, the corresponding distinction in the first order case is far from trivial as we now explain.

Define the "lift" of $\chi_1(\cdot;\mathbb{Z})$ to be the function $\widetilde{\mathcal{X}}_1(X)\colon \Gamma \to HH_1(\mathbb{Z}G)$ which takes γ to $T_1(\widetilde{\partial} \otimes \widetilde{D}^\gamma)$, the homology class of the cycle $\operatorname{trace}(\widetilde{\partial} \otimes \widetilde{D}^\gamma)$. The proof of Proposition 5.1 shows that this trace is independent of the choice of F^γ representing γ.

There is a right action of $Z(G)$ on $HH_*(\mathbb{Z}G)$. At the level of chains it is defined by

$$(g_1 \otimes \cdots \otimes g_n \otimes m) \cdot \omega = g_1 \otimes \cdots \otimes g_n \otimes (m\omega^{-1})$$

where $\omega \in Z(G)$. One easily checks that this action is compatible with d and hence makes $HH_*(\mathbb{Z}G)$ into a right $Z(G)$-module. The summand $HH_*(\mathbb{Z}G)_C$ is taken by the right action of ω isomorphically onto the summand $HH_*(\mathbb{Z}G)_{C\omega^{-1}}$ where $C\omega^{-1}$ is the conjugacy class $\{g\omega^{-1} \mid g \in C\}$.

Since η maps Γ into $Z(G)$, η defines a right action of Γ on $C_*(\mathbb{Z}G, \mathbb{Z}G)$ and on $HH_1(\mathbb{Z}G)$. By considering lifts of homotopies, we clearly get:

PROPOSITION 5.5. *When* $HH_1(\mathbb{Z}G)$ *is regarded as a right* Γ*-module,* $\widetilde{\mathcal{X}}_1(X)$ *becomes a (right) derivation*; i.e. $\widetilde{\mathcal{X}}_1(X)(\gamma_1\gamma_2) = \widetilde{\mathcal{X}}_1(X)(\gamma_1) + \widetilde{\mathcal{X}}_1(X)(\gamma_2) \cdot \gamma_1$.

Derivations modulo inner derivations yield one-dimensional cohomology; in particular, $\widetilde{\mathcal{X}}_1(X)$ defines a cohomology class in $H^1(\Gamma, HH_1(\mathbb{Z}G))$.

DEFINITION. $\widetilde{\chi}_1(X) = [\widetilde{\mathcal{X}}_1(X)] \in H^1(\Gamma, HH_1(\mathbb{Z}G))$.

This $\widetilde{\chi}_1(X)$ is a refinement of $\chi_1(X)$; see Proposition 5.7 below. $\widetilde{\mathcal{X}}_1(X)(\gamma) \equiv T_1(\widetilde{\partial} \otimes \widetilde{D}^\gamma)$ is the "deeper trace" referred to in §3.

The derivation $\widetilde{\mathcal{X}}_1(X)$ depends on the choice of lifts $\tilde{e}$ of the cells e of X. However, we have:

PROPOSITION 5.6. *Up to inner derivations,* $\widetilde{\mathcal{X}}_1(X)$ *is independent of the choice cell orientations and of the choice of lifts. Hence* $\widetilde{\chi}_1(X)$ *is a well-defined cohomology class.*

One can view Definition A_1 as defining a cohomology class

$$\chi_1(X) \in H^1(\Gamma, H_1(G)) \cong \operatorname{Hom}(\Gamma, H_1(G)).$$

Clearly we have:

PROPOSITION 5.7. *Under the homomorphism induced by* $\varepsilon_*\colon HH_1(\mathbb{Z}G) \to H_1(G)$, $\widetilde{\chi}_1(X)$ *is taken to* $\chi_1(X)$. *Thus Definition* A_1 *is independent of the choice of lifts.*

Despite Propositions 5.1 and 5.7, the formula in Definition A_1 might appear to depend on the CW structure of X. However, we have:

THEOREM 5.8. *The cohomology classes* $\widetilde{\chi}_1(X)$ *and* $\chi_1(X)$ *are homotopy invariants.*

Theorem 1.2 follows from this.

REMARK 5.9. Observe that $T_0(\widetilde{I}) = \chi(X)C(1)$. This is mapped by the homomorphism $\varepsilon_*\colon HH_0(\mathbb{Z}G) \to H_0(G) \cong \mathbb{Z}$ to $\chi(X)$. This the sense in which Definition A_1 is analogous to Definition A_0. For a development of this analogy, see [**GN1**].

6. Equivalence of Definitions A_1 and C_1

Our method of proving the equivalence of Definitions A_1 and B_1 is to prove that both are equivalent to Definition C_1 in the oriented manifold case, and then to extend to other complexes via regular neighborhoods. A direct algebraic proof that Definitions A_1 and B_1 are equivalent under the hypotheses of Theorem 1.1(i) might be of some interest.

This section is different in flavor from the rest of the paper. Our aim is to give the interested reader an outline of the geometry which relates Definitions A_1 and C_1. (To see the point, the reader should think through the classical geometrical proof that Definitions A_0 and C_0 are equivalent; the situation here is analogous but more complicated.) The full details of our argument can be found in §6(B) of [**GN1**]. Some readers may wish to skip this section: its methods will not be used subsequently.

Let M be a compact connected oriented PL or smooth n-manifold with boundary (as well as being the underlying simplicial complex of a compatible triangulation). We are to show that $\chi_1(M)(\gamma) = -\theta(\gamma)$; the case of other coefficient rings R will then follow immediately.

Let $J\colon M \times I \to M$ be a homotopy from id_M to a map j, such that the graph of $J\big|_{M\times[1/2,1]}$ meets the graph of $p\big|_{M\times[1/2,1]}$ transversely in $\big|\chi(M)\big|$ arcs; this can be achieved by classical techniques of cancelling unnecessary pairs of fixed points. Note that j will then have precisely $\big|\chi(M)\big|$ fixed points, all transverse and having the same fixed point index. Denote by $\bar{F}^\gamma$, the concatenated homotopy $J^{-1}\star F^\gamma \star J$. It is clear that, using Definition A_1, $\mathrm{trace}(\widetilde{\partial}_{k+1}D_k^\gamma) = \mathrm{trace}(\widetilde{\partial}_{k+1}\bar{D}_k^\gamma)$, since the new contributions cancel one another. By perturbing rel $M \times \{0,1\}$, we may assume that the graph of $\bar{F}^\gamma$ meets the graph of p transversely, and that $\mathrm{Fix}(\bar{F}^\gamma)$ consists of circles in $\mathring{M} \times (0,1)$ and $\big|\chi(M)\big|$ arcs in $\mathring{M} \times I$ joining $\mathring{M} \times \{0\}$ to $\mathring{M} \times \{1\}$.

In defining $\theta(\gamma)$ in §1, we oriented the 1-manifold of fixed points by the "intersection orientation". We must make this precise. We may assume, adjusting $\bar{F}^\gamma$ if necessary, that (in addition to transverse intersection of the graphs) $\mathrm{Fix}(\bar{F}^\gamma) \cap X \times \{t\}$ is finite for all t. Recall that an isolated fixed point x of $\bar{F}_t^\gamma\colon X \to X$ has a *fixed point index* $\iota(\bar{F}_t^\gamma, x)$; it is defined to be the degree of $(\mathrm{id} - h\bar{F}_t^\gamma h^{-1})\colon N_\rho\big(h(x)\big) - \big\{h(x)\big\} \to \mathbb{R}^n - \{0\}$ where h is a homeomorphism from a neighborhood of $x \in X$ to a neighborhood $h(x) \in \mathbb{R}^n$ and N_ρ is a ball of radius $\rho > 0$ for suitably small ρ. Our assumptions imply that for all but finitely points $(x,t) \in \mathrm{Fix}(\bar{F}^\gamma)$, $\iota(\bar{F}_t^\gamma, x) = \pm 1$. The *intersection orientation* can be found for each component S of $\mathrm{Fix}(\bar{F}^\gamma)$ by picking any $(x,t) \in S$ at which $\iota(\bar{F}_t^\gamma, x) = 1$ and orienting S near (x,t) in the direction of increasing time. It is shown in [**DG**, §8 and §11] that this is consistent. If there should be no such points (x,t) then S is an arc projecting homeomorphically onto I and is to be oriented in the direction of decreasing time. It is a fact that the same orientation is obtained by giving $M \times I$ and $M \times I \times M$ the product orientations, giving the graphs of p and $\bar{F}^\gamma$ the orientations induced by projection onto $M \times I$, giving their intersection the induced orientation with respect to the order $(p, \bar{F}^\gamma)$ and projecting the latter orientation onto $\mathrm{Fix}(\bar{F}^\gamma)$.

We first discuss the case where $\chi(M) = 0$. Then there are no arcs in $\mathrm{Fix}(\bar{F}^\gamma)$. Our plan is to take a transverse circle of fixed points of $\bar{F}^\gamma$, to analyze what it contributes to $\widetilde{\mathcal{X}}_1(X)(\gamma) \equiv T_1(\widetilde{\partial} \otimes \widetilde{\bar{D}}^\gamma) \in HH_1(\mathbb{Z}G)$, to show that all contributions to $\widetilde{\mathcal{X}}_1(X)(\gamma)$ arise in this way, and that the sum of the contributions (which define $\chi_1(X)(\gamma)$ via Proposition 4.8) is precisely $-\theta(\gamma)$.

Let J be a subdivision of I, L a subdivision of M, and K a subdivision of L. Let $K \times J$ denote a product triangulation of $M \times I$ using orderings of the vertices of K and J. For suitable choices of J, K and L we may assume that $\bar{F}^\gamma \colon K \times J \to L$ is simplicial, that $\mathrm{Fix}(\bar{F}^\gamma)$ is a PL 1-manifold and the following conditions hold:

(a) For each $(n-1)$-simplex σ^{n-1} of K, $\mathrm{Fix}(\bar{F}^\gamma)$ is transverse to $\sigma^{n-1} \times I$ and each point of $\mathrm{Fix}(\bar{F}^\gamma) \cap (\sigma^{n-1} \times I)$ lies in the interior of either:
  (i) an n-simplex α^n of the subcomplex $\sigma^{n-1} \times J$ of $K \times J$; or
  (ii) an $(n-1)$-simplex β^{n-1} of $\sigma^{n-1} \times J$ such that β^{n-1} projects onto σ^{n-1} and β^{n-1} is neither $\sigma^{n-1} \times \{0\}$ nor $\sigma^{n-1} \times \{1\}$.

(b) For each simplex σ^k of K with $k < n-1$ or with $k = n-1$ and $\sigma \in \partial K$, $\mathrm{Fix}(\bar{F}^\gamma) \cap (\sigma^k \times I) = \varnothing$.

(c) For each $t \in I$, $\mathrm{Fix}(\bar{F}^\gamma) \cap (X \times \{t\})$ is a finite set.

Recall that a simplex μ is the *carrier* of a simplex σ if and only if $\mathring{\sigma} \subset \mathring{\mu}$. Orient the n-simplices of L compatibly with the given manifold orientation on X; orient the other simplices of L arbitrarily. If a k-simplex of K has a k-dimensional carrier in L, let it inherit its orientation from its carrier; orient the other simplices of K arbitrarily. Orient the 1-simplices of J in the positive direction and orient the vertices of J by $+1$ (an orientation on a vertex is a choice of $+1$ or -1). If a k-simplex of $K \times J$ lies in $\big[(k-1)\text{-simplex of } K\big] \times I$, let it inherit the product orientation. If a k-simplex β of $K \times J$ projects onto a k-simplex of K, orient β so that this projection preserves orientation.

Viewed as a map $\bar{F}^\gamma \colon |K \times J| \to |K|$, $\bar{F}^\gamma$ is cellular. The cellular chain map $\bar{F}^\gamma_* \colon C_*(K \times J) \to C_*(K)$ is the composite:

$$C_*(K \times J) \xrightarrow{\bar{F}^\gamma_*} C_*(L) \xrightarrow{\text{subdivision}} C_*(K).$$

Choose a lift $\widetilde{\mu} \in \widetilde{L}$ for each simplex μ of L. Choose a lift $\widetilde{\sigma} \in \widetilde{K}$ for each simplex of σ of K so that if μ is the carrier of σ then $\widetilde{\mu}$ is the carrier of $\widetilde{\sigma}$. Lifts of simplices of $\sigma \times J \subset K \times J$ are to be chosen in $\widetilde{\sigma} \times J \subset \widetilde{K} \times J$. Orient all lifts of simplices of L, K, and $K \times J$ so that the covering projections preserve orientation.

Index all the simplices of K by natural numbers so that if $i < j$ then $\dim \sigma_i \leq \dim \sigma_j$. The (i,j)-entry of $\widetilde{\partial} \equiv [\oplus_p \widetilde{\partial}_p]$ can only be nonzero when σ_i is a codimension one face of σ_j. Denote the (i,j)-entries of $\widetilde{\partial}$ and $\widetilde{\bar{D}^\gamma}$ by $\widetilde{\partial}_{ij}$ and $\widetilde{\bar{D}^\gamma}_{ij}$ respectively. We analyze $\mathrm{trace}(\widetilde{\partial} \otimes \widetilde{\bar{D}^\gamma}) = \sum_{i,j} \widetilde{\partial}_{ij} \otimes \widetilde{\bar{D}^\gamma}_{ji}$.

The following propositions are stated to give a general idea of what is involved; see [**GN1**, §6(B)] for details.

Proposition 6.1. *If $\widetilde{\partial}_{ij} \otimes \widetilde{\bar{D}^\gamma}_{ji} \neq 0$ then $\dim \sigma_i = n-1$, $\dim \sigma_j = n$, σ_i^{n-1} is a face of σ_j^n, and $\bar{F}^\gamma$ has a fixed point in $\sigma_i^{n-1} \times I$. There is exactly one $k \neq j$ such that $\sigma_i^{n-1} = \sigma_j^n \cap \sigma_k^n$.*

Proposition 6.2. *Let $\mu \in L$ be the carrier of $\sigma_i^{n-1} \in K$. If $\dim \mu = n$ then all points of* $\mathrm{Fix}(\bar{F}^\gamma) \cap (\sigma_i^{n-1} \times I)$ *are of the type* (a)(i). *If $\dim \mu = n-1$ then all points of* $\mathrm{Fix}(\bar{F}^\gamma) \cap (\sigma_i^{n-1} \times I)$ *are of the type* (a)(ii).

Let i and j be such that $\widetilde{\partial}_{ij} \otimes \widetilde{\bar{D}^\gamma}_{ji} \neq 0$. Then Proposition 6.1 gives us information about σ_i^{n-1}, σ_j^n and a simplex σ_k^n. We now examine $\widetilde{\partial}_{ij} \otimes \widetilde{\bar{D}^\gamma}_{ji} + \widetilde{\partial}_{ik} \otimes$

$\widetilde{\bar{D}^\gamma}_{ki}$ in the two cases appearing in Proposition 6.2.

PROPOSITION 6.3. *If* $\dim \mu = n$ *then* $\widetilde{\bar{D}^\gamma}_{ji} = \widetilde{\bar{D}^\gamma}_{ki}$ *and* $\widetilde{\partial}_{ij} = -\widetilde{\partial}_{ik}$. *Hence* $\widetilde{\partial}_{ij} \otimes \widetilde{\bar{D}^\gamma}_{ji} + \widetilde{\partial}_{ik} \otimes \widetilde{\bar{D}^\gamma}_{ki} = 0$.

Let $g_{ij} \in G$ be such that $\widetilde{\sigma}_i^{n-1} g_{ij}$ is a face of $\widetilde{\sigma}_j^n$.

PROPOSITION 6.4. *If* $\dim \mu = n-1$ *then* $\widetilde{\partial}_{ij} \otimes \widetilde{\bar{D}^\gamma}_{ji} + \widetilde{\partial}_{ik} \otimes \widetilde{\bar{D}^\gamma}_{ki}$ *is a sum of terms of the form* $\pm(g_{ij} \otimes g_{ij}^{-1} g - g_{ik} \otimes g_{ik}^{-1} g)$ *where* g *marks the fixed point class of a fixed point of* $\bar{F}^\gamma$ *in* $\sigma_i^{n-1} \times I$.

For analysis of the sign in Proposition 6.4, we need some notation. Let $\bar{F}^\gamma$ have a fixed point $(x,t) \in \sigma_i^{n-1} \times I$. By Proposition 6.2, $(x,t) \in \mathring{\beta}^{n-1}$ where β is an $(n-1)$-simplex of $\sigma_i^{n-1} \times J$ which projects to σ_i^{n-1} and is not $\sigma_i^{n-1} \times \{0\}$ or $\sigma_i^{n-1} \times \{1\}$. This β is a face of exactly two n-simplices α^n and δ^n in $\sigma_i^{n-1} \times J$. We have $\sigma_i^{n-1} = \sigma_j^n \cap \sigma_k^n$ and $\mu = \nu_j \cap \nu_k$ where ν_j and ν_k are the carriers of σ_j and σ_k respectively. Then $\widetilde{\nu}_j g_{ij}^{-1} \cap \widetilde{\nu}_k g_{ik}^{-1} = \widetilde{\mu}$. Since $x \in \mathring{\sigma}_i^{n-1}$, $x = \bar{F}^\gamma(x,t) \in \mathring{\mu}$, and so $\bar{F}^\gamma$ maps β isomorphically onto μ. Since (x,t) is an isolated point of $\mathrm{Fix}(\bar{F}^\gamma) \cap \sigma_i^{n-1} \times I$, $\bar{F}^\gamma(\alpha) = \nu_j$ or ν_k and $\bar{F}^\gamma(\delta) = \nu_j$ or ν_k. It can be shown that the nonzero contributions arise only when $\bar{F}^\gamma(\alpha) \neq \bar{F}^\gamma(\delta)$, and that the sign is $\epsilon\eta$ where η is the incidence number of σ_j^n with σ_i^{n-1}, and ϵ is the degree of the embedding $\bar{F}^\gamma \colon \alpha \to |K|$. To formulate the desired result, we must adopt a convention for specifying σ_j^n and α: *let* α *be the* n*-simplex having* β *as a face whose other vertex is separated from* $\sigma_i^{n-1} \times \{1\}$ *by* β (i.e. "α is on the $\sigma_i^{n-1} \times \{0\}$ side of β"), *and let* $\nu_j^n = \bar{F}^\gamma(\alpha)$. This and our orientation convention together imply that β has incidence number $(-1)^{n+1}$ with α, and incidence number $(-1)^n$ with δ.

Let $(x,t) \in \mathring{\beta}^{n-1}$, and let $\widetilde{\bar{F}^\gamma}(\tilde{x},t) = \tilde{x}g$ where $g \in G$. As noted above, we can assume $\bar{F}^\gamma(\alpha) \neq \bar{F}^\gamma(\delta)$. Near (x,t), $\mathrm{Fix}(\bar{F}^\gamma)$ consists of two straight arcs A_j and A_k, intersecting at (x,t), labelled so that the arcs $p(A_j)$ and $p(A_k)$ lie in ν_j and ν_k respectively. The intersection orientation of $\mathrm{Fix}(\bar{F}^\gamma)$ orients $A_j \cup A_k$.

PROPOSITION 6.5. *The contribution of* (x,t) *to* $\widetilde{\partial}_{ij} \otimes \widetilde{\bar{D}^\gamma}_{ji} + \widetilde{\partial}_{ik} \otimes \widetilde{\bar{D}^\gamma}_{ki}$ *is* $g_{ij} \otimes g_{ij}^{-1} g - g_{ik} \otimes g_{ik}^{-1} g$ *if and only if the arc* $p(A_j) \cup p(A_k)$ *with intersection orientation points from* ν_j *to* ν_k.

We are now ready to analyze $\widetilde{\mathcal{X}}_1(X)(\gamma)$ and $\chi_1(X)(\gamma)$. Let S be a circle of fixed points in $\mathrm{Fix}(\bar{F}^\gamma)$. By assumption, S misses $|K^{n-2} \times J|$, and meets $|K^{n-1} \times J|$ transversely in a finite set of points which we list in cyclic order with respect to the intersection orientation of S: $(x_1,t_1), \dots, (x_m,t_m)$. For each $1 \le q \le m$, define $g_q \in G$ by $\widetilde{\bar{F}^\gamma}(\tilde{x}_q,t_q) = \tilde{x}_q g_q$. We have $x_q \in \mathring{\sigma}_{i_q}^{n-1}$ for some $\sigma_{i_q}^{n-1} \in K$. The simplex σ_{i_q} is a face of two n-simplices which, until now, we have called $\sigma_{j_q}^n$ and $\sigma_{k_q}^n$. The indices j_q and k_q are natural numbers. Without altering our indexing of the simplices of K, it is convenient to use a different notation. Let $\{a_q, b_q\} = \{j_q, k_q\}$ with the convention that *the oriented circle* S *crosses* $\sigma_{i_q}^{n-1} \times I$ *from the* $\sigma_{a_q}^n$*-side to the* $\sigma_{b_q}^n$*-side.*

PROPOSITION 6.6. *The total contribution of* S *to* $\mathrm{trace}(\widetilde{\partial} \otimes \widetilde{\bar{D}^\gamma})$ *is*

$$T_S = \sum_{q=1}^{m} g_{i_q a_q} \otimes g_{i_q a_q}^{-1} g_q - g_{i_q b_q} \otimes g_{i_q b_q}^{-1} g_q.$$

This is homologous to

$$\bar{g} \otimes \bar{g}^{-1}(g_{i_m b_m}^{-1} g_m g_{i_m b_m}),$$

where $\bar{g} = g_{i_m b_m}^{-1} g_{i_m, a_m} \cdots g_{i_1 b_1}^{-1} g_{i_1, a_1}$.

By Corollary 4.2, we conclude:

PROPOSITION 6.7. *The contribution of* S *to* $\chi_1(X)(\gamma)$ *is* $\{\bar{g}\} \in H_1(G)$.

One can also analyze the contribution of S to the cycle $U(\gamma)$ of §1 (recall that $U(\gamma)$ defines $\theta(\gamma)$). One looks at the lifts of the $(n-1)$-simplexes and n-simplexes mentioned above to the universal cover and, by covering space theory, one finds:

PROPOSITION 6.8. *The oriented circle* S *lifts to a path starting at* $\widetilde{\sigma}_{i_1}^{n-1}$ *and ending at* $\widetilde{\sigma}_{i_1}^{n-1} g_{i_1 a_1} \bar{g} g_{i_1 a_1}^{-1}$. *Hence the homology class of* S *is* $\{\bar{g}^{-1}\} = -\{\bar{g}\}$.

Propositions 6.7 and 6.8 establish the desired equivalence when $\chi(M) = 0$ because of the fundamental "one-parameter Nielsen-Wecken fixed point theorem". Theorem 1.8 of [**GN1**], which in the present context implies:

PROPOSITION 6.9. *All contributions to* $\operatorname{trace}(\widetilde{\partial} \otimes \widetilde{\widetilde{D}^{\gamma}})$ *arise from circles of fixed points as in Proposition* 6.6.

When $\chi(M) \neq 0$, more delicate arguments are needed to show that an arc of fixed points does not contribute; see [**GN4**] for more details.

Theorem 1.1(ii) is proved by replacing X with a finite simplicial complex (which may be done), embedding in a Euclidean space and using the methods described here applied to a regular neighborhood (see [**GN4**]).

REMARK. We have just outlined the proof that $\chi_1(X)(\gamma) \equiv A\left(\operatorname{trace}(\widetilde{\partial} D^{\gamma})\right) \in H_1(G)$ is given by the intersection class $-\theta(\gamma)$. The corresponding "universal cover" statement is a description of

$$\widetilde{\mathcal{X}}_1(X)(\gamma) \equiv T_1(\widetilde{\partial} \otimes \widetilde{\widetilde{D}^{\gamma}}) \in HH_1(\mathbb{Z}G) = \bigoplus_{C \in G_1} H_1(Z(g_C)).$$

Pick a path μ as in §3 from $(v, 0)$ to a point of one of the circles, S_1, and a path ν_i as in §3 joining S_1 to each of the other circles S_i. Using these paths, each circle S_i defines a loop based at $(v, 0)$. These loops lift to loops in the covering space corresponding to $Z(g_C)$ where $g_C = \left[(p \circ \mu)(F \circ \mu)^{-1} \tau^{-1}\right]$. Thus, together, they define an element $\theta^C(\gamma) \in H_1(Z(g_C))$. It is shown in §6(B) of [**GN1**] that $\widetilde{\mathcal{X}}_1(X)(\gamma) = -\sum_{C \in G_1} \theta^C(\gamma)$.

7. Equivalence of Definitions $\mathbf{B_1}$ and $\mathbf{C_1}$

This involves more standard algebraic topology so our outline will be brief.

The trace formula in Definition B_1 was introduced by Knill in [**Kn**]. As we remarked in §1, it is independent of basis. Moreover, it is a straightforward exercise to show that it is a homotopy invariant. In [**Kn**] there also appears an "intersection class", whose definition we now recall. Actually, the context in [**Kn**] is much more general: we only extract what we need.

Throughout this section, all homology and cohomology groups will have coefficients in the principal ideal domain R.

HYPOTHESIS 7.1. Let M be a compatibly oriented, compact, codimension 0, PL submanifold $i\colon M \hookrightarrow \mathbb{R}^n$. Let $F\colon M \times S^1 \to M$ be such that $\operatorname{Fix}(F) \cap \partial M \times S^1 = \varnothing$.

Let $[M \times S^1] \in H_{n+1}(M \times S^1, \partial M \times S^1)$ be the fundamental class of $M \times S^1$ and let $[\mathbb{R}^n]$ be the generator of $H_n(\mathbb{R}^n, \mathbb{R}^n - \{0\})$ determined by the orientation. Following Leray [**Le**] and Dold [**D**], Knill defines the intersection class of F to be the image, $I_R(F)$, of $[M \times S^1]$ under the following composition:

$$H_{n+1}(M, \partial M) \times S^1) \to H_{n+1}(M \times S^1, M \times S^1 - \operatorname{Fix}(F)) \xrightarrow{(iop - ioF, F)_*} H_{n+1}\Big((\mathbb{R}^n, \mathbb{R}^n - \{0\}) \times M\Big) \xrightarrow{\cong} H_1(M)$$

where $p\colon M \times S^1 \to M$ is projection and $H_{n+1}\Big((\mathbb{R}^n, \mathbb{R}^n - \{0\}) \times M\Big) \xrightarrow{\cong} H_1(M)$ is the inverse of the isomorphism $H_1(M) \xrightarrow{\cong} H_{n+1}\Big((\mathbb{R}^n, \mathbb{R}^n - \{0\}) \times M\Big)$, $y \mapsto [\mathbb{R}^n] \times y$.

THEOREM 7.2 ([**Kn**]). *Suppose $H_*(M)$ is a free R-module. Then*

$$-I_R(F) = \sum_{k \geq 0} (-1)^{k+1} \sum_j \bar{b}_j^k \cap F_*(b_j^k \times [S^1])$$

where $[S^1] \in H_1(S^1)$ is the fundamental class and where for each $k \geq 0$, $\{b_j^k\}$ is a basis for $H_k(X)$ with corresponding dual basis, $\{\bar{b}_j^k\}$, for $H^k(X)$.

We define another "intersection class" of a map $F\colon M \times S^1 \to M$, $\theta'_R(F) \in H_1(M)$, as follows (here M may be any compact oriented manifold with boundary and $\operatorname{Fix}(F) \cap \partial M \times S^1$ need not be empty). Let $(Y; \partial_1 Y, \partial_2 Y)$ be the manifold triad $Y = M \times S^1 \times M$, $\partial_1 Y = (\partial M) \times S^1 \times M$ and $\partial_2 Y = M \times S^1 \times \partial M$. The oriented graph of p defines a relative cycle representing $A \in H_{n+1}(Y, \partial_2 Y)$ and the oriented graph of F defines a relative cycle representing $B \in H_{n+1}(Y, \partial_1 Y)$. Let $\delta_1\colon H^n(Y, \partial_1 Y) \to H_{n+1}(Y, \partial_2 Y)$ and $\delta_2\colon H^n(Y, \partial_2 Y) \to H_{n+1}(Y, \partial_1 Y)$ be the Poincaré duality isomorphisms given by cap product with the fundamental class, $[Y]$, of Y. Then $\theta'_R(F) = (p_1)_*\big((\delta_2^{-1}(B) \cup \delta_1^{-1}(A)) \cap [Y]\big)$ where $p_1\colon M \times S^1 \times M \to M$ is projection to the first factor.

HYPOTHESIS 7.3. Let $F\colon M \times S^1 \to M$ and $z \in S^1$. Let $F_z\colon M \to M$ be given by $F_z(m) = F(m, z)$. Assume $(F_z)_*\colon H_1(M) \to H_1(M)$ is the identity. (Note that this condition is independent of the choice of z).

PROPOSITION 7.4. *Let the manifold M and the map $F\colon M \times S^1 \to M$ satisfy Hypotheses* 7.1 *and* 7.3. *Then $I_R(F) = \theta'_R(F)$.*

For a general position map Ψ^γ as in §1, we have $\theta_R(\gamma) = \theta'_R(\Psi^\gamma)$, and Hypothesis 7.3 is satisfied. So for manifolds M as in Hypothesis 7.1, we have $I_R(\Psi^\gamma) = \theta'_R(\Psi^\gamma) = \theta_R(\Psi^\gamma)$ as desired. Since regular neighborhoods of compact polyhedra in Euclidean space satisfy Hypothesis 7.1, the equivalence of Definitions B_1 and C_1 follows. This, together with the previously proved Theorem 1.1(ii) completes the proof of Theorem 1.1(i).

8. Sample calculations

We have seen that when X is aspherical, $\Gamma = Z(G)$, so $\chi_1(X)\colon \Gamma \to H_1(X)$ is trivial unless $\chi(X) = 0$, and similarly for $\widetilde{\chi}_1(X)$. Details of the following calculations can be found in [**GN4**].

EXAMPLE 8.1. Take $X = S^1$. Then Γ is generated by $[R]$, where R is positive rotation through 2π. Let $t \in \pi_1(S^1, v)$ be represented by the "clockwise" loop

$u \mapsto e^{-2\pi \imath u}$. Then $\widetilde{\mathcal{X}}_1(S^1)([R])$ is represented by the Hochschild cycle, in canonical form, $t \otimes t^{-1}t$. By Corollary 4.2 we conclude that $\chi_1([R]) = \{t\} \in H_1(S^1)$.

In the next two examples, X will be a finite 2-complex with one vertex,v, and one 2-cell, e^2 arising from a group presentation. In both of these examples, the element of the free group defined by the attaching map of the 2-cell is not a proper power and so by Lyndon's theorem [**Ly**], X is aspherical.

EXAMPLE 8.2. $G = \langle x, y \mid yx^m y^{-1} x^{-m} \rangle$, $m \geq 2$. Here, $\Gamma \cong Z(G)$ is generated by x^m. $\widetilde{\mathcal{X}}_1(X)(x^m)$ is represented by the cycle, in canonical form:

$$x^{-1} \otimes x(\sum_{i=1}^{m-1} x^{-i}) + x^{m-1} \otimes x^{-(m-1)} x^{-m}.$$

By Corollary 4.2, $\chi_1(X) = 0$. It can be shown that $\widetilde{\mathcal{X}}_1(X)$ is not an inner derivation. In particular, this is an example where the first order Euler characteristic is zero while $\widetilde{\chi}_1(X) \neq 0$.

EXAMPLE 8.3. $G = \langle x, y \mid x^m y^n \rangle$, $m \neq 0$ and $n \neq 0$. (If m and n are relatively prime, then G is the group of the $(m, -n)$ torus knot.) Here, $\Gamma \cong Z(G)$ is generated by $x^m = y^{-n}$. $\widetilde{\mathcal{X}}_1(X)(x^m)$ is represented by the cycle, in canonical form:

$$\sum_{i=1}^{m-1} (x^{-1} \otimes x x^{-i}) + \sum_{i=1}^{n-1} (y \otimes y^{-1} y^i) + x^{m-1} \otimes x^{-(m-1)} x^{-m} + y \otimes y^{-1} 1.$$

By Corollary 4.2, $\chi_1(X)(x^m) = -mA(x) = nA(y)$ (here, $A\colon G \to H_1(G)$ is Abelianization).

Next, we consider a family of non-aspherical examples. Let (p, q) be a pair of relatively prime positive integers with $p > 1$. The lens space $L(p, q)$ is the orbit space of the action of the cyclic group $\mathbb{Z}/p = \langle x \mid x^p = 1 \rangle$ on the 3-sphere $S^3 = \{(z_0, z_1) \in \mathbb{C}^2 \mid |z_0|^2 + |z_1|^2 = 1\}$ defined by $x(z_0, z_1) = (e^{2\pi \imath/p} z_0, e^{2\pi \imath q/p} z_1)$. The point in $L(p, q)$ determined by the orbit of $(z_0, z_1) \in S^3$ will be denoted $[z_0, z_1]$.

For any pair of integers (m, n) such that $m = n \bmod p$ define a smooth S^1 action $\gamma_{m,n}\colon S^1 \times L(p, q) \to L(p, q)$ by $e^{2\pi \imath \theta}[z_0, z_1] = [e^{2\pi \imath \theta m/p} z_0, e^{2\pi \imath \theta n q/p} z_1]$. These actions represent elements of $\Gamma = \pi_1\Big(\mathcal{E}\big(L(p, q)\big), \mathrm{id}\Big)$.

The group $HH_1(\mathbb{Z}[\mathbb{Z}/p])$ is isomorphic to a direct sum of p copies of $\mathbb{Z}/p$; furthermore, the Hochschild 1-cycles $\{x \otimes x^{-1-k} \mid k = 0, \ldots, p-1\}$ project to a set of generators for $HH_1(\mathbb{Z}[\mathbb{Z}/p])$. Define $c_i, d_i \in \mathbb{Z}$ for $0 \leq i \leq p-1$ by $m - i - 1 = (c_i - 1)p + b_i$ and $nq - i - 1 = (d_i - 1)p + b'_i$ where $0 \leq b_i, b'_i \leq p - 1$. Let $s_k = c_{k-1} + r d_{kq-1}$, where the indices are interpreted mod p and $rq = 1 \bmod p$.

There is a natural cell structure on the universal cover, S^3, of $L(p, q)$ (see [**GN1**, §5(B)]). Using this cell structure, [**GN1**, Lemma 5.3] asserts:

PROPOSITION 8.4. *$\widetilde{\mathcal{X}}_1\big(L(p, q)\big)([\gamma_{m,n}]) \in HH_1(\mathbb{Z}[\mathbb{Z}/p])$ is represented by the Hochschild cycle $\sum_{k=0}^{p-1} s_k x \otimes x^{-1-k}$.*

By Corollary 4.2, the homomorphism $\varepsilon\colon HH_1(\mathbb{Z}[\mathbb{Z}/p]) \to H_1(\mathbb{Z}/p)$ takes the generators $\{x \otimes x^{-1-k}\}$ to the same generator, α, of $H_1(\mathbb{Z}/p)$. From the proof of [**GN1**, Corollary 5.5], we deduce:

PROPOSITION 8.5. $\chi_1(L(p,q))([\gamma_{m,n}]) = (m+n)\alpha$.

REMARK 8.6. If p is odd then Propositions 8.4 and 8.5 give complete computations of $\widetilde{\chi}_1(L(p,q))$ and $\chi_1(L(p,q))$ respectively because the $[\gamma_{m,n}]$'s generate Γ; indeed, by [**GN1**, Proposition 5.7], for odd p, Γ is cyclic of order $2p^2$. The proof there also shows that $2[\gamma_{1,1}]$ is of order p^2 and that $p[\gamma_{0,p}]$ is of order 2 in Γ, so $[\gamma_{2,2+p^2}]$ generates Γ.

9. S^1-fibrations

In this section we discuss the first order Euler characteristic of the total space of an orientable Serre fibration with S^1-fiber.

Let $S^1 \longrightarrow X \xrightarrow{\pi} B$ be an orientable Serre fibration where B is a (not necessarily finite) connected CW complex and X has the homotopy type of a finite complex. By classical obstruction theory, fiber homotopy equivalence classes of orientable S^1-fibrations over a *CW* complex B are classified by the integral cohomology group $H^2(B;\mathbb{Z})$. Given an element $e \in H^2(B;\mathbb{Z}) \cong [B, \mathbb{C}P^\infty]$ one obtains a principal $U(1)$-bundle over B by pulling back, via a continuous map $B \rightarrow \mathbb{C}P^\infty$ representing e, the $U(1)$-bundle associated to the canonical complex line bundle over the infinite dimensional complex projective space $\mathbb{C}P^\infty$. Thus we can assume, without loss of generality, that $S^1 \longrightarrow X \xrightarrow{\pi} B$ is a principal $U(1)$-bundle. In particular, there is a free $U(1)$-action on X which we will write as $\Phi\colon X \times S^1 \longrightarrow X$. Let $\tau \in \Gamma \equiv \pi_1(\mathcal{E}(X), 1)$ be the element represented by Φ ($\Phi = \Phi^\tau$ in the notation of §1). For any coefficient ring F, let $\{\tau\} \in H_1(X;F)$ denote the image of τ under the composite:

$$\Gamma \xrightarrow{\eta} \pi_1(X) \longrightarrow H_1(X) \longrightarrow H_1(X;F).$$

Also, let e_F be the image of the element $e \in H^2(B;\mathbb{Z})$ which classifies $S^1 \longrightarrow X \xrightarrow{\pi} B$ under the homomorphism $H^2(B;\mathbb{Z}) \longrightarrow H^2(B;F)$.

LEMMA 9.1. *If F is a field, then $\{\tau\} \in H_1(X;F)$ is non-zero if and only if $e_F = 0$.*

THEOREM 9.2. *Let F be a field. If $e_F \neq 0$ then $\chi_1(X;F)(\tau) = 0$. If $e_F = 0$ then $H_*(B;F)$ is finite dimensional over F and $\chi_1(X;F)(\tau) = -\chi(B;F)\{\tau\}$ where $\chi(B;F) = \sum_{i\geq 0}(-1)^i \dim_F H_i(B;F)$.*

Next, we consider integer coefficients. Suppose that $S^1 \longrightarrow X \xrightarrow{\pi} B$ is a smooth orientable $U(1)$-bundle over a smooth, closed, oriented manifold B. Let λ be the one dimensional subbundle of the tangent bundle of X consisting of vectors which are tangent to the circle fibers and let be ν be a complementary bundle to λ. Then $\nu \cong \pi^*(T_B)$ where T_B is the tangent bundle of B. Let $[B] \in H_n(B;\mathbb{Z})$ be the fundamental class of B where $n = \dim B$. The Euler class, $\mathrm{Eul}(\nu) \in H^n(X;\mathbb{Z})$, is given by

$$\mathrm{Eul}(\nu) = \mathrm{Eul}(\pi^*(T_B)) = \pi^*(\mathrm{Eul}(T_B)) = \chi(B)\pi^*([B]^*)$$

where $[B]^* \in H^n(B;\mathbb{Z})$ is the generator determined by the condition $[B]^*([B]) = 1$; see [**MS**, Corollary 11.12]. The Gysin homology sequence for $S^1 \longrightarrow X \xrightarrow{\pi} B$ determines a fundamental class for X; $[X] \in H_{n+1}(X)$ is the image of $[B]$ under the homomorphism $\theta_n\colon H_n(B;\mathbb{Z}) \longrightarrow H_{n+1}(X;\mathbb{Z})$. In our paper [**GN3**] we investigate the algebraic topology and K-theory of flows. In particular, we express the Euler class as a trace. An oriented S^1-bundle supports a flow (rotation along the fibers), so Theorem 3.1 of [**GN3**] in this context yields:

THEOREM 9.3. $\chi_1(X)(\tau) = -\operatorname{Eul}(\nu) \cap [X]$ (i.e. $\chi_1(X)(\tau)$ *is the Poincaré dual of* $-\operatorname{Eul}(\nu)$).

Using this we deduce:

THEOREM 9.4. *Under the above hypotheses,* $\chi_1(X)(\tau) = -\chi(B)\{\tau\}$.

EXAMPLE 9.5. Let Σ_g be a closed oriented surface of genus $g > 1$ and let L_n be a complex line bundle over Σ_g with Chern number n. Let $M_{n,g}$ be the total space of the $U(1)$-bundle associated to L_n. Then $M_{n,g}$ is a closed oriented aspherical 3-manifold which fibers over Σ_g. The center of $\pi_1(M_{n,g})$ is the infinite cyclic group generated by τ (represented by a circle fiber); the image, $\{\tau\}$, of τ in $H_1(M_{n,g}) \cong \mathbb{Z}^{2g} \oplus \mathbb{Z}/n$ generates the $\mathbb{Z}/n$ summand. By Theorem 9.4, $\chi_1(M_{n,g})\colon \mathbb{Z} \to H_1(M_{n,g})$ is given by $\chi_1(M_{n,g})(\tau) = (2g-2)\{\tau\}$.

Let T^n, where $n > 1$, be the n-torus (i.e. the n-fold product of copies of $U(1)$). Let X be a closed oriented smooth manifold and let $\rho\colon T^n \times X \to X$ be a smooth free action of T^n. This action defines a homomorphism $\bar{\rho}\colon T^n \to \operatorname{Diff}(X)$ where $\operatorname{Diff}(X)$ is the diffeomorphism group of X. Let $\Gamma_\rho \subset \Gamma$ be the image of the composite:

$$\pi_1(T^n, 1) \xrightarrow{\bar{\rho}_\#} \pi_1(\operatorname{Diff}(X), \mathrm{id}) \longrightarrow \pi_1(\mathcal{E}(X), \mathrm{id}) = \Gamma.$$

PROPOSITION 9.6. *The restriction of* $\chi_1(X)\colon \Gamma \to H_1(X)$ *to* Γ_ρ *is the zero homomorphism.*

COROLLARY 9.7. *If* $n > 1$ *then* $\chi_1(T^n)\colon \mathbb{Z}^n \to \mathbb{Z}^n$ *is zero.*

10. Mapping tori

Let $f\colon Z \to Z$ be a cellular homotopy equivalence of a (not necessarily finite) complex. The *mapping torus* of f, denoted by $T(Z,f)$, is the space obtained from $Z \times [0,1]$ by identifying $(z,1)$ with $(f(z),0)$ for each $z \in Z$. In this section we discuss the first order Euler characteristic of $T(Z,f)$. In [**GN4**], mapping tori of more general maps f are treated.

The image of $(z,u) \in Z \times [0,1]$ in $T(Z,f)$ will be denoted by $[z,u]$. Choose a basepath σ from v to $f(v)$ and let $\theta\colon H \to H$ be the automorphism of $H \equiv \pi_1(Z,v)$ determined by f and σ.

Let $X = T(Z,f)$. Choose $w = [v,0]$ as a basepoint for X and let $G = \pi_1(X,w)$. There is a canonical map of X to the standard circle S^1 (realized as complex numbers of unit modulus) given by: $p_f\colon X \to S^1$, $p_f([z,s]) = e^{2\pi i s}$. Let $i\colon Z \hookrightarrow X$ be the inclusion $z \mapsto [z,0]$.

Recall that $\Gamma = \pi_1(\mathcal{E}(X), \mathrm{id})$. Let $\Gamma_{S^1} = \pi_1(\mathcal{E}(S^1), \mathrm{id})$. Let $\bar{\gamma}\colon I \to X$ be the path $\bar{\gamma}(u) = [v,u]$ and let $\gamma_0\colon I \to X$ be the path $\gamma_0 = \bar{\gamma}(i \circ \sigma)^{-1}$. Define a continuous map $P\colon X^X \to (S^1)^{S^1}$ by $P(g)(e^{2\pi i u}) = p_f(g(\gamma_0(u)))$. Then P induces a homomorphism $P_*\colon \Gamma \to \Gamma_{S^1}$. We define an identification $\Gamma_{S^1} \xrightarrow{\cong} \mathbb{Z}$ by sending the generator $[s \mapsto (e^{2\pi i u} \mapsto e^{2\pi i(u+s)})] \in \Gamma_{S^1}$ to $1 \in \mathbb{Z}$. The *rotation degree* of $\gamma \in \Gamma$ is the integer $P_*(\gamma)$.

The group G is a semidirect product of H with $T \equiv \pi_1(S^1, 1)$; there is an exact sequence: $H \rightarrowtail G \twoheadrightarrow T$ where $H \rightarrowtail G$ is induced by the inclusion $i\colon Z \hookrightarrow X$ and $G \twoheadrightarrow T$ is induced by p_f. We write $t = [\gamma_0]^{-1} \in G$, projecting to a generator of T, so that $\theta\colon H \to H$ is given by $h \mapsto tht^{-1}$.

The homotopy equivalence f is a *periodic homotopy idempotent* if for some $q > 0$, f^q is homotopic to id_Z, and f is *coherent* if there is a homotopy $N\colon \mathrm{id} \simeq f^q$

such that the following diagram commutes up to homotopy rel $Z \times \{0,1\} \times I$:

$$\begin{array}{ccc} Z \times I & \xrightarrow{f \times \mathrm{id}} & Z \times I \\ \big\downarrow N & & \big\downarrow N \\ Z & \xrightarrow{f} & Z. \end{array}$$

Given f, the least $q > 0$ for which such an N exists (assuming that it exists at all) is the *period* of f. Homotopy theoretic arguments yield the following fundamental fact:

PROPOSITION 10.1. *Let $f\colon Z \to Z$ be a homotopy equivalence for which the rotation degree homomorphism $P_*\colon \Gamma \to \mathbb{Z}$ is non-zero, and let q be the least positive element of $P_*(\Gamma)$. Then f is a coherent periodic homotopy idempotent of period q.*

The automorphism $\theta\colon H \to H$ induces an automorphism $\theta_{\mathrm{ab}}\colon G_{\mathrm{ab}} \to G_{\mathrm{ab}}$. We identify G_{ab} with $\mathrm{coker}(\mathrm{id} - \theta_{\mathrm{ab}}) \times \mathbb{Z}$ by sending $ht^n \in G$ to $(\{h\}, -n)$. When Z, and hence also X, is finite we can obtain from Proposition 10.1:

THEOREM 10.2. *Let $f\colon Z \to Z$ be a cellular homotopy equivalence and let $X = T(Z, f)$.*

1. *If f is not a coherent periodic homotopy idempotent then $\chi_1(X)(\gamma) = 0$ and $\chi_1(X;\mathbb{Q})(\gamma) = 0$ for all $\gamma \in \Gamma$.*
2. *If f is a coherent periodic homotopy idempotent of period q, and $\gamma \in \Gamma$ has rotation degree $\nu q > 0$ then*

$$\chi_1(X)(\gamma) = \left(\sum_{n \geq 0} \sum_{i=0}^{\nu q - 1} (-1)^n A\left(\mathrm{trace}([\tilde{f}_n][f_n^i])\right), -\sum_{i=0}^{\nu q-1} L(f^i) \right) \quad \textit{and}$$

$$\chi_1(X;\mathbb{Q})(\gamma) = \left(0, -\sum_{i=0}^{\nu q-1} L(f^i) \right).$$

These formulas are determined by the rotation degree of γ.

REMARK. The above formulas are deduced using Proposition 5.7 from formulas for the "lifted" invariants $\widetilde{\chi}_1(X)$ and $\widetilde{\chi}_1(X;\mathbb{Q})$; see [**GN4**]. The corresponding expression for $\widetilde{\chi}_1(X;\mathbb{Q})$ is closely connected to the "non-commutative" theory of zeta functions developed in our paper [**GN3**].

EXAMPLE 10.3. Let $f = \mathrm{id}_Z$. Then $X = T(Z, \mathrm{id}_Z) = Z \times S^1$. Let $\nu \geq 0$. The "rotation along the S^1 factor", $\mathcal{R}_\nu \equiv \mathrm{id} \times R^\nu$ (see Example 8.1), represents an element of $\Gamma = \pi_1(\mathcal{E}(Z \times S^1), \mathrm{id})$ of rotation degree ν. Then

$$\widetilde{\mathcal{X}}_1(Z \times S^1)([\mathcal{R}_\nu]) = \chi(Z) T_1\left(t \otimes \frac{1 - t^\nu}{1 - t} \right).$$

This formula also holds for $\nu < 0$. It follows that $\chi_1(Z \times S^1)([\mathcal{R}_\nu]) = (0, -\chi(Z)\nu) = \chi(Z)\nu\{t\}$ where $\{t\} \in H_1(Z \times S^1) \cong H_1(Z) \oplus H_1(S^1)$ is the generator of the $H_1(S^1)$ summand determined by t.

Now we will apply these results to the case where Z, and hence also $X = T(Z, f)$, is aspherical. Some preliminary group theory is needed.

Let H be an arbitrary group, let $\theta\colon H \to H$ be an automorphism, and let G be the semidirect product $\langle H, t \mid tht^{-1} = \theta(h) \text{ for all } h \in H \rangle$. Write $\mathrm{Fix}(\theta) = \{h \in H \mid \theta(h) = h\}$ and write $\langle x \rangle$ for the cyclic subgroup generated by $x \in G$.

Let $\mathrm{Out}(H) = \mathrm{Aut}(H)/\mathrm{Inn}(H)$ be the group of outer automorphisms of H, i.e. the quotient of the group, $\mathrm{Aut}(H)$, of automorphisms of H by the normal subgroup $\mathrm{Inn}(H)$ of inner automorphisms.

PROPOSITION 10.4. *If θ has infinite order in* $\mathrm{Out}(H)$, *then* $Z(G) = Z(H) \cap \mathrm{Fix}(\theta)$. *If θ has finite order r in* $\mathrm{Out}(H)$, *and $h_0 \in H$ is such that* $\theta^r(\cdot) = h_0(\cdot)h_0^{-1}$, *there are two cases:*

1. *if no positive power of h_0 lies in $Z(H)\,\mathrm{Fix}(\theta)$ then $Z(G) = Z(H) \cap \mathrm{Fix}(\theta)$,*
2. *if some positive power of h_0 lies in $Z(H)\,\mathrm{Fix}(\theta)$ then*

$$Z(G) = \big(Z(H) \cap \mathrm{Fix}(\theta)\big)\langle u h_0^{-p} t^{rp}\rangle$$

where p is the smallest positive integer such that $h_0^{-p} \in Z(H)\,\mathrm{Fix}(\theta)$ and $u \in Z(H)$ is such that $u h_0^{-p} \in \mathrm{Fix}(\theta)$.

DEFINITION 10.5. Let $\theta\colon H \to H$ be an automorphism which has finite order r in $\mathrm{Out}(H)$, and let $h_0 \in H$ be such that $\theta^r(\cdot) = h_0(\cdot)h_0^{-1}$. If some positive power of h_0 lies in $Z(H)\,\mathrm{Fix}(\theta)$ we say θ is *special*. The *period* of a special automorphism is the integer $q = pr$ where p is the least positive integer such $h_0^{-p} \in Z(H)\,\mathrm{Fix}(\theta)$.

REMARK. θ is special if and only if the image of h_0 has finite order in the quotient group $N_H\big(\mathrm{Fix}(\theta)\big)/Z(H)\,\mathrm{Fix}(\theta)$ where $N_H\big(\mathrm{Fix}(\theta)\big)$ is the normalizer of $\mathrm{Fix}(\theta)$ in H. In particular, if $\theta^m = \mathrm{id}$ (with $m > 0$) then θ is special because $h_0^p \in Z(H)$ where $p = m/r$.

Here are some other conditions which imply that θ is special:

PROPOSITION 10.6. *Suppose $\theta\colon H \to H$ has finite order r in* $\mathrm{Out}(H)$, *the restriction of θ to $Z(H)$ is the identity, and $Z(H)$ is torsion free. Then θ is special with period r.*

PROPOSITION 10.7. *Suppose $\theta\colon H \to H$ has finite odd order r in* $\mathrm{Out}(H)$ *and the restriction of θ to $Z(H)$ is given by $h \mapsto h^{-1}$. Then θ is special with period r.*

COROLLARY 10.8. *Suppose $Z(H)$ is infinite cyclic and the automorphism θ has finite odd order r in* $\mathrm{Out}(H)$. *Then θ is special with period r.*

Let Z be a (not necessarily finite) $K(H,1)$ complex and let $f\colon Z \to Z$ be a continuous map which induces θ (after choosing a basepoint and basepath). The homomorphism $(p_f)_*\colon G \to \mathbb{Z}$ is identified with $ht^m \mapsto -m$. Since $\Gamma = \pi_1\big(\mathcal{E}(Z), \mathrm{id}\big) \cong Z(G)$, the rotation degree homomorphism, $P_*\colon Z(G) \to \mathbb{Z}$ is just the restriction of $(p_f)_*$. Thus we have:

PROPOSITION 10.9. *There is an exact sequence*

$$0 \to Z(H) \cap \mathrm{Fix}(\theta) \to Z(G) \xrightarrow{P_*} \mathbb{Z}.$$

If θ is not special then $P_* = 0$. If θ is special then $P_*\big(Z(G)\big) = q\mathbb{Z}$ where q is the period of θ.

Proposition 10.1 yields:

COROLLARY 10.10. *The map f is an eventually coherent periodic homotopy idempotent of period $q > 0$ if and only if θ is a special automorphism of period $q > 0$.*

Now suppose H is of type $\mathcal{F}$ so that we may take Z to be a finite $K(H,1)$ complex. Then $X = T(Z,f)$ is a finite $K(G,1)$ complex. By Theorem 10.2 and Corollary 10.10, we get:

THEOREM 10.11. *If θ is a special automorphism of period $q > 0$ then*

$$\chi_1(G)(ht^{\nu q}) = \left(\sum_{n\geq 0} \sum_{i=0}^{\nu q - 1} (-1)^n A\left(\operatorname{trace}([\tilde{f}_n][f_n^i])\right), -(q/r)\nu \sum_{i=0}^{r-1} L(f^i) \right)$$

and

$$\chi_1(G;\mathbb{Q})(ht^{\nu q}) = \left(0, -(q/r)\nu \sum_{i=0}^{r-1} L(f^i)\right) = (q/r)\nu \sum_{i=0}^{r-1} L(f^i)\{t\}$$

where r is the order of θ in $\operatorname{Out}(H)$ *and* $h \in \operatorname{Fix}(\theta) \cap h_0^{-\nu q/r} Z(H)$. *If θ is not special then $\chi_1(G) = 0$ and $\chi_1(G;\mathbb{Q}) = 0$.*

For the corresponding formulas for $\widetilde{\mathcal{X}}_1$, see [**GN4**].

11. A higher analog of Gottlieb's theorem

Let X be a finite aspherical complex. Gottlieb's theorem (see Corollary 2.2 and Proposition 5.3) asserts that if $\chi(X) \neq 0$ then the center of $G \equiv \pi_1(X)$ is trivial. We have an analogous theorem for $\chi_1(X;F)$ where F is a field: if $\chi_1(X;F) \neq 0$ then the center of G is infinite cyclic provided G satisfies an extra hypothesis (explained below) related to the Bass Conjecture.

Let R be a commutative ground ring and let S be any associative R-algebra with unit. The Hochschild homology group $HH_0(S)$ is the R-module $S/[S,S]$ where $[S,S]$ is the R-submodule of S generated by $\{ab - ba \mid a, b \in S\}$; see §4. The *Hattori-Stallings* trace $T_0\colon K_0(S) \to HH_0(S)$ is defined as follows. Let $A\colon M \to M$ be an idempotent endomorphism of a free, finitely generated right S-module M representing $x \in K_0(S)$. If $[A]$ is the matrix of A with respect to a given basis for M then $T_0(x)$ is defined to be $T_0([A]) \in HH_0(S)$; see §4.

Consider the groupring, RG, of a group G over R. Then $HH_0(RG)$ is naturally isomorphic to the free R-module generated by G_1, the set of conjugacy classes of G (see §4 for an explanation in the case $R = \mathbb{Z}$). Recall that for $g \in G$ we write $C(g) \in G_1$ for the conjugacy class of g, $HH_0(RG)_{C(g)}$ for the summand of $HH_0(RG)$ corresponding to $C(g)$ and $x_{C(g)}$ for the $C(g)$-component of $x \in HH_0(RG)$. Also write $HH_0(RG) = HH_0(RG)_{C(1)} \oplus HH_0(RG)'$ where $1 \in G$ is the identity element of G, and $HH_0(RG)'$ is the direct sum of the remaining factors. The augmentation homomorphism $\varepsilon\colon RG \to R$ induces a homomorphism $\varepsilon_*\colon HH_0(RG) \to HH_0(R) = R$.

PROPERTY A. We say that the group G has *Property A over R* if the image of $T_0\colon K_0(RG) \to HH_0(RG)$ lies in the $HH_0(RG)_{C(1)}$ summand.

PROPERTY B. We say that the group G has *Property B over R* if the composite $K_0(RG) \xrightarrow{T_0} HH_0(RG) \xrightarrow{\text{projection}} HH_0(RG)' \xrightarrow{\varepsilon_*} R$ is zero.

Clearly, if G has Property A over R then it also has Property B over R. The following two conjectures have become known as the Strong Bass Conjecture and Weak Bass Conjecture respectively; see [**Bass**] and [**Stall**, §4.1]. .

CONJECTURE A (STRONG CONJECTURE). Every group has Property A over $\mathbb{Z}$.

CONJECTURE B (WEAK CONJECTURE). Every group has Property B over $\mathbb{Z}$.

The corresponding conjectures are false over $\mathbb{Q}$ for a group which has nontrivial torsion; instead, one could conjecture:

CONJECTURE A′ (STRONG CONJECTURE). Every torsion free group has Property A over $\mathbb{Q}$.

CONJECTURE B′ (WEAK CONJECTURE). Every torsion free group has Property B over $\mathbb{Q}$.

Each element of the center of G, $Z(G)$, makes up its own conjugacy class. Given a subgroup N of $Z(G)$, let $HH_0(RG)_N = \bigoplus_{C(g)\in c(N)} HH_0(RG)_{C(g)}$ where $c(N)$ is the set of conjugacy classes in G represented by elements of N. Then $HH_0(RG) = HH_0(RG)_N \oplus HH_0(RG)'_N$ where $HH_0(RG)'_N$ is the direct sum of the factors corresponding to the conjugacy classes not in $c(N)$.

PROPERTY C. We say that the group G has *Property C over R* if there exists a subgroup N of $Z(G)$ such that the composite

$$K_0(RG) \xrightarrow{T_0} HH_0(RG) \xrightarrow{\text{projection}} HH_0(RG)'_N \xrightarrow{\varepsilon_*} R$$

is zero.

By taking N to be the trivial subgroup of $Z(G)$ we see that if G has Property B over R then it also has Property C over R.

Recall that a group G is said to have finite cohomological dimension over R if there exists an integer N such that $H^k(G, M) = 0$ for all RG-modules M and for all $k > N$. Also, G is said to be of type FP^∞ over R if R has a resolution by finitely generated projective RG-modules.

The following proposition is derived from the techniques of [**Stall**, §3].

PROPOSITION 11.1. *Let R be a principal ideal domain of characteristic p. Suppose that G is of type FP^∞ over R and has finite cohomological dimension over R. Suppose that G has a subgroup H of finite index which has Property C over R; furthermore, if $p > 0$ assume that p does not divide $[G\colon H]$. If the Euler characteristic $\chi(G; R) \equiv \sum_{i\geq 0}(-1)^i \operatorname{rank}_R H_i(G, R)$ is non-zero modulo p then the center of G is finite.*

We will be interested in groups with the property that certain of their central quotients have Property C "virtually":

PROPERTY D. Let p be the characteristic of R. We say that the group G has *Property D over R* if the following condition holds. Given any element τ in the center of G with the property that the extension class $e_R \in H^2(G/\langle\tau\rangle; R)$ is zero (where $\langle\tau\rangle$ is the cyclic subgroup generated by τ), there is a finite index subgroup $H \subset G/\langle\tau\rangle$ such that H has Property C over R; moreover, if $p > 0$ we require that p does not divide $[G : H]$.

The next theorem is our "higher" analog of Gottlieb's theorem over a field of arbitrary characteristic; Theorem 11.4, below, is a sharper version over $\mathbb{Q}$.

THEOREM 11.2. *Let F be a field. Suppose G is a group of type $\mathcal{F}$ such that G has Property D over F. If $\chi_1(G; F) \neq 0$, then the center of G is infinite cyclic.*

In order to apply the above theorem in the case $R = \mathbb{Q}$, we need:

PROPOSITION 11.3. *Let G be a finitely generated group which has Property B over $\mathbb{Q}$. Then G has Property D over $\mathbb{Q}$.*

Combining Theorem 11.2 and Proposition 11.3 yields:

THEOREM 11.4. *Suppose that G is a group of type $\mathcal{F}$ and has Property B over $\mathbb{Q}$. If $\chi_1(G; \mathbb{Q}) \neq 0$, then the center of G is infinite cyclic.*

This motivates the following:

CONJECTURE B$''$. *Every group of type $\mathcal{F}$ has Property B over $\mathbb{Q}$.*

Of course, Conjecture B$'$ implies Conjecture B$''$.

A group G is a *linear group* if it is a subgroup of $GL(n,k)$ where k is a field of characteristic zero. Bass [**Bass**, Theorem 9.6] proved that a torsion free linear group has Property A over $\mathbb{C}$ (and thus has Property B over $\mathbb{Q}$); also see [**Eck**].

COROLLARY 11.5. *Suppose G is a linear group of type $\mathcal{F}$. If $\chi_1(G;\mathbb{Q}) \neq 0$, then the center of G is infinite cyclic.*

Eckmann [**Eck**] proved that a group of cohomological dimension 2 over $\mathbb{Q}$ has Property A over $\mathbb{Q}$. Consequently:

COROLLARY 11.6. *Suppose G is of type $\mathcal{F}$ and has cohomological dimension* 2 *over $\mathbb{Q}$. If $\chi_1(G;\mathbb{Q}) \neq 0$, then the center of G is infinite cyclic.*

12. Outer automorphisms of groups of type $\mathcal{F}$

An application of the previous sections is the following theorem which relates the algebraic topology of a special automorphism (see Definition 10.5) of a group of type $\mathcal{F}$ to the fixed group of the automorphism.

THEOREM 12.1. *Let H be a group of type $\mathcal{F}$ which has Property B over $\mathbb{Q}$* (i.e. *H satisfies Bass's Weak Conjecture over $\mathbb{Q}$). Suppose that $\theta\colon H \to H$ is a special automorphism whose order in* $\mathrm{Out}(H)$ *is $r > 1$. If the sum of the Lefschetz numbers $\sum_{i=1}^{r-1} L(\theta^i)$ is non-zero then $Z(H) \cap \mathrm{Fix}(\theta) = (1)$.*

PROOF. By Proposition 5.3 and Corollary 2.2, we may assume $\chi(H) = L(\theta^0) = 0$. Then $\sum_{i=0}^{r-1} L(\theta^i) = \sum_{i=1}^{r-1} L(\theta^i) \neq 0$. Let G be the semidirect product $G = H \times_\theta T$ where T is infinite cyclic. It can be shown ([**GN4**]) that G also has Property B over $\mathbb{Q}$. Applying Theorem 10.11 to G, we have that $\chi_1(G;\mathbb{Q}) \neq 0$. By Theorem 11.4, $Z(G)$ is infinite cyclic. By Proposition 10.9 , there is an exact sequence $1 \to Z(H) \cap \mathrm{Fix}(\theta) \to Z(G) \xrightarrow{P_*} q\mathbb{Z} \to 1$ where the period of θ, q, is positive. It follows that $Z(H) \cap \mathrm{Fix}(\theta) = (1)$. $\square$

COROLLARY 12.2. *Let H be a group of type $\mathcal{F}$ which has Property B over $\mathbb{Q}$. Suppose that $\theta\colon H \to H$ is a special automorphism of order* 2 *in* $\mathrm{Out}(H)$. *If $L(\theta) \neq 0$ then $Z(H) \cap \mathrm{Fix}(\theta) = (1)$.*

COROLLARY 12.3. *Let H be a group of type $\mathcal{F}$ which has Property B over $\mathbb{Q}$. Suppose $Z(H) \neq (1)$, the automorphism $\theta\colon H \to H$ has finite order r in* $\mathrm{Out}(H)$ *and the restriction of θ to $Z(H)$ is the identity. Then $\sum_{i=1}^{r-1} L(\theta^i) = 0$.*

A special automorphism may have infinite order in $\mathrm{Aut}(H)$. If we make the assumption that θ is of finite order in $\mathrm{Aut}(H)$ then θ is necessarily special (see §10) and the Bass Conjecture hypothesis can be dispensed with in Theorem 12.1 and Corollary 12.2:

PROPOSITION 12.3. *Let H be a group of type $\mathcal{F}$. Suppose that $\theta\colon H \to H$ has finite order in* $\mathrm{Aut}(H)$ *and $L(\theta) \neq 0$. Then $Z(H) \cap \mathrm{Fix}(\theta) = (1)$.*

Note that $\sum_{i=1}^{r-1} L(\theta^i) \neq 0$ implies one of the $L(\theta^i)$'s is non-zero. Since $\mathrm{Fix}(\theta) \subset \mathrm{Fix}(\theta^i)$ for $i \geq 0$, we recover Theorem 12.1 (but without the Bass Conjecture hypothesis) in the special case where θ has finite order in $\mathrm{Aut}(H)$.

References

[Bass] H. Bass, *Euler characteristics and characters of discrete groups*, Invent. Math. **35** (1976), 155–196.

[B] K. S. Brown, *Cohomology of Groups*, Springer-Verlag, New York, 1982.

[DG] D. Dimovski and R. Geoghegan, *One-parameter fixed point theory*, Forum Math. **2** (1990), 125–154.

[D] A. Dold, *Fixed point index and fixed point theorem for Euclidean neighborhood retracts*, Topology **4** (1965), 1–8.

[Eck] B. Eckmann, *Cyclic homology of groups and the Bass conjecture*, Comment. Math. Helv. **61** (1986), 193–202.

[GN1] R. Geoghegan and A. Nicas, *Parametrized Lefschetz-Nielsen fixed point theory and Hochschild homology traces*, Amer. J. Math. **116** (1994).

[GN2] ——, *Lefschetz Trace Formulae, Zeta Functions and Torsion in Dynamics*, Proceedings of the Joint Summer Research Conference on Nielsen Theory and Dynamical Systems, Contemporary Math., vol. 152, 1993, pp. 141–157.

[GN3] ——, *Trace and torsion in the theory of flows*, Topology (to appear).

[GN4] ——, *Higher Euler characteristics* (preprint).

[Got] D. H. Gottlieb, *A certain subgroup of the fundamental group*, Amer. J. Math. **87** (1965), 840–856.

[Kn] R. J. Knill, *On the homology of a fixed point set*, Bull. Amer. Math. Soc. **77** (1971), 184–190.

[Le] J. Leray, *Sur les équations et les transformations*, J. Math. Pures Appl. (9) **24** (1945), 201–248.

[Ly] R. C. Lyndon, *Cohomology theory of groups with a single defining relation*, Annals of Math. **52** (1950), 650–665.

[MS] J. Milnor and J. Stasheff, *Characteristic Classes*, Ann. of Math. Studies, No. 76, Princeton Univ. Press, Princeton, NJ, 1974.

[Stall] J. Stallings, *Centerless groups—An algebraic formulation of Gottlieb's theorem*, Topology **4** (1965), 129–134.

Department of Mathematics, SUNY Binghamton, Binghamton, NY 13902–6000, USA

E-mail address: ross@math.binghamton.edu

Department of Mathematics, McMaster University, Hamilton, Ontario L8S 4K1, Canada

E-mail address: nicas@mcmail.cis.mcmaster.ca

Centre de Recherches Mathématiques
CRM Proceedings and Lecture Notes
Volume **6**, 1994

On Fibrewise Homotopy Theory

Ioan M. James

The aim of this article is to give some account of recent work on fibrewise homotopy theory by members and former members of my research group at Oxford. To a considerable extent the results obtained are fibrewise versions of results in ordinary homotopy theory obtained by Hilton and his associates many years ago.

Some of this is fairly routine work but much of it is not and there are some unsolved problems which may be of interest. The main topics discussed are fibrewise Hopf spaces, fibrewise coHopf spaces and fibrewise category. A few results are stated without proof because of space limitations.

1. Introduction

Ordinary homotopy theory is normally concerned with pointed spaces, i.e. spaces with basepoint. While playing an essential role the basepoint does not receive much attention. Every non-empty space admits a basepoint. In a path-connected space it generally does not matter which basepoint is chosen although if it is necessary for the space to be non-degenerate or well-pointed the choice may be more restricted. Throughout this article the terms space, map, homotopy etc. are used in the pointed sense. Subspaces must always contain the basepoint which is assumed to be closed.

Equivariant homotopy theory is not normally concerned with pointed spaces. One reason is that there may be no fixed points. Throughout this article, however, we use the terms G-space, G-map, G-homotopy etc. in the pointed sense, where G is a topological group. Thus we do not regard the sphere S^{n-1} as an $O(n)$-space, since there are no fixed points. We do regard the sphere S^n as an $O(n)$-space, choosing one of the poles as basepoint, but we do not regard S^n as a suspension, in the equivariant sense.

Fibrewise homotopy theory is normally concerned with fibrewise pointed spaces, i.e. fibrewise spaces with sections. It is usually convenient to require the sections to be closed and we do so here. By definition every sectionable fibrewise space admits a section. In a vertically-connected fibrewise space it often does not matter which section is chosen but if it is necessary for the fibrewise space to be fibrewise non-degenerate or fibrewise well-pointed the choice may be more restricted. In any case one obviously does not wish to be restricted to vertically-connected fibrewise spaces. Throughout this article the terms fibrewise space, fibrewise homotopy etc. are used in the pointed sense. Subspaces must always contain the section.

1991 *Mathematics Subject Classification.* Primary: 55P99; Secondary: 55R10.

This is the final form of the paper.

There is an important link between equivariant homotopy theory and fibrewise homotopy theory, as follows. Let G be a compact topological group and let P be a principal G-bundle over the base space B. For each G-space A the associated bundle $P_{\#}A$ is defined, as a fibrewise space over B and similarly with G-maps and G-homotopies. We refer to $P_{\#}$ as the *associated bundle functor*.

In what follows we adopt the terminology and notation of [**19**] but with some simplifications. Because we are only concerned with fibrewise pointed spaces, for example, it seems safe to use the term fibrewise suspension, and the notation Σ_B, since we are only concerned with the reduced fibrewise suspension.

The methodology of fibrewise homotopy theory may be outlined as follows. Once the fibrewise version of a result in ordinary homotopy theory has been correctly formulated there may be more than one way of proving it, depending on what restrictions are acceptable. Ideally one constructs, step by step, a fibrewise version of one of the proofs used to establish the ordinary version of the result. Not all proofs may be suitable for this treatment. But a useful alternative is to apply Dold's "local-global" theorem. The appropriate form of this theorem is the section-preserving version[1] given by Eggar in [**9**], which we restate here for the convenience of the reader.

THEOREM 1.1 (DOLD-EGGAR). *Let ϕ: $X \to Y$ be a fibrewise map, where X and Y are fibrewise spaces over B. Suppose that B admits a numerable covering such that ϕ_V is a fibrewise homotopy equivalence over V for each member V of the covering. Then ϕ is a fibrewise homotopy equivalence over B.*

In the applications it is usually a corollary of Theorem 1.1 which is used, as follows. Recall that a subspace V of B is said to be *categorical* if V is contractible in B. We say that B is *numerably categorical* if B admits a numerable covering by categorical subsets. Then we have

COROLLARY 1.2. *Let B be numerably categorical. Let ϕ: $X \to Y$ be a fibrewise map, where X and Y are fibrant fibrewise spaces over B. Suppose that the restriction of ϕ to the fibre is a homotopy equivalence. Then ϕ is a fibrewise homotopy equivalence over B.*

For some results we need to restrict ourselves to the class of *well-sectioned* fibrewise spaces, i.e. those for which the section is a *fibrewise cofibration*, as defined in §22 of [**19**] (see also (1.3) of [**5**]). Of course B is always well-sectioned, as a fibrewise space over itself. Also the fibrewise collapse $X/_B X'$ is well-sectioned, for any fibrewise cofibred pair over B.

The associated bundle functor $P_{\#}$ transforms G-cofibrations into fibrewise cofibrations over B, as shown in §24 of [**19**], for each G-bundle P over B. In particular $P_{\#}$ transforms well-pointed G-spaces into well-sectioned fibrewise spaces.

2. Fibrewise Hopf and coHopf spaces

Hopf spaces and coHopf spaces are discussed in many places in the literature (see §12 of [**5**], for example). The corresponding concepts in equivariant homotopy theory have also been considered. In what follows it is the fibrewise versions of these concepts which we shall be concerned with. Although these have been discussed in several places in the literature (see [**21**] and [**28**], for example) they are less well-known and so it seems best to begin with an outline of the basic theory.

[1] Eggar assumes B to be paracompact but this can be avoided by using (7.24) of [**18**] instead.

As before we work over a base space B. Consider a fibrewise space X over B. By a *fibrewise multiplication* on X we mean a fibrewise map

$$m\colon X \times_B X \to X.$$

By a *fibrewise Hopf structure* on X we mean a fibrewise multiplication m such that

$$m \circ (id \times c) \circ \Delta \simeq id_X \simeq m \circ (c \times id) \circ \Delta$$

by fibrewise homotopies, as shown below

$$X \xrightarrow{\Delta} X \times_B X \underset{c\times id}{\xrightarrow{id\times c}} X \times_B X \xrightarrow{m} X.$$

Here Δ denotes the diagonal and c the fibrewise constant. Given such a structure on X we describe X as a *fibrewise Hopf space.*

By a *fibrewise comultiplication* on X we mean a fibrewise map

$$m\colon X \to X \vee_B X.$$

By a *fibrewise coHopf structure* on X we mean a fibrewise comultiplication m such that

$$\nabla \circ (id \vee c) \circ m \simeq id_X \simeq \nabla \circ (c \vee id) \circ m$$

by fibrewise homotopies, as shown below

$$X \xrightarrow{m} X \vee_B X \underset{c\vee id}{\xrightarrow{id\vee c}} X \vee_B X \xrightarrow{\nabla} X.$$

Here ∇ denotes the codiagonal and c, as before, the fibrewise constant. Given such a structure on X we describe X as a *fibrewise coHopf space.*

For example, let T be a (pointed) space. Then $B \times T$, with the section determined by the basepoint, is fibrewise Hopf if T is Hopf and fibrewise coHopf if T is coHopf.

Sectionable fibrewise spaces may admit fibrewise Hopf structure with one choice of section but not with another. Thus consider the product $S \times S$, for some (pointed) S. If S is a Hopf space, with basepoint e, then $S \times S$ is a fibrewise Hopf space with the axial section $S \times \{e\}$, as we have just observed. Suppose, however, that we replace the axial section by the diagonal. Then a fibrewise Hopf structure on $S \times S$ would determine a map of $S \times S \times S$ into S which satisfies the Hopf condition on $S \times S \times \{e\}$ and maps the diagonal of $S \times S \times S$ identically onto S. When S is a sphere (say $S = S^1$) this is a homological impossibility. Similar examples show that sectionable fibrewise spaces may admit fibrewise coHopf structure with one choice of section but not with another.

For fibrant fibrewise spaces the following result is useful.

PROPOSITION 2.1. *Let B be numerably categorical. Let X be a fibrant fibrewise space over B.*

(i) *If a Hopf structure on the fibre of X can be extended to a fibrewise multiplication on X then it can be extended to a fibrewise Hopf structure.*

(ii) *If a coHopf structure on the fibre of X can be extended to a fibrewise comultiplication on X then it can be extended to a fibrewise coHopf structure.*

Here we mean extend up to homotopy, rather than extend as a map. To prove (i) let $m\colon X \times_B X \to X$ be a fibrewise multiplication on X extending a Hopf structure on the fibre. Let $\sigma_j\colon X \to X \times_B X$ $(j = 1, 2)$ be the axial sections of the fibrewise product. Then the restriction of the fibrewise map $m \circ \sigma_j\colon X \to X$ to the fibre is homotopic to the identity. Hence $m \circ \sigma_j$ is a fibrewise homotopy equivalence, by Theorem 1.1. Let $\alpha_j\colon X \to X$ be a fibrewise homotopy inverse of $m \circ \sigma_j$. Then

$$m \circ (\alpha_1 \times \alpha_2)\colon X \times_B X \to X$$

is a fibrewise Hopf structure on X which extends the given Hopf structure on the fibre. This proves (i) and the proof of (ii) is similar.

For fibrewise multiplications and fibrewise comultiplications the terms *fibrewise homotopy-commutative* and *fibrewise homotopy-associative* are self-explanatory, also the terms *left* and *right fibrewise homotopy inverse.* When fibrewise homotopy-associativity holds the left and right fibrewise homotopy inverses, if they exist, are equivalent, up to fibrewise homotopy.

Fibrewise homotopy-associative fibrewise Hopf structures which admit fibrewise homotopy inverses are said to be *fibrewise group-like.* For example the fibrewise topological groups of [**24**] are fibrewise group-like. Also fibrewise loop-spaces are fibrewise group-like.

Fibrewise homotopy-associative fibrewise coHopf structures which admit fibrewise homotopy inverses are said to be *fibrewise cogroup-like.* For example fibrewise suspensions are fibrewise cogroup-like.

By the distributive law for the fibrewise smash product, a fibrewise comultiplication on X determines a fibrewise comultiplication on $X \wedge_B Y$ for all fibrewise spaces Y. If the former is fibrewise coHopf then so is the latter, and similarly with the other conditions. Thus $X \wedge_B Y$ is fibrewise cogroup-like whenever X is fibrewise cogroup-like.

Given fibrewise spaces X and Y the fibrewise mapping-space $\operatorname{map}_B(X, Y)$ obtains a fibrewise Hopf structure if either (i) X is fibrewise coHopf or (ii) Y is fibrewise Hopf and X is fibrewise regular. If both (i) and (ii) hold the fibrewise Hopf structures on $\operatorname{map}_B(X, Y)$ which arise are equivalent, in the sense of fibrewise homotopy, and are fibrewise homotopy-commutative.

So far the two theories we have been considering have been completely dual, in the sense of Eckmann-Hilton, but now differences begin to emerge. Let X be a fibrewise space over B. Recall (see [**22**]) that a subspace V of X is said to be *fibrewise categorical* if V is fibrewise contractible in X. We say that X is *numerably fibrewise categorical* if X admits a numerable covering by fibrewise categorical subsets.

PROPOSITION 2.2. *Let X be a numerably fibrewise categorical well-sectioned fibrewise Hopf space over B. Then X admits fibrewise homotopy inverses on each side.*

For let $m\colon X \times_B X \to X$ be the fibrewise Hopf structure. By using the fibrewise homotopy extension property we may suppose, with no real loss of generality, that the section $s\colon B \to X$ is a strict neutral section for m, in the sense that

$$m \circ (c \times id) \circ \Delta = id = m \circ (id \times c) \circ \Delta,$$

where $c = s \circ p$ as before. Regard $X \times_B X$ as a fibrewise space over X using the first projection and the section given by $(c \times id) \circ \Delta$. Then the fibrewise shearing

map
$$k\colon X \times_B X \to X \times_B X$$
is fibrewise over X, where the components of k are given by $\pi_1 \circ k = \pi_2$, $\pi_2 \circ k = m$. By Theorem 1.1 k is a fibrewise homotopy equivalence. Hence the composition
$$X \xrightarrow{u} X \times_B X \xrightarrow{\ell} X \times_B X \xrightarrow{\pi_2} X$$
provides a right fibrewise homotopy inverse for m, where $u = (id \times c) \circ \Delta$ and ℓ is a fibrewise homotopy inverse of k. Similarly m admits a left fibrewise homotopy inverse.

One would hope for a similar result to be true in the dual case. However, even for ordinary coHopf spaces there is a problem here, and the best way to proceed seems to be to prove

PROPOSITION 2.3. *Let B be numerably categorical. Let X be a fibrant fibrewise coHopf space over B. Suppose that the coHopf structure on the fibre admits a homotopy inverse on either side. Then the fibrewise coHopf structure on X admits a fibrewise homotopy inverse on the same side.*

For suppose that the fibrewise coHopf structure m, restricted to the fibre, admits a homotopy inverse on the right. Consider the fibrewise "coshearing" map
$$k\colon X \vee_B X \to X \vee_B X$$
given by the identity on the first summand, by m on the second. By hypothesis k is a homotopy equivalence on the fibre and hence, by (1.2), a fibrewise homotopy equivalence. Now the composition
$$X \xrightarrow{\sigma_2} X \vee_B X \xrightarrow{\ell} X \vee_B X \xrightarrow{u} X$$
provides a right fibrewise homotopy inverse for m, where $u = \nabla \circ (id \vee c)$ and ℓ is a fibrewise homotopy inverse of k.

The existence of homotopy inverses on each side for ordinary coHopf structures has been discussed by Ganea [**14**] who shows (using coshearing maps) that it is sufficient for the fibres to be simply-connected.

For each fibrewise space X over B the right adjoint of the identity on $\Sigma_B X$ is a fibrewise map $u\colon X \to \Omega_B \Sigma_B X$ while the left adjoint of the identity on $\Omega_B X$ is a fibrewise map $v\colon \Sigma_B \Omega_B X \to X$. Following the practice of Ganea [**14**] and others in the ordinary theory let us describe a left inverse of u, up to fibrewise homotopy, as a *fibrewise retraction*, and a right inverse of v, up to fibrewise homotopy, as a *fibrewise coretraction*. When X is well-sectioned the qualification "up to fibrewise homotopy" is unnecessary.

Clearly if X is a fibrewise retract of $\Omega_B \Sigma_B X$ then X is a fibrewise Hopf space, while if X is a fibrewise coretract of $\Sigma_B \Omega_B X$ then X is a fibrewise coHopf space. Under certain conditions the converse implications hold.

PROPOSITION 2.4. *Let B be numerably categorical. Let X be a vertically-connected fibrant fibrewise Hopf space over B. Then X is a fibrewise retract of the fibrewise loop-space $\Omega_B \Sigma_B X$ on the fibrewise suspension of X.*

For consider the *fibrewise reduced product space* $J_B(X)$ of X, i.e. the fibrewise free monoid generated by X with topology as in [**20**]. The fibrewise Hopf structure on X determines a fibrewise retraction of $J_B(X)$ onto X by extending multiplicatively. Now it is shown[2] in [**20**], under the hypotheses of Proposition 2.4, that $J_B(X)$ has the same fibrewise homotopy type as $\Omega_B\Sigma_B X$, hence the result.

PROPOSITION 2.5. *Let X be a well-sectioned fibrewise coHopf space over B. Then X is a fibrewise coretract of the fibrewise suspension $\Sigma_B\Omega_B X$ of the fibrewise loop-space on X.*

For each fibrewise space Y we have a standard fibrewise coHopf structure

$$k\colon \Sigma_B Y \to \Sigma_B Y \vee_B \Sigma_B Y,$$

in particular we have this when $Y = \Omega_B X$. Consider the diagram shown below where v is as before and j is the inclusion

$$\begin{array}{ccc} \Sigma_B\Omega_B X & \xrightarrow{\ v\ } & X \\ {\scriptstyle (v\vee v)\circ k}\downarrow & & \downarrow{\scriptstyle \Delta} \\ X \vee_B X & \xrightarrow[\ j\ ]{} & X \times_B X \end{array}$$

It is not difficult to show (see (6.63) and (6.65) of [**18**]) that the fibrewise homotopy fibres of v and j are equivalent, in the sense of fibrewise homotopy type, moreover the verticals in the diagram induce a fibrewise homotopy equivalence of the fibrewise homotopy fibres. It follows that fibrewise coHopf structures $X \to X \vee_B X$ correspond precisely to fibrewise coretractions $X \to \Sigma_B\Omega_B X$.

In the ordinary case this result is due to Ganea [**14**] who works in the category of CW-spaces. Sunderland [**29**] gives a fibrewise version of Ganea's proof which is similar to the one indicated here. Sunderland [**30**] uses Dold's theorem and Ganea's result instead.

It is hard to believe that there is not a proof of Proposition 2.4 on similar lines but I am not aware of one. Almost certainly the assumptions needed to make the proof we have given work are unnecessarily restrictive.

In the ordinary theory the existence of Hopf structure is equivalent to the vanishing of the Whitehead square, on the one hand, and to the existence of maps of Hopf invariant one on the other. Fibrewise versions of these results can be developed (see [**11**], for example), but these are somewhat routine. Again one can construct the fibrewise projective plane, from a fibrewise multiplication, and analyse its fibrewise cohomology. For reasons of space we shall not pursue these matters here, in full generality, although some special cases will be discussed later on.

3. Sectioned sphere-bundles

We turn now from the general theory to an important class of special cases, the sectioned sphere-bundles. As we shall see there is plenty of interest in this range of examples. We may begin with a (real) n-plane bundle ξ over B. The fibrewise compactification ξ_B^+ is a sectioned n-sphere bundle, the section being the complement of ξ, and every sectioned n-sphere bundle can be represented in this way. To look at the situation in another way, ξ_B^+ can be obtained from the sphere S^n, as a (pointed) $O(n)$-space, by applying the associated bundle functor

[2] Admittedly the assumptions are somewhat different but the same argument applies.

$P_\#$ where P is a principal $O(n)$-bundle over B. Let us begin our discussion with some observations about $O(n)$-spaces.

Points of S^n are represented in the form

$$(\cos\alpha, x\sin\alpha) \quad (x \in S^{n-1}, 0 \le \alpha \le \pi)$$

with $e = (1, 0)$ as basepoint. We regard S^n as an $O(n)$-space so that $g \in O(n)$ transforms $(\cos\alpha, x\sin\alpha)$ into $(\cos\alpha, g.x\sin\alpha)$. For each integer k we have an $O(n)$-map $\theta_k \colon S^n \to S^n$ of degree k, where

$$\theta_k(\cos\alpha, x\sin\alpha) = (\cos k\alpha, x\sin k\alpha).$$

Note that $\theta_{k_1k_2} = \theta_{k_1}\theta_{k_2}$.

We regard $S^n \times S^n$ as an $O(n)$-space using the diagonal action. There is a well-known $O(n)$-map

$$\phi\colon S^n \times S^n \to S^n$$

of bidegree $\bigl(1 + (-1)^{n+1}, 1\bigr)$, which sends each pair (y, z) into the reflection of $\theta_{-1}z$ in the hyperplane orthogonal to y.

The coproduct $S^n \vee S^n$ may be identified with the invariant subspace

$$S^n \times \{e\} \cup \{e\} \times S^n \subset S^n \times S^n$$

Another $O(n)$-map

$$\psi\colon S^n \to S^n \vee S^n$$

of bidegree $\bigl(1 + (-1)^{n+1}, 1\bigr)$ is defined which sends $(\cos\alpha, x\sin\alpha)$ into

$$\begin{array}{lll} (0) & \bigl((\cos 3\alpha, x\sin 3\alpha), e\bigr) & (0 \le \alpha \le 2\pi/3) \\ (1) & \bigl(e, (\cos 3\alpha, x\sin 3\alpha)\bigr) & (2\pi/3 \le \alpha \le \pi). \end{array}$$

We use these $O(n)$-maps, as described above, to construct fibrewise maps of the associated sectioned n-sphere bundles.[3] Specifically let ξ, as before, be an n-plane bundle with fibrewise compactification ξ_B^+. Then for each integer k we have a fibrewise map $\Theta_k \colon \xi_B^+ \to \xi_B^+$ of degree k on the fibre. We also have the fibrewise multiplication

$$\Phi\colon \xi_B^+ \times_B \xi_B^+ \to \xi_B^+$$

of bidegree $\bigl(1 + (-1)^{n+1}, 1\bigr)$ on the fibres, and the fibrewise comultiplication

$$\Psi\colon \xi_B^+ \to \xi_B^+ \vee_B \xi_B^+$$

of the same bidegree on the fibres.

PROPOSITION 3.1. *Let B be numerably categorical. Let ξ be an n-plane bundle over B, where n is odd. Suppose that the sectioned n-sphere bundle ξ_B^+ admits a fibrewise multiplication of bidegree $(2k + 1, 2l + 1)$ on the fibres, where k and l are integers. Then ξ_B^+ admits fibrewise Hopf structure.*

[3] Although we do not need to use Noakes pioneering work in this area a courtesy reference to **[27]** here is surely appropriate.

For let m be a fibrewise multiplication on ξ_B^+ of bidegree $(2k+1, 2l+1)$ on the fibres. Replace m by m', where

$$m'(y,z) = \Phi\Big(\Theta_{-k}y, \Phi(\Theta_{-l}z, m(y,z))\Big)$$

for y, z in the same fibre of ξ_B^+. Since Φ has bidegree $(2,1)$ on each fibre we find that m' has bidegree $(1,1)$ on the fibres. By Proposition 2.1, therefore, ξ_B^+ admits fibrewise Hopf structure. Similarly using Ψ instead of Φ we obtain

PROPOSITION 3.2. *Let B be numerably categorical. Let ξ be an n-plane bundle over B, where n is odd. Suppose that the sectioned n-sphere bundle ξ_B^+ admits a fibrewise comultiplication of bidegree $(2k+1, 2l+1)$ on the fibres, where k, l are integers. Then ξ_B^+ admits fibrewise coHopf structure.*

As we shall see later the conclusion no longer holds for even values of n (it is unnecessary to say this in the case of Proposition 3.1 since S^n cannot admit a multiplication of bidegree $(2k+1, 2l+1)$ when n is even). For odd values of n, however, these two results show that the existence of fibrewise Hopf structure and the existence of fibrewise coHopf structure on sectioned n-sphere bundles are 2-local problems. By using the machinery of fibrewise localization (see [**25**], for example) one can describe the situation with greater precision but the message is that odd primes are irrelevant.

As we have already remarked, there is a relation between fibrewise multiplications

$$m\colon X \times_B X \to X$$

where $X = \xi_B^+$, and the corresponding fibrewise maps

$$c(m)\colon X *_B X \to \Sigma_B X,$$

obtained by the fibrewise version of the Hopf construction. Clearly if m determines a Hopf structure on the fibre S^n then $c(m)$ determines a map $S^{2n+1} \to S^{n+1}$ of Hopf invariant one, and conversely. In fact one can go further and show that the existence of a fibrewise map

$$h\colon X *_B X \to \Sigma_B X$$

of Hopf invariant one on the fibres implies the existence of a fibrewise map

$$m\colon X \times_B X \to X$$

of bidegree $(1,1)$ on the fibres and hence the existence of fibrewise Hopf structure, by Proposition 2.1.

For which n-plane bundles ξ over B does ξ_B^+ admit fibrewise Hopf structure? The question only arises when $n = 0$, 1, 3 or 7, since otherwise the fibre does not admit Hopf structure. Ignoring the trivial case $n = 0$ we consider the other three cases in turn.

The classical Hopf structure on S^1, arising from complex multiplication, is $O(1)$-equivariant. Hence every line bundle ξ satisfies the condition and there is no more to be said.

The classical Hopf structure on S^3, arising from quaternionic multiplication, is $SO(3)$-equivariant. Hence every orientable 3-plane bundle ξ satisfies the condition.

The classical Hopf structure on S^7, arising from Cayley multiplication, is G_2-equivariant, where G_2 is the exceptional Lie group. Hence a 7-plane bundle ξ satisfies the condition if its structure group can be reduced to G_2.

What can be said in the other direction? It is only to be expected that the existence of fibrewise Hopf structure has implications for the characteristic classes. We find

PROPOSITION 3.3. *Let B be a CW-complex. A necessary condition on the n-plane bundle ξ for ξ_B^+ to admit fibrewise Hopf structure is that $w_i(\xi) = 0$ whenever $n+1-i$ is not a power of two.*

This was proved by Cook in his thesis through an analysis of the cohomology of the fibrewise projective plane determined by a fibrewise Hopf structure; a simpler proof is given by Cook and Crabb in [**4**].

The result shows, in particular that if $n = 3$ then $w_1(\xi) = 0$, so that ξ is orientable, while if $n = 7$ then $w_1(\xi) = 0$ and $w_2(\xi) = 0$ so that ξ admits spin structure. Thus we see that orientability is both necessary and sufficient when $n = 3$. Provided B is a finite complex and dim $B < 8$ we see that the existence of spin structure is both necessary and sufficient, since if the structural group is reducible to spin(7) it is also reducible to G_2.

Turning now to the dual problem, we recall that the Thom space B^ξ of the n-plane bundle ξ is equivalent to the space obtained from ξ_B^+ by collapsing the section. If ξ_B^+ admits fibrewise coHopf structure then B^ξ admits coHopf structure. This implies, in particular, that products are trivial in the cohomology ring of B^ξ. Since the Euler class $e(\xi)$ of ξ maps into the square of the Thom class under the Thom isomorphism

$$H^n(B) \to H^{2n}(B^\xi)$$

we conclude that $e(\xi) = 0$ when ξ_B^+ admits fibrewise Hopf structure.

This result is due to Sunderland [**29**] who established fibrewise versions of many of the results about coHopf spaces proved by Berstein, Ganea and Hilton. For example Sunderland obtains conditions for a fibrewise coHopf space to be a fibrewise suspension. Full details will be given in his forthcoming paper [**30**] but here is just one illustration.

Take B to be a smooth $(n+1)$-manifold M with vanishing Euler characteristic. Then the tangent n-sphere bundle $T(M)$ is sectionable. Can the section be chosen so that $T(M)$, as a sectioned n-sphere bundle, admits fibrewise coHopf structure? Certainly this is so if M admits a field of linearly independent 2-frames since then $T(M)$ with one vector (say the first) in each 2-frame providing the section is a fibrewise suspension. In fact Sunderland shows that this sufficient condition is also necessary for M simply-connected and $n \geq 4$. Necessary and sufficient conditions for M to admit a 2-field are given in [**31**] for the case when M is compact. It would be interesting to know more about the class of "fibrewise coHopf" manifolds, in this sense.

4. Sectioned sphere-bundles over spheres

Let us turn now to the case where the base space is a sphere. Specifically, consider the sectioned oriented n-sphere bundle X_α over S^m $(m > 1)$ corresponding, in the standard classification, to the element $\alpha \in \pi_{m-1}SO(n)$. For which α does X_α admit fibrewise Hopf structure or fibrewise coHopf structure? We take the second question first.

PROPOSITION 4.1. *The fibrewise pointed space X_α admits a fibrewise coHopf structure if and only if $\Sigma^n_* \rho_* \alpha = 0$.*

Here Σ_* denotes the suspension operator and ρ_* is induced by the evaluation map $\rho\colon SO(n) \to S^{n-1}$. We have $\rho_*\alpha = 0$ if and only if X_α admits a second section, orthogonal to the first, in which case X_α is the fibrewise suspension of a sectioned $(n-1)$-sphere bundle. In the metastable range, when $m < 2n-2$, the iterated suspension Σ^n_* is an isomorphism and so the existence of a fibrewise coHopf structure implies that $\rho_*\alpha = 0$ and hence that X_α is a fibrewise suspension. By taking $\alpha = 4\gamma$, where γ generates the infinite cyclic group $\pi_3 SO(3)$, we obtain an example where X_α is fibrewise coHopf, but not a fibrewise suspension, since the Thom space is the Berstein-Hilton non-suspension. However for even values of n it follows from the exactness of the EHP sequence that $\Sigma^n_* J\alpha = HJ\alpha = 0$ implies that $J\alpha \in \Sigma_* \pi_{m+n-1}(S^{n-1})$ and hence the existence of a fibrewise coHopf structure on X_α implies that X_a is a fibrewise suspension. By taking α so that $\rho_*\alpha$ has order 3 in $\pi_6(S^3) = \rho_* \pi_6 SO(4)$ we see that Propsition 3.2 above breaks down for even values of n.

To prove Proposition 4.1 we use the analysis of the homotopy theory of sphere-bundles over spheres developed by J. H. C. Whitehead and myself in [**23**]. Writing $X_\alpha = X$, for simplicity, we have the direct sum decomposition

$$\pi_t(S^m) \oplus \pi_t(S^n) \approx \pi_t(X) \quad (t = 2, 3, \dots),$$

where the isomorphism is given by the section $\iota_m \in \pi_m(X)$ on the first summand, by the inclusion $\iota_n \in \pi_n(X)$ of the fibre on the second. Recall from [**23**] that with appropriate sign conventions the relation

$$[\iota_m, \iota_n] = \iota_n \circ J\alpha$$

holds in $\pi_{m+n-1}(X)$, where square brackets denote the Whitehead product.

Similarly with the associated bundle $Z = X \vee_B X$ with fibre the coproduct $S^n_1 \vee S^n_2$ of copies of S^n. We have the direct sum decomposition

$$\pi_t(S^m) \oplus \pi_t(S^n_1 \vee S^n_2) \approx \pi_t(Z)$$

and the relations

$$[\iota_m, \iota^k_n] = \iota^k_n \circ J\alpha \quad (k = 1, 2)$$

where $\iota^k_n \in \pi_n(Z)$ is the class of the kth insertion of the coproduct.

Suppose that X admits a fibrewise coHopf structure $f\colon X \to Z$. Then we have a commutative diagram as shown below, where $g\colon S^n \to S^n_1 \vee S^n_2$ is the induced coHopf structure on the fibre.

$$\begin{array}{ccc} \pi_t(S^m) \oplus \pi_t(S^n) & \approx & \pi_t(X) \\ \Big\downarrow {\scriptstyle id\,|\,g_*} & & \Big\downarrow {\scriptstyle f_*} \\ \pi_t(S^m) \oplus \pi_t(S^n_1 \vee S^n_2) & \approx & \pi_t(Z) \end{array}$$

From the first relation stated above we obtain that

$$[\iota_m, \iota^1_n + \iota^2_n] = (\iota^1_n + \iota^2_n) \circ J\alpha$$

in $\pi_{m+n-1}(Z)$. Using the second relation and the left distributive law this reduces to

$$[\iota_n^1, \iota_n^2] \circ HJ\alpha = 0,$$

where H denotes the generalized Hopf invariant. However $[\iota_n^1, \iota_n^2] \in \pi_{2n-1}(Z)$ is the image of the basic Whitehead product of $\pi_{2n-1}(S_1^n \vee S_2^n)$ and so it follows that $HJ\alpha = 0$. Since $HJ\alpha = \pm\Sigma_*^n \rho_* \alpha$, this proves that the condition in Proposition 4.1 is necessary.

Conversely suppose that $\Sigma_*^n \rho_* \alpha = 0$ and hence $HJ\alpha = 0$. Since the difference $[\iota_m, \iota_n] - \iota_n \circ J\alpha$ is the class of the attaching map of the $(m+n)$-cell $X \backslash (S^m \vee S^n)$ of X we find, by reversing the above argument, that there exists a fibrewise map $f\colon X \to Z$ extending the coHopf structure on S^n. Hence X admits fibrewise coHopf structure by Proposition 2.1.

Let us turn now to the problem of the existence of fibrewise Hopf structure. Instead of dealing with this directly we first consider a related problem: does there exist a pointed map

$$X *_B X \to \Sigma_B X$$

which is given, on each fibre, by a map of which the homotopy class $\beta \in \pi_{2n+1}(S^{n+1})$ has Hopf invariant one? Such questions are answered in [23] where it is shown that such a fibrewise map exists if and only if

$$\Sigma_* J\alpha \circ \Sigma_*^m \beta = (2\beta - w_{n+1}) \circ \Sigma_*^{n+1} J\alpha. \tag{3}$$

Here $w_{n+1} = [\iota_{n+1}, \iota_{n+1}] \in \pi_{2n+1}(S^{n+1})$ is the Whitehead square. We recall, incidentally, that $2\beta - w_{n+1} \in \Sigma_* \pi_{2n}(S^n)$. We see, therefore, that X admits fibrewise Hopf structure if and only if β can be chosen with Hopf invariant one so that (3) is satisfied.

In particular take $n = 7$ and $m = 8$. Then $\pi_7 SO(7)$ is cyclic infinite and $\pi_{14}(S^7) = J\pi_7 SO(7)$ is cyclic of order 120 for $n = 7$. Let σ denote the Hopf class in $\pi_{15}(S^8)$. Then $w_8 = 2\sigma + \Sigma_* \sigma'$, where σ' generates $\pi_{14}(S^7)$. Now $\beta = \sigma + k\Sigma_* \sigma'$ and $J\alpha = l\sigma'$, for some integers k, l. The relation in (3) reduces to

$$(3 - 2k) l \Sigma_* \sigma' \circ \Sigma_*^8 \sigma = 0.$$

However $\sigma' \circ \Sigma_*^7 \sigma$ has order 24 in $\pi_{21}(S^7)$ and so k can be found to satisfy the relation if and only $l \equiv 0 \bmod 8$. Therefore the sectioned 7-sphere bundles over S^8 which admit fibrewise Hopf structure are precisely those which arise from elements of the subgroup $8\pi_7 SO(7)$ of $\pi_7 SO(7)$ by means of the clutching construction.

5. Fibrewise category

Discussion of fibrewise coHopf spaces leads on naturally to the idea of fibrewise (Lusternik-Schnirelmann) category. This may be defined, as in [**22**], in terms of coverings by fibrewise categorical subsets. Alternatively it may be defined, as we do here, by using a fibrewise version of the characterization of ordinary category due to G.W. Whitehead.

For any fibrewise space X over B consider the fibrewise topological products $\Pi_B^n X$ $(n = 1, 2, \dots)$ of X with itself. Let $\Pi_B^n(X, B) \subset \Pi_B^n X$ denote the subspace such that the fibre over each point b of B is the "fat wedge" subspace of $\Pi^n X_b$. Thus $\Pi_B^n X$ contains the diagonal ΔX of X while $\Pi_B^n(X, B)$ contains the diagonal ΔB

of B. In other words the pair $(\Pi_B^n X, \Pi_B^n(X,B))$ contains the diagonal $\Delta(X,B) = (\Delta X, \Delta B)$ of the pair (X,B).

DEFINITION 5.1. Let X be a fibrewise space over B. The fibrewise category $\operatorname{cat}_B X$ of X is the least number n such that the diagonal

$$\Delta \colon X \to \Pi_B^n X$$

can be compressed into $\Pi_B^n(X,B)$ by a fibrewise homotopy.

If no such number exists we simply say that the fibrewise category is infinite. As (6.1) of [**22**] shows there is no difference, for a large class of fibrewise spaces, between fibrewise category as defined here and fibrewise category as defined in [**22**].

We see, for example, that $\operatorname{cat}_B X = 1$ if and only if X is fibrewise contractible, while $\operatorname{cat}_B X \leq 2$ if and only if X is fibrewise coHopf.

Clearly fibrewise category is an invariant of fibrewise homotopy type. Note

$$\operatorname{cat}_B(X) \geq \operatorname{cat}(X/B)$$

from which various lower bounds can be obtained. It turns out, however, that many of the lower bounds for fibrewise category are also lower bounds for weak fibrewise category, and so we proceed at once to the definition of the latter.

Recall that the n-fold fibrewise smash product $\Lambda_B^n X$ can be obtained from $\Pi_B^n X$ by fibrewise collapsing $\Pi_B^n(X,B)$; we denote the natural projection by ρ.

DEFINITION 5.2. Let X be a fibrewise space over B. The weak fibrewise category $w\operatorname{cat}_B X$ of X is the least number n such that the composition

$$X \xrightarrow{\Delta} \Pi_B^n X \xrightarrow{\rho} \Lambda_B^n X$$

is fibrewise nulhomotopic.

If no such number exists we simply say that the weak fibrewise category is infinite. Clearly fibrewise category is never less than weak fibrewise category. Both are numerical invariants of fibrewise homotopy type. Lower bounds for the fibrewise category are usually also lower bounds for the weak fibrewise category. For example let H_B^* be a multiplicative fibrewise cohomology theory in the sense of Dold [**7**]. Then (cf. (3.3) of [**22**]) the nilpotency index of the reduced fibrewise cohomology ring $\tilde{H}_B^*(X)$ is a lower bound for $w\operatorname{cat}_B(X)$. This shows that even for sectioned sphere-bundles the weak fibrewise category can be as large as we please. For example, take $B = P(\mathbb{R}^n)$, the real projective $(n-1)$-space, and take ξ to be the Hopf line bundle; then $w\operatorname{cat}_B \xi_B^+ \geq n$.

Another type of nilpotency theorem is due to G. W. Whitehead [**33**]. We prove a fibrewise version in the following form.

PROPOSITION 5.3. *Let A be a fibrewise space over B. Then for each sectioned sphere-bundle X over B we have*

$$\operatorname{nil} \pi_B(\Sigma_B X, A) < w\operatorname{cat}_B X.$$

Here the left-hand side of the inequality denotes the index of nilpotency of the group of fibrewise homotopy classes of fibrewise maps of $\Sigma_B X$ into A, the group operation being determined by the standard fibrewise coHopf structure on the fibrewise suspension. Of course this group is isomorphic to $\pi_B(X, \Omega_B A)$, by taking adjoints; in a more general formulation $\Omega_B A$ is replaced by any fibrewise group-like space but in fact this can be deduced from the special case, using Proposition 2.4, provided the restrictions of Proposition 2.4 are satisfied.

Eckmann-Hilton [**8**] have given an elegant proof of the ordinary version of the Whitehead nilpotency theorem.[4] A large part of their argument is formulated in general terms which apply straightaway to our situation. However there remain two "topological lemmas" which Eckmann-Hilton prove, in the ordinary case, on the assumption that the spaces concerned are CW-complexes. This is not a suitable assumption in the case of the fibrewise version and so we proceed as follows instead.

The lemmas in question are formulated for CW-spaces in [**8**]. In fact it is only necessary to assume that all the pairs which occur in the argument are cofibred. Furthermore the equivariant form of each lemma holds under the assumption that all the pairs are equivariantly cofibred. After applying the associated bundle functor we arrive at fibrewise versions of the two lemmas, as follows. As in Proposition 5.3 let A be a fibrewise space over B and let X be a sectioned sphere-bundle over B. Then:

LEMMA 5.4. *The functor $\pi_B(\cdot, A)$ is exact on the fibrewise cofibration*

$$B \to \Sigma_B \Pi^n_B(X, B) \to \Sigma_B \Pi^n_B X \to \Sigma_B \Lambda^n_B X \to B.$$

LEMMA 5.5. *Let $f\colon \Sigma_B \Pi^n_B X \to A$ be a fibrewise map. Suppose that f is fibrewise nulhomotopic on the fibrewise suspension of each of the n subspaces which make up the fibrewise fat wedge $\Pi^n_B(X, B)$. Then f is fibrewise nulhomotopic.*

Now the remainder of the proof of Proposition 5.3 is purely formal.

Of course Lusternik-Schnirelmann introduced their eponymous numerical invariant because it provides a lower bound for the number of critical points of a smooth real-valued function on a manifold. No fibrewise version of this result has been found as yet. Although some recent work on critical sets by Igusa and others may be relevant it is not at all clear that fibrewise category in our sense has any connection with this.

Finally, is there a satisfactory dual of the definition of fibrewise category? Ganea [**12**] has given a definition of cocategory with various good properties, for example the Hopf spaces are precisely the spaces with cocategory ≤ 2. Hopkins [**15**] has given another definition, with similar properties, but it is not known whether his invariant is equal to Ganea's except in special cases. Fibrewise versions of both invariants need to be developed and this is work in progress.

References

1. I. Berstein and P. J. Hilton, *Category and generalized Hopf invariants*, Illinois J. Math. **4** (1960), 437–451.
2. ——, *On suspensions and comultiplications*, Topology **2** (1963), 73–82.
3. A. L. Cook, *Fibrewise Hopf spaces*, Ph.D. thesis, Oxford, 1991.
4. A. L. Cook and M. C. Crabb, *Fibrewise Hopf structures on sphere-bundles*, J. London Math. Soc. **48** (1993) 365–384.
5. T. tom Dieck, K. H. Kamps, and D. Puppe, *Homotopietheorie*, Lecture Notes in Math., vol. 57, Springer Verlag, 1970.
6. A. Dold, *Partitions of unity in the theory of fibrations*, Ann. of Math. **78** (1963), 223–255.
7. A. Dold, *Chern classes in general cohomology*, Inst. Naz. di Alta Mat. Symposia Mat. **5** (1970), 385–410.
8. B. Eckmann and P. J. Hilton, *A natural transformation in homotopy theory and a theorem of G. W. Whitehead*, Math. Zeitschr. **82** (1963), 115–124.
9. M. G. Eggar, *The piecing comparison theorem*, Indag. Math. **35** (1973), 320–330.

[4]Category, in [**8**], is normalized so that contractible spaces have category zero.

10. ———, *On structure-preserving maps between spaces with cross-sections*, J. London Math. Soc. (2) **7** (1973), 303–311.
11. ———, *Ex-homotopy theory*, Comp. Math. **27** (1973), 185–195.
12. T. Ganea, *Lusternik-Schnirelmann category and cocategory*, Proc. London Math. Soc. **10** (1960), 623–639.
13. ———, *On the homotopy suspension*, Comment. Math. Helv. **43** (1968), 225–234.
14. ———, *Cogroups and suspensions*, Invent. Math. **9** (1970), 185-197.
15. M. J. Hopkins, *Formulations of cocategory and the iterated suspension*, Astérisque **113/4** (1984), 212-226.
16. I. M. James, *On the suspension triad of a sphere*, Ann. of Math. **63** (1956), 407-429.
17. ———, *On category, in the sense of Lusternik-Schnirelmann*, Topology **17** (1978), 331-348.
18. ———, *General topology and homotopy theory*, Springer Verlag, 1985
19. ———, *Fibrewise topology*, Cambridge Univ. Press, 1988.
20. ———, *Fibrewise reduced product spaces*, Adams Memorial Symposium on Algebraic Topology I, Cambridge Univ. Press, 1992, pp. 179-185.
21. ———, *Fibrewise coHopf spaces*, Glasnik Math. **27** (1992), 183-190.
22. I. M. James and J. R. Morris, *Fibrewise category*, Proc. Roy. Soc. Edinburgh **119A** (1991), 171–190.
23. I. M. James and J. H. C. Whitehead, *The homotopy theory of sphere-bundles over spheres*, Proc. London Math. Soc. **4** (1954), 196–218.
24. I. M. James and P. Zhang, *Fibrewise transformation groups and fibrewise fibre bundles*, Northeast Chinese Math. J. **8** (1992), 263–274.
25. J. P. May, *Fibrewise localization and completion*, Trans. Amer. Math. Soc. **258** (1980), 127–146.
26. J. R. Morris, *Fibrewise Lusternik-Schnirelmann category*, M.Sc. thesis, Oxford, 1989.
27. J. L. Noakes, *Self-maps of sphere-bundles* I, J. Pure Appl. Algebra **10** (1977), 95–99.
28. H. Scheerer, *On H-spaces over a base*, Collect. Math. **36** (1985), 219–226.
29. A. M. Sunderland, *Fibrewise coHopf spaces*, Ph.D. thesis, Oxford, 1992.
30. ———, *Fibrewise coHopf spaces* (in preparation).
31. E. Thomas, *Vector fields on manifolds*, Bull. Amer. Math. Soc. **75** (1969), 643–683.
32. G. W. Whitehead, *A generalization of the Hopf invariant*, Ann. of Math. **5** (1950), 192–237.
33. ———, *On mappings into group-like spaces*, Comment. Math. Helv. **28** (1954), 320–328.

Mathematical Institute, University of Oxford, 24–29 St. Giles, Oxford OX1 3LB, England

E-mail address: imj@vax.oxford.ac.uk

Centre de Recherches Mathématiques
CRM Proceedings and Lecture Notes
Volume 6, 1994

The Mislin Genus of a Space

Charles A. McGibbon

ABSTRACT. Let X be a nilpotent space of finite type. The Mislin genus of X is defined to be the set of all homotopy types $[Y]$ of finite type spaces Y such that the localizations $Y_{(p)}$ and $X_{(p)}$ are homotopy equivalent for every prime p. This paper is a survey of results and open questions about the Mislin genus.

Introduction

Given a nilpotent space X of finite type, let $\mathcal{G}(X)$ denote the set of all homotopy types $[Y]$, of nilpotent finite type spaces Y, such that for every prime p, the localizations $Y_{(p)}$ and $X_{(p)}$ are homotopy equivalent. $\mathcal{G}(X)$ is called the Mislin genus of X, after G. Mislin who first introduced this notion into the homotopy category in [**65**]. In general, $\mathcal{G}(X)$ is only a filtered set[1] with a distinguished member, namely $[X]$. It is not a functor and it is not well behaved with respect to products or direct limits. Nevertheless, it is an interesting gadget that, in some sense, measures the complexity of a space up to homotopy. More precisely, it measures the extent to which the p-local symmetries of a space, for the different primes p, fail to match up in the rational homotopy type. This goal of this paper is to describe some of the features of $\mathcal{G}(X)$ that have been discovered in the past 20 years and to also point out some questions about the Mislin genus that remain to be solved.

1. The extended genus

The reader may wonder, in the definition of $\mathcal{G}(X)$, if it is really necessary to require the spaces Y to also have finite type. The answer is yes if one wants only finite type spaces represented in $\mathcal{G}(X)$[2]. The reason is evident in the algebra. Consider, for example, the additive subgroup of the rational numbers generated by the fractions $1/p$ for all primes p. This is a group A which is not finitely generated and yet $A_{(p)} \approx \mathbb{Z}_{(p)}$ for all p. It turns out that there are uncountably many isomorphism classes of Abelian groups with this property, [**41**]. Of course, only one of them is finitely generated. Define the *extended genus* of a space X to be the localization genus of X without the finite type restriction on the Y's. The following example shows what can happen in this case.

1991 *Mathematics Subject Classification.* Primary: 55P60; Secondary: 55P15.
This is the final form of the paper.

[1]Define $F_n\mathcal{G}(X) = \{[Y] \in \mathcal{G}(X) \mid Y \text{ has the same } n-\text{type as } X\}$. This filtration is not necessarily complete when X is infinite dimensional.

[2]This point was overlooked in [**65**] and consequently some of the claims made there are false without the finite type restriction on the other members of $\mathcal{G}(X)$.

EXAMPLE A. *If X is a sphere S^n or an Eilenberg-MacLane space $K(\mathbb{Z}, n)$, where $n \geq 1$, then the Mislin genus of X has just one element but the extended genus of X is uncountably large.*

PROOF. Let A denote one of the non-finitely generated Abelian groups which is locally isomorphic to $\mathbb{Z}$. Then A has rank 1 and can be represented as a direct limit

$$\mathbb{Z} \xrightarrow{d_1} \mathbb{Z} \xrightarrow{d_2} \cdots$$

Since homology commutes with direct limits, the Moore space $M(A, n)$ can be constructed as the infinite telescope of the sequence of maps

$$S^n \xrightarrow{d_1} S^n \xrightarrow{d_2} \cdots .$$

The inclusion of S^n in the left end of the telescope is a homotopy equivalence at each prime p and thus $M(A, n)$ is in the extended genus of the sphere. The other claims made in the example are obvious. $\square$

The example just given could be generalized to finite bouquets of spheres and finite products of $K(\mathbb{Z}, n)$'s. These spaces are, in some sense, the most elementary examples of co-H-spaces and H-spaces. Thus the cardinality of the Mislin genus of a co-H-space or an H-space might be viewed as a measure of the complexity of its homotopy type. However, for other spaces, such as Grassmann manifolds (see §5), this interpretation seems somewhat flawed. At any rate, the following result shows that in many interesting cases this measure is finite.

2. Wilkerson's finiteness theorem

THEOREM 1. *Suppose that X is a simply connected CW-complex of finite type with $H_n(X; \mathbb{Z}) = 0$ for n sufficiently large (or $\pi_n X = 0$ for $n \gg 0$). Then the Mislin genus $\mathcal{G}(X)$ is a finite set.*

Although Wilkerson only stated this theorem for finite complexes in [**100**], it is clear from his proof that this more general result is true. Pickel had previously proved a result like this for nilpotent groups,[**86**]. For a finitely generated nilpotent group N, he established the finiteness of the set $\widehat{G}_0(N)$, of isomorphism classes of finitely generated nilpotent groups M whose rationalization $M_0 \approx N_0$ and whose profinite completion $\widehat{M} \approx \widehat{N}$. In doing this he used the double coset formula

$$\widehat{G}_0(N) \approx \operatorname{Aut}(\widehat{N}) \backslash \operatorname{Aut}\big((\widehat{N})_0\big) / \operatorname{Aut}(N_0).$$

Here $\operatorname{Aut}(N_0)$ is a linear algebraic group $\mathcal{G}_{\mathbb{Q}}$, by the Mal'cev correspondence between divisible nilpotent groups and nilpotent Lie algebras over $\mathbb{Q}$. Pickel reduces his proof to a special case (the lattice nilpotent groups) where this coset space can be identified with $\mathcal{G}_{\widehat{\mathbb{Z}}} \backslash \mathcal{G}_{A_f} / \mathcal{G}_{\mathbb{Q}}$. The finiteness of this coset space is then a result about adele groups due to Borel, [**10**]. Pickel goes on to show that if one drops the requirement that $M_0 \approx N_0$ in the above definition of $\widehat{G}_0(N)$, the resulting genus set is still finite.

In his proof of Theorem 1, Wilkerson uses $\widehat{G}_0(X)$ the set consisting of all homotopy types $[Y]$ of finite type nilpotent spaces Y whose profinite completion $\widehat{Y}$ is homotopy equivalent to $\widehat{X}$ and whose rationalization $Y_0 \simeq X_0$. It follows that $\mathcal{G}(X) \subseteq \widehat{G}_0(X)$ because $\widehat{X} \simeq \prod_p X_{\hat{p}}$ and $X_{\hat{p}}$, the p-adic completion of X, is homotopy equivalent to the p-adic completion of $X_{(p)}$[3]. He then establishes the double coset formula

$$\widehat{G}_0(X) \approx r_* \operatorname{Aut}(\widehat{X}) \backslash \operatorname{Caut}(\widehat{X}_0)/fc_* \operatorname{Aut}(X_0).$$

In this formula r denotes the rationalization functor, $\operatorname{Caut}(\widehat{X}_0)$ denotes the group of homotopy classes of those self equivalences f which induce $\mathbb{Q} \otimes \widehat{\mathbb{Z}}$ module maps on $\pi_*(\widehat{X}_0)$, and fc denotes Sullivan's formal completion.

In establishing the finiteness of this double coset space, Wilkerson shows that most of Pickel's proof can be adapted to the automorphism groups which arise in homotopy theory. However, the adaptation is not a straightforward one. First the set of homotopy classes $[X, X]$, containing $\operatorname{Aut}(X)$, is replaced by the set of loop homotopy classes $[\Omega X, \Omega X]$. Then the loop space ΩX is replaced by Kan's free simplicial group GX. Kan's results on minimal simplicial groups are then extended to include localizations and completions of GX. By using minimal simplicial groups, the problem of classification up to homotopy type is reduced to one of classification up to isomorphism. He then shows that the finiteness conditions on X allow one, in the analysis of $[GX, GX]$, to replace GX by a large nilpotent quotient of itself. After this step, the relevance of Pickel's work becomes apparent.

Two groups G and H are said to be *commensurable* if they contain subgroups of finite index, $\widetilde{G}$ and $\widetilde{H}$, with $\widetilde{G} \approx \widetilde{H}$. This is the definition Pickel uses and it plays an important role in his paper. In Wilkerson's paper the term has a slightly different meaning; there commensurability is the equivalence relation generated by homomorphisms whose kernels and cokernels are finite. Using this notion he shows that if X and Y are two 1-connected finite CW-complexes with the same rational homotopy type then the groups $\operatorname{Aut}(X)$ and $\operatorname{Aut}(Y)$ are commensurable. He also shows that $\operatorname{Aut}(X_0)$ is a linear algebraic group and that $\operatorname{Aut}(X)$ is commensurable with an arithmetic subgroup of $\operatorname{Aut}(X_0)$ and hence is finitely presented. These results remain among the deepest ones yet obtained on $\operatorname{Aut}(X)$.

The completion genus $\widehat{G}(X)$ is defined to be the set homotopy types $[Y]$ of finite type Y such that the profinite completions $\widehat{X} \simeq \widehat{Y}$. Wilkerson also showed that this third genus set is finite under the conditions of Theorem 1. In general there are inclusions $\mathcal{G}(X) \subseteq \widehat{G}_0(X) \subseteq \widehat{G}(X)$. For examples that show these inclusions are proper see Belfi and Wilkerson, [**6**].

3. Genus and non-cancellation phenomena

The first examples of non-cancellation in the homotopy category were given by Freyd, [**29**], in the stable range. In nonstable homotopy theory the following is one of the most famous and also one of the simplest examples of non-cancellation.

EXAMPLE B. *If $E_{5\omega}$ denotes the Hilton-Roitberg criminal, then $E_{5\omega} \times S^3 \simeq \operatorname{Sp}(2) \times S^3$.*

The H-space $E_{5\omega}$ was the first example of a finite H-space not of the homotopy type of a compact Lie group, or S^7, or some product of the two[4], [**48**]. It can be

[3]Unfortunately, Wilkerson claims that the two genus sets are equal when X is rationally an H-space. This is false as Example F of [**63**] shows. However, Belfi and Wilkerson, [**6**], show that his claim is true if one also requires that X be a finite complex or that $\pi_n X = 0$ for n sufficiently large.

[4]In retrospect it seems odd that this example wasn't discovered at least 10 years sooner. Its construction and the proof that it is an H-space certainly require no elaborate machinery.

obtained as a pullback of the principal Sp(1)-bundle in the diagram

$$\begin{array}{ccc} & & \mathrm{Sp}(2) \\ & & \downarrow \pi \\ S^7 & \xrightarrow{5} & S^7 \end{array} .$$

In Sp(2) the attaching map ω for the 7-cell has order 12. This fact and the above description imply that $E_{5\omega}$ is a nontrivial member of $\mathcal{G}(\mathrm{Sp}(2))$. Indeed by comparing the 7-skeletons and noting that $\omega \neq \pm 5\omega$ it follows that $E_{5\omega}$ and Sp(2) cannot have the same homotopy type. Away from the primes 2 and 3 they both are equivalent to $S^3 \times S^7$. The map between them in the pullback diagram above is an equivalence at every prime $\neq 5$. Thus the two spaces are in the same genus.

PROOF OF EXAMPLE B. I will write down a homotopy equivalence between the two products. To this end choose homology generators for $E_{5\omega}$ and Sp(2) and let $d\colon [E_{5\omega}, \mathrm{Sp}(2)] \to \mathbb{Z}^2$ record the degrees of the induced homomorphism in homology in dimensions 3 and 7. We have just seen a map f with $d(f) = (1, 5)$. It is easy to find another, g, with $d(g) = (0, 12)$. In the group $[E_{5\omega}, \mathrm{Sp}(2)]$ let $h = 5f - 2g$, so that $d(h) = (5, 1)$. Now let

$$\phi_1 : E_{5\omega} \to \mathrm{Sp}(2) \times S^3$$

be a map which projects, on the first factor to h, and, on the second factor, to an extension of the degree 12 map on S^3. Let

$$\phi_2 \colon S^3 \to \mathrm{Sp}(2) \times S^3$$

project to 3 times a generator of $\pi_3 \mathrm{Sp}(2)$ on the first factor and to 7 times the identity on the second. Since the matrix

$$\begin{pmatrix} 5 & 3 \\ 12 & 7 \end{pmatrix}$$

is invertible over the integers, the product map $\mu(\phi_1 \times \phi_2)$ is then seen to induce an isomorphism in integral cohomology and hence it is a homotopy equivalence. For a different proof, see [**47**, p. 139]. $\square$

The following theorem is beautiful illustration of the connection between genus and non-cancellation. The possibility of such a result was first raised by Mislin in [**65**]. Later he proved the equivalence of i) and iii) in [**66**]. About the same time, Zabrodsky showed that i) implies ii) and iii) in [**103**]. Wilkerson then proved ii) implies i) in [**99**].

THEOREM 2. *Let X be a finite H-space with the rational homotopy type of $S = S^{n_1} \times \cdots \times S^{n_r}$. Then for any other space Y the following statements are equivalent*:

i) $Y \in \mathcal{G}(X)$.

ii) *The k-fold products X^k and Y^k are homotopy equivalent for some positive k.*

iii) $X \times S \simeq Y \times S$.

The main ingredient in Wilkerson's proof is the following p-local result. It is a unique factorization result which allows one to cancel in certain p-local situations.

THEOREM 3. *If X is a 1-connected H-space with finitely generated homology then $X_{(p)} \simeq X_1 \times \cdots \times X_t$ where the X_i are irreducible p-local H-spaces and are unique up to order.*

Wilkerson proved this theorem by a Postnikov tower argument and so it applies equally well to p-local H-spaces with only finitely many nonzero homotopy groups. When combined with results from **[6]** and **[98]** it implies that one also has cancellation among p-complete H-spaces of finite type. There are also Eckmann-Hilton dual versions of Theorems 2 and 3 in terms of finite co-H-spaces and $\vee$-sums.

The following example, due to Bokor, **[8]**, shows that in special integral cases it is sometimes possible to cancel up to homotopy.

EXAMPLE C. *Let $X = S^{2n} \cup_f e^{4n}$ and $Y = S^{2n} \cup_g e^{4n}$ where $n \geq 1$ and the attaching maps f and g have infinite order. Then the following statements are equivalent:*

i) $X \simeq Y$.

ii) $X \vee S^{2n} \vee S^{4n} \simeq Y \vee S^{2n} \vee S^{4n}$.

iii) $\overset{k}{\bigvee} X \simeq \overset{k}{\bigvee} Y$ *for some positive integer k.*

The book **[47]** contains extensive results on non-cancellation phenomena among the total spaces of principal G-bundles. Other papers dealing with non-cancellation in homotopy theory include **[9]**, **[57]**, and **[73]**.

4. Generic properties

A homotopy theoretic property of a space X is said to be *generic* if it is one had by every member of $\mathcal{G}(X)$. Some homotopy invariants such as the Euler characteristic of X (if defined) or the graded Abelian group $H_*(X;\mathbb{Z})$, up to isomorphism, are obvious generic invariants (remember, we are assuming here that X has finite type). The property of being homotopy equivalent to a Poincaré duality complex is likewise a generic property (**[47]**, page 108). The following short quiz deals with a number of other properties.

Quiz

PROBLEM 1. Given a nilpotent space X of finite type, consider the following five possibilities. Which of these properties are generic?

The space X has the homotopy type of:

1.1) a finite complex;

1.2) a 1-connected smooth closed manifold;

1.3) an H-space;

1.4) a stunted complex projective space;

1.5) a finite loop space with a maximal torus.

PROBLEM 2. Which of the following homotopy invariants of a space X are generic invariants?

2.1) the graded ring $H^*(X;\mathbb{Z})$, up to isomorphism;

2.2) the Lusternik-Schnirelmann category of X;

2.3) the isomorphism class of $\mathrm{Aut}(X)$, the group of homotopy classes of self homotopy equivalences of X.

Answers

1.1. The correct answer is no; having the homotopy type of a finite complex is not a generic property. Mislin demonstrated this in [**68**] with the following example. First identify S^3 with $SU(2)$; then Q_8, the quaternion group of order 8, is a subgroup and the coset space S^3/Q_8 is a compact 3-manifold. Mislin showed that $\mathcal{G}(S^3/Q_8)$ is the two element set $\{S^3/Q_8, V\}$ where V is a complex whose Wall obstruction $\widetilde{w}(V)$ has order 2. Hence V does not have the homotopy type of a finite complex by [**94**]. Of course this example also shows that having the homotopy type of a closed manifold is not a generic property.

1.2. The answer is again no. An example can be found in the genus of the quaternionic projective plane $\mathbb{H}P^2$. Up to homotopy, there are 6 different simply connected closed 8-manifolds whose integral cohomology rings are isomorphic to $H^*\mathbb{H}P^2$. A detailed study of the various TOP, PL, and smooth structures on these 6 homotopy types was made by Eells and Kuiper, [**26**]. See also Kahn, [**54**]. In Kahn's notation, the 6 homotopy types are represented by the manifolds $W(h)$, where $h = 1, \ldots, 6$ with $\mathbb{H}P^2 \simeq W(1)$. Three of the six admit smooth structures ($h = 1$, 4 and 5)—the other three do not. The genus of $\mathbb{H}P^2$ has four members (see Example F) and so the answer follows. In fact, the four members can be represented as $W(h)$ where $h = 1$, 3, 4 and 6. Thus only two of the four homotopy types in $\mathcal{G}(\mathbb{H}P^2)$ admit smooth structures.

1.3. The answer, in general, is unknown. However some important special cases are known—if $\pi_n X = 0$ or if $H_n X = 0$ for n sufficiently large, then the answer is yes. See [**47**, p. 115] or [**105**, p. 141]. Moreover Mislin has shown in [**69**] that if X is a finite H-space, then so is every member of $\mathcal{G}(X)$. In other words, the Wall obstruction is always zero in the genus of a finite H-space.

1.4. More generally one could ask if having the *stable* homotopy type of a stunted complex projective space is a generic property. Davis raised this question in [**20**]. Let CP_n^k denote the quotient $\mathbb{C}P^{n+k}/\mathbb{C}P^{n-1}$. The first two cases (where k is 0 or 1) are trivial since here $\mathcal{G}(CP_n^k) = *$. The first nontrivial case is $k = 2$: the CP_n^2's fall into 14 different stable homotopy types and some of them have nontrivial genera. However Davis shows that these stable genera contain only stunted complex projective spaces.

The first counterexample occurs when $k = 3$ and I am grateful to Davis for pointing it out to me. According to Feder and Gitler, [**28**], CP_m^3 and CP_n^3 have the same stable homotopy type if and only if $m - n \equiv 0 \bmod 24$. Based on primary cohomology operations alone, it follows that if $\mathbb{C}P^4$ and CP_n^3 are in the same stable genus then $n \equiv 1 \bmod 6$. Notice that if $\mathbb{C}P^4$ and CP_7^3 are in the same stable genus then, by restriction, $\mathbb{C}P^3$ and CP_7^2 are either stably equivalent or in the same stable genus. But Theorem 1.3 of [**20**] rules out both possibilities. It follows that the stable genus of $\mathbb{C}P^4$ contains at most 2 other stunted projective spaces. However, as will be seen in Example G, the stable genus of $\mathbb{C}P^4$ has 4 members. Thus having the stable homotopy type of a stunted complex projective space is not a generic property.

1.5. The answer is no. A relevant example involves the Lie group Sp(2). It has a maximal torus, of course, but in its genus lies the Hilton-Roitberg criminal, $E_{5\omega}$, and I contend that the latter has no maximal torus. More precisely I claim that there is no space B such that

i) $\Omega B \simeq E_{5\omega}$, and

ii) $\exists f\colon BT^2 \to B$ whose fiber is a finite complex.

The proof uses the Adams e-invariant, [**2**]. To begin, one can choose K-theory generators so that e'_R applied to the 8-skeleton of $\Sigma\,\mathrm{Sp}(2)$ takes the value $-1/12$ and e'_R applied to the 8-skeleton of $\Sigma E_{5\omega}$ equals $-5/12$. The second value follows from the first by naturality. One way to obtain the first value is to use the maps,

$$\Sigma\,\mathrm{Sp}(2) \xrightarrow{i} B\,\mathrm{Sp}(2) \xrightarrow{j} BT^2.$$

Here i is the standard inclusion and j represents a maximal torus. Recall that $K(BT^2) \approx \mathbb{Z}\big[[x_1, x_2]\big]$ where $\eta_i = x_i + 1$ is the pullback of the Hopf line bundle over $\mathbb{C}P^\infty$ via projection on the ith factor. The K-theory of $B\,\mathrm{Sp}(2)$ embeds in that of BT^2 as the ring of invariants of the Weyl group W of $\mathrm{Sp}(2)$. The group W is generated by the involutions $x_i \mapsto \Psi^{-1}x_i$ and the interchange of x_1 and x_2. Consider then, the W-invariant class

$$z = x_1 + \Psi^{-1}(x_1) + x_2 + \Psi^{-1}(x_2).$$

The Chern character of this class is

$$\begin{aligned} ch(z) &= e^{y_1} - 1 + e^{-y_1} - 1 + e^{y_2} - 1 + e^{-y_2} - 1 \\ &= (y_1^2 + y_2^2) + \frac{2}{4!}(y_1^4 + y_2^4) + \text{higher terms} \\ &= u + \frac{2}{4!}(u^2 - 2w) + \text{higher terms.} \end{aligned}$$

Here y_i denotes the Chern class $c_1(x_i)$, while

$$u = y_1^2 + y_2^2 \text{ and } v = y_1^2 y_2^2.$$

It is well known that $H^*B\,\mathrm{Sp}(2)$ can be identified, using j^*, with $\mathbb{Z}[u, v]$. In K-theory, apply the map i^* to the class in $K^*B\,\mathrm{Sp}(2)$ that maps to z. Since $i^*(u^2), = 0$, the claim $e'_R\big(\Sigma\,\mathrm{Sp}(2)_8\big) = -1/12$ then follows from Adams's description of e'_R in terms of the Chern character.

Now suppose that $E_{5\omega}$ had a loop structure B with a maximal torus. Then for each prime p, the p-completion of this integral maximal torus would be a maximal torus of a p-compact group in the sense of Dwyer and Wilkerson, [**25**]. According to their results, this p-complete maximal torus, for each prime p, would have a Weyl group say G_p of order 8, and the image of $H^*(B; \mathbb{Z}/p)$, under f_p^*, would coincide with the ring of invariants $H^*(BT^2; \mathbb{Z}/p)^{G_p}$. However, from the Clark-Ewing paper, [**19**], it follows that G_p must equal W, the Weyl group of $\mathrm{Sp}(2)$ for all $p \equiv -1 \bmod 4$ (there are no other possibilities). Up to isomorphism, there is only one representation of W in $GL(2, \mathbb{Z}_p)$ as reflection group. It follows that for all such primes the mod p reduction of $f^*H^*(B; \mathbb{Z})$ coincides with the mod p reduction of $H^*(BT^2; \mathbb{Z})^W$. Since this is true for infinitely many primes, the two subrings must coincide integrally. The Chern character is a monomorphism in this torsion-free situation, and so the image f^*K^*B must likewise coincide with the ring of invariants of W acting on the K-theory of BT^2. Using this information, one can then calculate the value e'_R on $\Sigma E_{5\omega}$ in the same manner as was done above for $\mathrm{Sp}(2)$—and arrive at the same value, $-1/12$. Since this does not agree the earlier value of $-5/12$ it follows that $E_{5\omega}$ has no loop structure which contains a maximal torus.[5]

2.1. The cohomology ring is in general not a generic invariant. Here is an example, taken from the theory of quadratic forms, [**7**]. Let p and q be two distinct primes and let X_1 and X_2 be two cell complexes of the form

$$X_i = (S^2 \vee S^2) \cup_{\Phi_i} e^4, \text{ for } i = 1, 2$$

where

$$\begin{aligned}\Phi_1 &= [\iota_1, \iota_1] + pq[\iota_2, \iota_2]\\ \Phi_2 &= p[\iota_1, \iota_1] + q[\iota_2, \iota_2].\end{aligned}$$

Consider the map from X_1 to X_2 induced by the commutative diagram

$$\begin{array}{ccc} S^3 & \xrightarrow{p} & S^3 \\ \downarrow{\scriptstyle \Phi_1} & & \downarrow{\scriptstyle \Phi_2} \\ S^2 \vee S^2 & \xrightarrow{\left(\begin{smallmatrix} p & 0 \\ 0 & 1 \end{smallmatrix}\right)} & S^2 \vee S^2 \end{array}.$$

This map is an equivalence at every prime $\ell \neq p$. Similarly, the map induced by the commutative diagram

$$\begin{array}{ccc} S^3 & \xrightarrow{q} & S^3 \\ \downarrow{\scriptstyle \Phi_1} & & \downarrow{\scriptstyle \Phi_2} \\ S^2 \vee S^2 & \xrightarrow{\left(\begin{smallmatrix} 0 & 1 \\ q & 0 \end{smallmatrix}\right)} & S^2 \vee S^2 \end{array}$$

induces an equivalence at every prime $\ell \neq q$. Thus X_1 and X_2 are in the same genus. However, their integral cohomology rings are not isomorphic. Indeed suppose that f was an isomorphism from $H^*(X_2; \mathbb{Z})$ to $H^*(X_1; \mathbb{Z})$. In degree 2, with respect to the obvious choice of basis, it could be represented by a 2×2 matrix with integer entries a, b, c, d, and in degree 4 the homomorphism f would have degree ± 1. It would follow that the integers a and b must satisfy the equation $pa^2 + qb^2 = \pm 1$, which is clearly impossible.

2.2. The answer is unknown. At first glance a yes answer seems plausible, but after considering some delicate calculations of $\text{cat}(X)$, such as in [**90**], a no answer does not seem so unreasonable.

2.3. In general $\text{Aut}(X)$ is not a generic invariant. Let $B = \mathbb{H}P^\infty \times \mathbb{H}P^\infty$; it is clear that $\text{Aut}(B)$ contains an involution that interchanges the two factors. However, Møller has constructed a space $Y \in \mathcal{G}(B)$ with $\text{Aut}(Y) = *$, ([**71**], Example 3.7). The answer to 2.3 for finite complexes is unknown.

[5] Notice that the claim proved here is stronger (at least in this special case) than the results of Notbohm and Smith on the existence of maximal tori. For a discussion of their results, see Section 9.

5. On the genus of generalized flag manifolds

Let M denote the homogeneous space G/H where $G = U(n_1 + \cdots + n_k)$ and where $H = U(n_1) \times \cdots \times U(n_k)$ is embedded in G in the obvious way. If N denotes the normalizer of H in G, then the finite group N/H acts on M and, in fact, is faithfully represented in $\mathrm{Aut}(M_0)$. It is also known that in $\mathrm{Aut}(M_0)$ there is a central subgroup, call it $\Psi(M_0)$, consisting of maps ψ^r (one for each nonzero $r \in \mathbb{Q}$) such that for each n, the map ψ^r induces multiplication by r^n on $H^{2n}(M;\mathbb{Q})$. The following conjecture says that these two sources account for all the self equivalences of M_0.

CONJECTURE 4. Let $M = G/H$ be a generalized flag manifold as just described. Then every map $f \in \mathrm{Aut}(M_0)$ can be expressed in the form $f = \psi^r \sigma$, where $\psi^r \in \Psi(M_0)$ and $\sigma \in N/H$.

This conjecture is known to be true for the classical flag manifolds $U(n)/T^n$, for the complex Grassmann manifolds $G_p(\mathbb{C}^{p+q})$ where $p \neq q$, and in a few other cases. Glover and Mislin proved the following theorem in [**32**].

THEOREM 5. *If M is a generalized complex flag manifold for which Conjecture 4 holds, then $\mathcal{G}(M) = *$.*

To prove this they first establish the double coset formula

$$\mathcal{G}(X_L) \approx \mathrm{Aut}(X_0)\backslash \mathrm{Aut}(X_0)^L \Big/ \prod_{p \in L} \mathrm{Aut}_p(X_0)$$

which is valid for any *finite* set of primes L and any nilpotent finite type space X. Here X_L denotes the localization of X at L. In the formula, $\mathrm{Aut}(X_0)^L$ denotes the set of functions of L into $\mathrm{Aut}(X_0)$ which, of course, is isomorphic to the $|L|$-fold product of $\mathrm{Aut}(X_0)$. The $\mathrm{Aut}(X_0)$ on the left is embedded diagonally in this product. On the right $\mathrm{Aut}_p(X_0)$ denotes the image $\{\mathrm{Aut}(X_0) \leftarrow \mathrm{Aut}(X_{(p)})\}$.

They then show that the surjection $\mathcal{G}(X) \to \mathcal{G}(X_L)$ is a bijection if for each $Y \in \mathcal{G}(X)$,

i) $Y_{L'} \simeq X_{L'}$, where L' denotes the complement of L , and

ii) each $f \in \mathrm{Aut}(Y_0)$ can be expressed as $f = f_1 f_2$ with $f_1 \in \mathrm{Aut}_L(Y_0)$ and $f_2 \in \mathrm{Aut}_{L'}(Y_0)$.

Now assume that Conjecture 4 holds for M and that L is a finite set primes, say $L = \{p_1, \ldots, p_n\}$. A typical element in $\phi \in \mathrm{Aut}(M_0)^L$ must then have the form

$$\phi = (\psi^{r_1}\sigma_1, \ldots, \psi^{r_n}\sigma_n).$$

Choose a number $\lambda \in \mathbb{Q}$ so that for each i, $r_i = \lambda s_i$ where s_i is a unit in $\mathbb{Z}_{(p_i)}$. A result of Friedlander [**31**] implies that ψ^{s_i} lifts to a self equivalence of $M_{(p_i)}$. Thus

$$\phi = \psi^\lambda(\psi^{s_1}\sigma_1, \cdots, \psi^{s_n}\sigma_n)$$

which is in the double coset of the identity and so $\mathcal{G}(M_L) = *$. Finally since M is a formal space there is, for any $N \in \mathcal{G}(M)$ a map from N to M which becomes a homotopy equivalence when rationalized. This together with the finiteness of the genus implies that $\mathcal{G}(M) = \mathcal{G}(M_L)$ for some finite set of primes and completes their proof.

For any compact Lie group G with maximal torus T, it is now known that the Mislin genus of G/T is also trivial. This is due to Papadima, [**82**], who computes $\mathrm{Aut}\big((G/T)_0\big)$ (it is isomorphic to the normalizer of the Weyl group W in $GL(\Gamma \otimes \mathbb{Q})$, where Γ denotes the integral lattice in the Lie algebra of T) and shows it has property ii) above. Since G/T is a formal space, the proof that $\mathcal{G}(G/T) = *$ then proceeds as above.

It is interesting to note that $\varinjlim_k G_n(\mathbb{C}^{n+k}) \simeq BU(n)$ and that the genus of $BU(n)$ is uncountably large when $n \geq 2$ (see Theorem 9). This examples illustrates how $\mathcal{G}(\ )$ fails to commute with direct limits.

6. Zabrodsky's presentation of the genus for an H_0-space

Recall that an H_0-space is a nilpotent space whose rationalization is an H-space. Thus H-spaces, odd dimensional spheres, and complex Stiefel manifolds are all examples of H_0-spaces while S^{2n}, for $n > 0$, and complex Grassmann manifolds are not. Let X denote a simply connected H_0-space of finite type and which has only finitely many nonzero homotopy groups. The main result in Zabrodsky's paper [**104**], is an exact sequence

$$\mathcal{E}_t(X) \xrightarrow{d} \left(\mathbb{Z}_t^*/\pm 1\right)^{\ell} \to \mathcal{G}(X) \to *$$

which is defined as follows. In the middle term, $\mathbb{Z}_t^*$ denotes the group of units in the ring of integers modulo t[6]. The exponent ℓ is the number of positive degrees k, in which the module of indecomposables in rational cohomology, $Q\widetilde{H}^k(X;\mathbb{Q}) \neq 0$. If there is more than one such degree k, order them $k_1 < \cdots < k_\ell$. The number t depends upon the space X. The prime divisors of t include those primes p, for which there is p-torsion in either the homotopy groups of X, or in the integral homology groups of X through degree k_ℓ, or in the cokernel of the Hurewicz homomorphism, through the same range. Zabrodsky gives a description of the smallest possible exponents $\nu_p(t)$, as well. However, for most purposes the exact value of t is unimportant; indeed given any one choice of t, any integer multiple of it works equally well in this sequence. The first term in the sequence, $\mathcal{E}_t(X)$, denotes the monoid (under composition) of homotopy classes of those self-maps of X, which are local equivalences at each prime divisor of t. Such maps are sometimes referred to as *t-equivalences* of X. The function d assigns to each such map a sequence of determinants—or rather the image of such a sequence in the middle group. The jth determinant here is that of the linear transformation on $QH^{k_j}(X;\mathbb{Q})$ induced by the map f. Zabrodsky shows that this image is a subgroup and that the quotient is isomorphic as an Abelian group to $\mathcal{G}(X)$.

Given a finite nilpotent CW-complex X, let $X^{(m)}$ denote its Postnikov approximation through dimension m. In other words, the homotopy groups of $X^{(m)}$ are naturally isomorphic to those of X in degrees $\leq m$ and are zero in higher degrees. If $m \geq \dim X$, then the function $[Y] \mapsto [Y^{(m)}]$ induces a bijection between $\mathcal{G}(X)$ and $\mathcal{G}(X^{(m)})$. The inverse function is induced by restriction to the n-skeleton. Of course, if X is an H_0-space, then so is $X^{(m)}$ for any m. Thus Zabrodsky's sequence applies equally well to 1-connected finite H_0-spaces. To each self map f of X there corresponds another, $f^{(m)}$ of $X^{(m)}$, that induces essentially the same endomorphism of $QH^*(X^{(m)};\mathbb{Q})$. It follows that for H_0-spaces X which satisfy a finiteness condition on their homotopy groups or one on their homology groups, the map $\mathcal{G}(X) \to \mathcal{G}(X^{(m)})$ is surjective for any m.

[6] In other places this will sometimes be denoted $(\mathbb{Z}/t)^*$.

The group structure on $\mathcal{G}(X)$ is induced as follows. Given $[Y]$ and $[Z]$ in $\mathcal{G}(X)$, partition the set of all primes into two nonempty sets, say A and B and then choose maps $\alpha\colon Y \to X$ and $\beta\colon Z \to X$ such that both maps are equivalences at every prime p dividing $t = t(X)$ and so that α is an equivalence at A and β is an at B. For H_0-spaces of the sort considered here, such maps always exist, [**104**]. Then define $[Y] + [Z]$ to be the homotopy type of the pullback of

$$\begin{array}{ccc} & & Z \\ & & \big\downarrow{\scriptstyle \beta.} \\ Y & \xrightarrow{\alpha} & X \end{array}$$

If X is a finite complex in the stable range (and hence a co-H-space), there is a similar presentation of $\mathcal{G}(X)$. This was first proved by Davis in [**20**]; a much shorter proof of this was subsequently given in [**60**].

Consider Zabrodsky's sequence in the case of Sp(2). Let f be a self map of Sp(2), and let $d_n(f)$ denote the degree of f^* on $H^n(\mathrm{Sp}(2);\mathbb{Q})$. It is known that $d_3(f) \equiv d_7(f)$ mod 12 and this is the only restriction on these degrees. The number t can be taken to be some integer multiple of 12. Disregard the action of $\{\pm 1\}$ on each of the factors for the moment and consider the following commutative diagram

$$\begin{array}{ccccc} \mathbb{Z}_t^* & \longrightarrow & (\mathbb{Z}_t^*)^2 & \xrightarrow{\pi_1} & \mathbb{Z}_t^* \\ \big\uparrow{\scriptstyle \delta_7|K} & & \big\uparrow{\scriptstyle \delta} & & \big\uparrow{\scriptstyle =} \\ K & \longrightarrow & \mathcal{E}_t(\mathrm{Sp}(2)) & \xrightarrow{\delta_3} & \mathbb{Z}_t^* \end{array}$$

where the top row is split exact and where δ_n denotes the mod t reduction of $d_n(\;)$. On the bottom row, the kernel K of δ_3 consists of maps f with $d_7(f) \equiv 1$ mod 12. Hence if one reduces the top row mod 12, it follows that $\mathrm{coker}(\delta) \approx \mathrm{coker}(\delta_7|K) \approx \mathbb{Z}_{12}^*$. Hence

$$\mathcal{G}(\mathrm{Sp}(2)) \approx \mathbb{Z}_{12}^*/\pm 1 = \{\bar{1}, \bar{5}\}.$$

The calculation of the genus of the exceptional Lie group G_2 is very similar; just replace 12 with 15. More examples involving Zabrodsky's sequence will be considered in the next section.

7. Calculating $\mathcal{G}(\mathbf{X})$—some examples

EXAMPLE D. *Let $X = S^m \cup_g e^n$ where the attaching map $g \in \pi_{n-1}S^m$ is a suspension class of order q. Then the cardinality of $\mathcal{G}(X)$ is $\frac{1}{2}\varphi(q)$, where φ denotes the Euler function.*

Examples of this sort were first studied by Molnar in [**73**]. It follows that for each λ relatively prime to the order of g, the mapping cone of λg is in $\mathcal{G}(X)$ and conversely every member of the genus can be described this way.

EXAMPLE E. *Let $X = (S^2 \vee S^2) \cup_\Phi e^4$ where $\Phi = [\iota_1, \iota_1] + d[\iota_2, \iota_2]$. If the coefficient d is the product of n distinct primes, where $n \geq 2$, then the cardinality of $\mathcal{G}(X)$ is at least $n+1$.*

Notice that this example shows that for a finite CW-complex X, the cardinality of $\mathcal{G}(X)$ is not bounded by some function of the dimension of X or the number of cells in X. Example E is similar to the one described in the answer Problem 2.1 in Section 4. Order the prime divisors of d. If p is the kth prime divisor, let X_k denote the mapping cone of $p[\iota_1, \iota_1] + \frac{d}{p}[\iota_2, \iota_2]$. It is left to the reader to check that X_k is in the genus of X and that the integral cohomology ring of X_k is not isomorphic to that of X or of X_i where $i \neq k$.

EXAMPLE F. *The Mislin genus of quaternionic projective n-space, $\mathbb{H}P^n$, is isomorphic to $\mathbb{Z}/2 \times \cdots \times \mathbb{Z}/2$, where the number of factors equals the number of primes p such that $2 \leq p \leq 2n-1$.*

This result was proved in [**60**] using Zabrodsky's sequence. Even though $\mathbb{H}P^n$ is not an H_0-space, notice that its Postnikov approximation $X^{(4n+1)}$ is an H_0-space and that the sets $\mathcal{G}(X)$ and $\mathcal{G}(X^{(4n+1)})$ are isomorphic. To calculate this genus using Zabrodsky's sequence one needs the following classification of certain self maps of $\mathbb{H}P^n$.

PROPOSITION F.1. *Assume that λ is a number not divisible by any prime $\leq 2n$. Then there is a self map of quaternionic projective n-space with degree λ on $H_4\mathbb{H}P^n$ if and only if $\lambda \equiv x^2 \bmod p$ for each odd prime $p < 2n$ and $\lambda \equiv 1 \bmod 8$ if $n \geq 2$.*

This result is essentially a description of the first term in Zabrodsky's sequence where t is some multiple of all the primes less than $2n$

$$\mathcal{E}_t(\mathbb{H}P^n) \xrightarrow{d} \mathbb{Z}_t^*/\pm 1 \longrightarrow \mathcal{G}(\mathbb{H}P^n).$$

The order of the middle term is, of course, $\frac{1}{2}\varphi(t)$. By the Chinese remainder theorem the image of d has order $\frac{1}{2}\varphi(t)2^{-k}$ where k is the number of primes less than $2n$. The result follows from these facts. Incidentally, it is interesting to note the the self maps of $\mathbb{H}P^n$, for $3 < n < \infty$ have yet to be classified up to homology. For more information on this problem see [**60**].

Since localization commutes with suspension it makes sense to take the direct limit of the sequence

$$\mathcal{G}(X) \to \mathcal{G}(\Sigma X) \to \mathcal{G}(\Sigma^2 X) \to \cdots,$$

when X is a finite complex. The result is called the stable genus of X. In the next example this limit is achieved in one suspension.

EXAMPLE G. *The stable genus of $\mathbb{C}P^n$ equals $\mathcal{G}(\Sigma\mathbb{C}P^n)$. It is trivial for $n \leq 2$ and for $n \geq 3$ has cardinality $\prod_{m=3}^{n}\big[\varphi(m!)/2\big]$.*

This result was proved in [**60**] using Zabrodsky's sequence and the classification, up to homology, of stable self maps of $\mathbb{C}P^n$. To describe that classification let f be a self map of $\Sigma^k\mathbb{C}P^n$ and let $\delta(f) = \big(\delta_1(f), \cdots, \delta_n(f)\big)$ where $\delta_q(f)$ denotes the degree of f_* on $H_{k+2q}(\Sigma^k\mathbb{C}P^n; \mathbb{Z})$. Then define

$$D_n = \big\{\delta(f) \mid f\colon \Sigma^k\mathbb{C}P^n \hookleftarrow \text{ where } k \geq 1\big\}.$$

Notice that D_n is a subring of $\mathbb{Z}^n$; addition is induced by the suspension co-H-structure and multiplication is induced by composition of self maps. Both operations are defined componentwise in $\mathbb{Z}^n$. The following classification was proved in [**60**] and describes D_n as a subgroup of $\mathbb{Z}^n$.

THEOREM G.1. *The stable self maps of $\mathbb{C}P^n$ have the following properties:*

i) *There is an epimorphism $D_{n+1} \to D_n$, for each n, induced by the restriction of self maps.*
ii) *The vectors $(1, 1, \ldots, 1)$, $(2, 4, \cdots, 2^n)$, $\ldots$, $(n, n^2, \ldots, n^n)$ form a basis for D_n.*

iii) *If* $(0, \ldots, 0, \lambda_m, \ldots, \lambda_n) \in D_n$, *then* $m!$ *divides* λ_i *for* $i \geq m$. *Among all such classes, the case* $\lambda_m = m!$ *occurs.*

Notice that the vectors in part ii) come from self maps of $\mathbb{C}P^n$. After one suspension it is possible to take linear combinations of these maps. The main point here is that, up to homology, these are the only maps—no others exist, no matter how many more times one suspends. This is why the stable genus of $\mathbb{C}P^n$ is obtained in one suspension. To see how the rest of Example G follows from this result, consider the group of units $(D_n \otimes \mathbb{Z}_t)^*$ where t is some integer multiple of $n!$. Define a filtration, for $k \geq 1$,

$$F_k = \{(\lambda_1, \cdots, \lambda_n) \in (D_n \otimes \mathbb{Z}_t)^* \mid \lambda_i = 1 \text{ for } i < k\}$$

and take the projection onto the kth coordinate, $\pi_k \colon F_k \to \mathbb{Z}_t^*$. Of course, the kernel of this projection is F_{k+1}. By part iii) of the previous theorem it follows that the image of π_k is contained in the kernel of the reduction map $\rho_k \colon \mathbb{Z}_t^* \to (\mathbb{Z}/k!)^*$. I will show that the image of π_k equals the kernel of ρ_k. Let P^r denote $\Sigma^j \mathbb{C}P^r$ where $j \geq 1$. It suffices to show that any t-equivalence, say g of P^{r-1} extends (up to homology) to a t-equivalence of P^r. By the first part of Theorem G.1 it may be assumed that g extends to some self map $\bar{g}$ of P^r. If γ denotes the attaching map for the top cell of P^r, it follows that $g_*(\gamma) = \lambda\gamma$ for some integer λ, which also equals the degree of $\bar{g}$ on the top cell of P^r. Since g was a t-equivalence and the order of γ divides t, it follows that $(\lambda, |\gamma|) = 1$. If $(\lambda, t) = 1$, then $\bar{g}$ is a suitable extension; if not, the degree of $\bar{g}$ can be altered on the top cell of P^r by any integer multiple of $|\gamma|$. Since λ and $|\gamma|$ are relatively prime an integer m can be chosen, by Dirichlet's theorem, so that $\lambda' = \lambda + m|\gamma|$ is a prime greater than t[7]. Thus there is an extension g' of g which is a t-equivalence.

The order of the kernel of the reduction map from $\mathbb{Z}_t^*$ to $(\mathbb{Z}/k!)^*$ is $\varphi(t)/\varphi(k!)$. Hence

$$\text{order } (D_n \otimes \mathbb{Z}_t)^* = \prod_{k=1}^{n} \text{order } (F_k/F_{k+1}) = \frac{\varphi(t)^n}{\varphi(3!) \cdots \varphi(n!)}.$$

Thus the order of the cokernel of $(D \otimes \mathbb{Z}_t)^* \hookrightarrow ((\mathbb{Z}/t)^n)^*$ is $\varphi(3!) \cdots \varphi(n!)$. After the action of $\mathbb{Z}/2$ on each coordinate is taken into account, the claim about the cardinality follows.

8. Zabrodsky's conjecture for the genus of $\mathbf{SU(n+1)}$

Some people might claim that, from a homotopy point of view, the most interesting questions about the classical groups have long since been answered. The following problem is, I believe, a notable counterexample to such a claim. Zabrodsky posed it in his book, ([**105**], page 152), and it is still open.

CONJECTURE 6. For $n \geq 3$, the genus of $SU(n+1)$ has order $\prod_{m=3}^{n} [\varphi(m!)/2]$.

It is not difficult to check that the genus of $SU(n)$ is trivial for $n \leq 3$. Let SU denote the infinite special unitary group and let SU_m denote the Postnikov approximation $SU^{(2m)}$. As noted earlier there is an epimorphism from the genus of $SU(n+1)$ to the genus of any Postnikov approximation of itself—in particular to that of SU_{n+1}. In his book Zabrodsky shows that the genus of the approximation SU_{n+1} has cardinality $(\varphi(3!)/2) \cdots (\varphi(n!)/2)$. He calculates this by showing that for each n the epimorphism $\mathcal{G}(SU_{n+1}) \to \mathcal{G}(SU_n)$ has kernel $(\mathbb{Z}/n!)^*/\pm 1$. To

[7]This argument corrects an error in the proof of Lemma 4.8 of [**60**]

establish this kernel he actually constructs the appropriate members in the genus of $SU(n+1)$, each of which fibers over S^{2n+1} with fiber $SU(n)$.

I gave a different proof of this lower bound in [**60**]. There I noticed that since $\mathbb{C}P^n$ is a stable retract of $SU(n+1)$, each self map of $SU(n+1)$ induces a stable self map of $\mathbb{C}P^n$. This implies that the set of stable self maps of $\mathbb{C}P^n$ serves as an upper bound on the set of realizable endomorphisms of $QH^*SU(n+1)$. Because of Zabrodsky's sequence and Example G, this means that the genus of $\Sigma\mathbb{C}P^n$ is a lower bound for (i.e., a quotient of) the genus of $SU(n+1)$.

Of course, everyone knows that $\Sigma\mathbb{C}P^n$ is a generating complex for $SU(n+1)$ but Zabrodsky's conjecture (if true) suggests that there is a more intimate connection between the two spaces than the one currently understood. It also implies the existence of self maps of $SU(n+1)$ unlike any currently known.

PROPOSITION 6.1. *Zabrodsky's conjecture for $\mathcal{G}\big(SU(n+1)\big)$ is true if and only if for each positive integer q there exists a self map ψ_q of $SU(n+1)$ which induces multiplication by q^k on $\pi_{2k+1}SU(n+1)$ for $k=1,\cdots,n$.*

PROOF. Given a self map f of $SU(n+1)$, let $\delta'(f)=\big(\delta'_1(f),\cdots,\delta'_n(f)\big)$ where $\delta'_q(f)$ denotes the degree of f_* on $\pi_{2q+1}SU(n+1)$ and let

$$E_n=\big\{\delta'(f)\mid f\colon SU(n+1)\hookleftarrow\big\}.$$

As noted above $E_n\subseteq D_n$. If the two groups are equal then Zabrodsky's conjecture is true. By part ii) of Theorem G.1, equality between the two groups is equivalent to the existence of the self maps ψ_q on $SU(n+1)$ for all positive q. Suppose that some of the ψ_q's are *not* contained in D'_n. Then E_n must be a proper subgroup of D_n. The index of E_n in $\mathbb{Z}^n$ must then be strictly larger than the index of D_n in $\mathbb{Z}^n$. The latter, of course, equals $(1!)(2!)\cdots(n!)$. Again let t be some fixed multiple of $n!$. It suffices to show that the cardinality of $(E_n\otimes\mathbb{Z}_t)^*$ is strictly less than that of $(D_n\otimes\mathbb{Z}_t)^*$. Take a basis for E_n similar to the one described in part iii) of Theorem G.1; that is, a basis which forms an upper triangular matrix (m_{ij}). The product of the diagonal entries in this matrix equals the index of E_n in $\mathbb{Z}^n$. Since $E_n\subseteq D_n$ it follows, again from part iii) that $k!$ divides each of the entries in the kth row. The order of $(E_n\otimes\mathbb{Z}_t)^*$ can be computed using a filtration argument like the one used on $(D_n\otimes\mathbb{Z}_t)^*$. It follows that

$$\text{order }(E_n\otimes\mathbb{Z}_t)^*\leq\frac{\varphi(t)^n}{\varphi(m_{11})\varphi(m_{22})\cdots\varphi(m_{nn})}.$$

Since each m_{kk} is a nonzero integer multiple of $k!$ and at least one m_{kk} is strictly greater than $k!$, the result follows. □

What is known about the self maps of $SU(n+1)$? Here is a list of the most obvious ones

1) Power maps—for each integer λ, the power map $x\mapsto x^\lambda$ induces the vector $(\lambda,\lambda,\cdots\lambda)$ in E_n.
2) Sending a matrix to its complex conjugate yields $\big(1,-1,\cdots(-1)^{n+1}\big)$ in E_n.
3) Other loop maps[8]—for each positive integer q which is relatively prime to $(n+1)!$, the loop map $\Omega\psi^q$ induces the vector $(q^2,q^3,\ldots q^{n+1})$ in E_n.

[8]The existence of these loop maps is due, independently, to Sullivan [**93**], Wilkerson [**97**], and Friedlander [**30**]. That there are no others follows from the work of Hubbuck [**50**] and Ishiguro [**51**].

4) The composition $SU(n+1) \to S^{2n+1} \to SU(n+1)$, where the first map is the principal $SU(n)$ fibration and the second represents a generator of π_{2n+1}, induces the vector $(0, \cdots 0, n!)$ in E_n.

It should be possible to give an upper bound on the genus of $SU(n+1)$ using the maps just listed, but as far as I know, this has yet to be done.

9. Infinite spaces

By an infinite space I mean one which has nonzero homology and nonzero homotopy groups in infinitely many dimensions. The classifying space of a compact Lie group and the loop space of a finite complex are two important examples of such spaces. The first explicit description of the genus of an infinite space was the following result.

THEOREM 7. *Let B be a space in the Mislin genus of $\mathbb{H}P^\infty$. For each prime p, there is a homotopy invariant $(B/p) \in \{\pm 1\}$ having the following properties:*

1) *The symbols (B/p) provide a complete set of homotopy classification invariants for members of $\mathcal{G}(\mathbb{H}P^\infty)$.*
2) *$(B/p) = 1$ for all p when $B = \mathbb{H}P^\infty$ and every combination of values for the (B/p) occurs for some B.*
3) *$(B/2)$ is defined on the 8-skeleton of B, and for odd primes (B/p) is defined on the $2p+2$-skeleton of B.*
4) *If B contains a maximal torus, then $B = \mathbb{H}P^\infty$.*

Almost all of this theorem is due to Rector, [**87**]. In part (4), however, he only showed that if B contains a maximal torus, then $(B/p) = 1$ for all odd primes p. Later I showed in [**58**] that this was also true for $p = 2$. To say that B *contains a maximal torus* here, means only that there is a map $\pi\colon \mathbb{C}P^\infty \to B$ whose homotopy theoretic fiber is homotopy equivalent to a finite complex. Incidentally, when Theorem 7 was first established it was not known if every loop structure on S^3 (i.e., a space X with $\Omega X \simeq S^3$) was in the genus of $\mathbb{H}P^\infty$. Later I showed this was stably true at odd primes in [**59**]. It is now known to be true unstably as well, at all primes, by the work of Dwyer, Miller, and Wilkerson, [**23**]. In other words, for each prime p, there is only one loop structure on $S^3_{(p)}$.

Rector's invariants (B/p) are defined in terms of a certain integral cohomology generator of degree 4. Note that the cohomology of B is a polynomial algebra on a class of degree 4, and so there are just two choices for this generator. Assume for the moment that the correct choice has been made; call it $[u]$. (In the case of $\mathbb{H}P^\infty$, choose $[u]$ so that $\pi^*[u] = y^2$, where y generates $H^2\mathbb{C}P^\infty$). For each odd prime p, let u denote the image of $[u]$ in $H^*(B;\mathbb{Z}/p)$ and note that the Adem relation $\mathcal{P}^1\mathcal{P}^1 = 2\mathcal{P}^2$ implies that

$$\mathcal{P}^1 u = \pm 2u^{(p+1)/2}.$$

Define (B/p) to be the sign that occurs in this equation.

It remains to define $(B/2)$ and to choose the orientation class $[u]$ in a manner consistent with the choice made in the classical case. Rector does this using the e-invariant from KO-theory. Here is an alternate proof.

PROPOSITION 7.1. *Let P denote the 8-skeleton of a member of $\mathcal{G}(\mathbb{H}P^\infty)$. After one suspension,*

$$\Sigma P \simeq S^5 \cup_{k\nu} e^9 \text{ where } k = 1, 5, 7, \text{ or } 11.$$

All four values of k occur and are well defined mod24. *When* $k = 1, \Sigma P \simeq \Sigma\mathbb{H}P^2$.

Assume for the moment this is true. Then given B, suspend its 8-skeleton once and define

$$(B/2) = \begin{cases} 1 & \text{if } k \equiv 1 \bmod 4 \\ -1 & \text{otherwise} \end{cases}$$

and

$$(B/3) = \begin{cases} 1 & \text{if } k \equiv 1 \bmod 3 \\ -1 & \text{otherwise.} \end{cases}$$

Having done this choose $[u]$ in $H^4(B;\mathbb{Z})$ so that the previous definition of $(B/3)$ agrees with the one just given. This is consistent with the earlier choice made for $\mathbb{H}P^\infty$.

PROOF OF PROPOSITION 7.1. By the results of [**61**] it suffices to consider the attaching maps of the 4 members of $\mathcal{G}(\mathbb{H}P^2)$. This has been done in [**8**]: the attaching maps for the 8-cells have the form $\gamma + mE\omega$ where γ is the quaternionic Hopf fibration, ω is the Blakers-Massey element, and $m = 0, 4, 6$, and 10. Using the relations $E\gamma = \nu$, $E^2\omega = -2\nu$, and $C_{\lambda\nu} \simeq C_{-\lambda\nu}$, the result follows. □

The next three theorems were inspired, in large part, by Theorem 7. The first thing one notices in Rector's theorem is that the genus of $BSU(2)$ must be uncountably large. Møller has shown that the same is true of BG for all compact connected Lie groups G except tori, [**71**]. This remains true even after inverting finitely many primes.

THEOREM 8. *Let G be any non-Abelian compact connected Lie group and let L be any cofinite set of primes. Then $\mathcal{G}(BG_L)$ is uncountably large.*

Suppose that X and Y are finite H-spaces in the same genus and that L is some finite set of primes. It is known then that L-equivalences between X and Y exist ([**47**], page 114). This is true, more generally, for H_0-spaces with a finiteness condition on their homology or their homotopy groups, [**104**, Proposition 1.5]. Thus, for such spaces localized at L, the genus $\mathcal{G}(X_L)$ is trivial. However, for infinite H_0-spaces this is not necessary true as Møller shows in the following result from [**71**].

THEOREM 9. *Let G be a compact connected Lie group whose whose universal covering is non-contractible and not homotopy equivalent to $SU(2)$. Then $\mathcal{G}(BG_L)$ is countably infinite for all finite sets of two or more primes.*

Notbohm and Smith have generalized the last part of Theorem 7, in the following manner, [**75**], [**76**].

THEOREM 10. *Let G be a simply connected compact Lie group and let $X \in \mathcal{G}(BG)$. Then the following statements are equivalent:*

i) *X has a maximal torus .*

ii) *$K(X) \approx K(BG)$ as λ-rings.*

iii) *$X \simeq BG$.*

Thus within the genus of such a classifying space there exist homotopy theoretic properties which distinguish the Lie multiplication from all others! If one removes the requirement that G be simply connected in Theorem 10, then the resulting statement is not true. The reason why involves the following curious fact about the unitary group $U(n)$: *for each $n \geq 3$ there are $\frac{1}{2}\varphi(n)$ different Lie groups G, each diffeomorphic to $S^1 \times SU(n)$, whose classifying spaces BG are distinct up to* *homotopy*[9]. These groups occur in Baum's study of the set of all compact connected Lie groups belonging to a fixed Lie algebra, [**5**], and they are considered in detail by Notbohm and Smith in [**76**]. They can be presented as quotients of $S^1 \times SU(n)$ as follows. Let k be a unit in $\mathbb{Z}/n$. Then the group $FU_k(n)$, in the notation of [**76**], fits into an exact sequence

$$\mathbb{Z}/n \xrightarrow{\zeta_k} S^1 \times SU(n) \to FU_k(n),$$

where ζ_k sends the generator $\bar{1}$ to $(\omega^{-k}, \omega I_n)$. Here $\omega = e^{2\pi i/n}$ and I_n is the identity $n \times n$ matrix. Of course, $FU_1(n) = U(n)$. The classifying space $BFU_k(n)$ can be obtained as a pullback

$$\begin{array}{ccc} & & BPU(n) \\ & & \downarrow \\ K(\mathbb{Z}, 2) & \xrightarrow[\psi^k \rho]{} & K(\mathbb{Z}/n, 2) \end{array}$$

where $PU(n)$ denotes the projective unitary group, ρ is induced by the quotient map $\mathbb{Z} \to \mathbb{Z}/n$ and ψ^k is induced by multiplication by k. Since k is relatively prime to n, it follows easily that $BFU_k(n)$ is in the genus of BU. Since the $FU(n)$'s are Lie groups they have maximal tori and it is this for reason that Theorem 10 fails to hold in the non-simply connected case. However, Notbohm and Smith show that these are the only exceptions; that is, they show that if X is a member of $\mathcal{G}(BU(n))$ with a maximal torus, then $X \simeq BFU_k(n)$ for some k. Moreover they show that the equivalence of ii) and iii) is still valid in $\mathcal{G}(BU(n))$; indeed they show that the different $BFU(n)$'s can be distinguished using K-theory.

The question of when there are rational equivalences between two members of the same genus is an interesting one. Adams once constructed an example $X \in \mathcal{G}(\mathbb{H}P^\infty)$ with the property that every nonzero element in $\widetilde{K}^o(X)$ has infinitely many nonzero Chern classes, ([**3**], page 79). It follows that there are no rational equivalences from his X to $\mathbb{H}P^\infty$. The following result, due to Ishiguro, Møller and Notbohm ,[**53**], greatly strengthens and generalizes his example.

THEOREM 11. *Let X and Y be members of $\mathcal{G}(BG)$ where G is a simple Lie group. If there exists an essential map between X and Y, then $X \simeq Y$.*

This is a remarkable result; it demonstrates a vast difference between infinite H_0-spaces and finite ones. As mentioned earlier one can always find a rational equivalence between any two finite H_0-spaces in the same genus. Along these same lines Møller has shown in [**72**] the following.

THEOREM 12. *Let G be a connected compact Lie group and let $X \in \mathcal{G}(BG)$. If there exists a rational equivalence between X and BG, then $X = BH$ for some Lie group H, and there are rational equivalences between BG and BH in both directions.*

Let X be a connected CW-space. Following Wilkerson, [**98**], let $SNT(X)$ denote the set of all homotopy types $[Y]$ such that the Postnikov approximations $X^{(n)} \simeq Y^{(n)}$ for all n[10]. In analyzing the genus of an infinite space X, one is forced

[9]Once again $\varphi(\ )$ denotes the Euler phi-function.

[10]For another approach to $SNT(X)$ see [**21**] and [**22**].

to deal with the set $SNT(X)$. Consider, for example, the function

$$\mathcal{G}(X) \xrightarrow{\Phi} \varprojlim \mathcal{G}(X^{(n)})$$

that sends $[Y]$ to the coherent sequence $([Y^{(1)}], [Y^{(2)}], \cdots)$. The kernel of Φ is $SNT(X) \cap \mathcal{G}(X)$ and, as the next theorem shows, this set sometimes contains more than just $[X]$. Members of this intersection are sometimes called *clones of* X. In other words, a clone of X is a space Y such that $Y_{(p)} \simeq X_{(p)}$ for all primes p and $Y^{(n)} \simeq X^{(n)}$ for all natural numbers n. The following strange[11] result appeared in [**62**].

EXAMPLE H. *Let $X = \mathbb{H}P^\infty$, the infinite quaternionic projective space and let S denote a set of primes. Then, up to homotopy, $X_{(S)}$ has nontrivial clones (uncountably many, in fact) if and only if S is an infinite subset of primes whose complement is also infinite. Moreover, any two clones of $X_{(S)}$ become homotopy equivalent when localized at any finite set of primes.*

The six theorems just mentioned indicate that real progress has been achieved in study of the Mislin genus of BG, when G is a compact Lie group. It should be noted, however, that in general, not every loop structure on G will necessarily be in Mislin genus of BG or even in the complete genus $\widehat{\mathcal{G}}(BG)$, [**27**]; nevertheless these genera represent a good starting point for studying the various multiplications on the homotopy type of a Lie group. In contrast, very little is known about the genus of ΩK, where K is a finite complex. It is not known, for example, if there exists a finite complex K for which $\mathcal{G}(\Omega K)$ is infinite. One of the few results in this area is the following, which Møller and I proved in [**63**].

THEOREM 13. *If K is a 1-connected finite H_0-space then the function $[Y] \mapsto [\Omega Y]$ induces a surjection of $\mathcal{G}(K)$ onto $\mathcal{G}(\Omega K)$. In particular, the genus of ΩK is finite in this case.*

The proof of this theorem involves two maps,

$$\mathcal{G}(K) \longrightarrow \varprojlim \mathcal{G}(\Omega K^{(n)}) \longleftarrow \mathcal{G}(\Omega K).$$

The first map comes from taking Postnikov approximations and showing that the function

$$\mathcal{G}(K^{(n)}) \longrightarrow \mathcal{G}(\Omega K^{(n)})$$

is an epimorphism for each n. This is quite easy using Zabrodsky's sequence. Since K is finite dimensional this gives an epimorphism of $\mathcal{G}(K)$ onto $\varprojlim \mathcal{G}(\Omega K^{(n)})$. The other map is the one introduced as Φ above. Injectivity for Φ in this case is established by showing that $SNT(Y) = *$ for every $Y \in \mathcal{G}(\Omega K)$; surjectivity follows by showing that $SNT(\Omega K_{(p)}) = *$ for each p.

In general one should not expect the loop space functor to induce a surjection $\mathcal{G}(X) \to \mathcal{G}(\Omega X)$, as the following example shows.

EXAMPLE I. *If X is the Grassmann manifold of complex n-planes in $\mathbb{C}^{n+k}$ where $n \geq 5$ and $2k > n^2$ then $\mathcal{G}(X) = *$ but $\mathcal{G}(\Omega X) \neq *$.*

That $\mathcal{G}(X) = *$ is the theorem of Glover and Mislin, described in Section 5. Let W denote the Stiefel manifold of n-frames in $\mathbb{C}^{n+k}$. Consider the principal fibration $U(n) \longrightarrow W \longrightarrow X$, given by sending an n-frame of vectors to the n-plane they span. Since the dimension of $U(n)$ is n^2 and the connectivity of W

[11]It seems strange because $SNT(\mathbb{H}P^\infty) = *$ by Theorem 4 of [**63**].

is $2k$, the condition $n^2 < 2k$ insures that the inclusion of the fiber $U(n) \to W$ is nullhomotopic. This in turn forces $\Omega X \simeq \Omega W \times U(n)$. It is easy to see that $\mathcal{G}(U(n)) \subseteq \mathcal{G}(\Omega X)$ and since $\mathcal{G}(U(n)) \neq *$ for $n \geq 5$ (see Section 7) the example follows.

There is one application of the Mislin genus I would like to mention and since it involves infinite spaces this may be a good place to do it. Recall the notation $QX = \varinjlim \Omega^n \Sigma^n X$. This is, of course, an infinite space for most X. Bruner, Cohen and I recently proved the following result, [**12**].

THEOREM 14. *Let X and Y be finite CW-complexes. Then $\Sigma^n X$ and $\Sigma^n Y$ are homotopy equivalent for some large n if and only if the spaces QX and QY are homotopy equivalent.*

This result appears to have nothing to do with the Mislin genus and yet its proof uses a number of ideas already discussed in this survey. Of course, the nontrivial part is showing that $QX \simeq QY$ implies that X and Y are stably equivalent. This is first established one prime at a time. Results of Wilkerson (essentially the Eckmann-Hilton dual of his Theorem 3 in this paper) are used to first express the stable p-local homotopy types of X and Y as bouquets of prime retracts. The same sort of analysis is then carried out, p-locally, on QX and QY through an appropriate range of dimensions. The assumption that QX and QY are homotopy equivalent then forces the stable prime retracts of $X_{(p)}$ to match up with those of $Y_{(p)}$. Since this holds for each prime, it follows that X and Y must have the same stable genus. Then for large n we consider the function

$$\mathcal{G}(\Sigma^n X) \to \mathcal{G}(QX)$$

that sends $[Y] \mapsto [\Omega^n QY]$. This function is shown to be injective using Zabrodsky's genus sequence and the result follows for the case where X and Y are connected. In the non-connected case, cancellation arguments applied to the p-completions of QX and QY, and the Segal conjecture are used to first establish that X and Y must have the same number of components. Then further cancellation, this time p-locally, allows X and Y to be replaced, each by the bouquet of its components, and the proof reverts to the connected case. Thus many of the ideas and results associated with the Mislin genus are crucial in the proof of Theorem 14.

10. On the genus of nilpotent groups

This is a big topic with an extensive literature. I am afraid that I have neither the expertise nor the time to do little more than mention a few of the highlights. The first highlight would, of course, be Pickel's paper, [**86**], which was briefly described in Section 2. The set $\hat{G}_0(N)$, defined there is sometimes called the Pickel genus of N in the literature[12]. For a finitely generated nilpotent group N, the Mislin genus, $\mathcal{G}(N)$ is defined to be the set of isomorphism classes of finitely generated nilpotent groups M such that $N_{(p)} \approx M_{(p)}$ for all primes p. This notion first appears in the seminal paper, [**67**]. It follows that the Mislin genus of the *group* N is isomorphic to the Mislin genus of the *space* $K(N,1)$.

[12]More precisely, two groups N and M are in the same Pickel genus if they have the same finite images and $N_0 \approx M_0$. For finitely generated nilpotent groups this definition is equivalent to the one given for $\widehat{G}_0(N)$ [**96**].

Let $\mathcal{N}_0$ denote the category of finitely generated nilpotent groups with finite commutator subgroups. Hilton and Mislin have shown that when $N \in \mathcal{N}_0$, the set $\mathcal{G}(N)$ can be given the structure of an Abelian group in a fairly natural way; indeed, there is an exact sequence,

$$\mathcal{E}_t(N) \longrightarrow \mathbb{Z}_t^* \longrightarrow \mathcal{G}(N) \longrightarrow O$$

similar to Zabrodsky's genus sequence for 1-connected H_0-spaces[13] . A number of specific calculations of $\mathcal{G}(N)$ have been made using it. In a series of papers [**37**], [**38**], [**39**], [**40**] Hilton studied the Mislin genus of nilpotent groups N admitting presentations

$$N = \langle x, y \colon x^n = 1, yxy^{-1} = x^u \rangle.$$

Let t be the order of u mod n. It turns out that $\mathcal{G}(N)$ is isomorphic to $\mathbb{Z}_t^*/\{\pm 1\}$; moreover, if m is a positive integer such that $[m] \in \mathbb{Z}_t^*/\{\pm 1\}$, then the corresponding element of $\mathcal{G}(N)$ is the isomorphism class of the group

$$N_m = \langle x, y \colon x^n = 1, yxy^{-1} = x^{u^m} \rangle$$

and so all elements of $\mathcal{G}(N)$ are represented in this way.

For a nilpotent group N, let TN denote its torsion subgroup and let FN denote the quotient N/TN. Casacuberta and Hilton generalized the preceding example by considering groups in $\mathcal{N}_0$ with TN Abelian and with N a semidirect product for an action $\omega \colon FN \to \operatorname{Aut}(TN)$ such that $\omega(FN)$ is in the center of $\operatorname{Aut}(TN)$. Among other things in [**15**], they showed that if TN is a p-group, then $\mathcal{G}(N) \approx \mathcal{G}(\times^k N)$. Schuck, a recent Ph.D. student of Hilton's, studied this same class of groups and proved that if FN is cyclic then $\mathcal{G}(\times^k N) \approx \mathcal{G}(N)/H$ where H is a certain elementary Abelian 2-group while if FN is not cyclic then $\mathcal{G}(\times^k N)$ is trivial for $k \geq 1$.

In [**67**] Mislin established a version of Theorem 3 for groups. He showed that if $G \times A \approx H \times A$ where G and H are nilpotent, A is Abelian, and all three groups are finitely generated, then G and H must be locally isomorphic at each prime. The converse holds for a suitable A, provided G and H are in the same Mislin genus and have finite commutator subgroups. In this event he also showed that $\times^n G \approx \times^n H$ for some positive integer n. Warfield then noticed that this result could be extended to other groups (not necessarily nilpotent ones) if one replaces localization by profinite completion. He proved the following in [**96**].

THEOREM 15. *If G and H are finitely generated groups with finite commutator subgroups then the following conditions are equivalent:*

i) *the profinite completions of G and H are isomorphic.*

ii) $G \times \mathbb{Z} \approx H \times \mathbb{Z}$.

iii) $\times^n G \approx \times^n H$, *for some positive integer n.*

Warfield also showed that for nilpotent group N with a finite commutator subgroup, the Mislin genus $\mathcal{G}(N)$ and the complete genus $\widehat{G}(N)$ are the same. Thus Theorem 15 includes and strengthens Mislin's result and also applies to some non-nilpotent groups. This is an example of nilpotent results leading to more general ones. Another example is the significant generalization of Pickel's theorem due to

[13]Notice that when $N \in \mathcal{N}_0$, the space $K(N, 1)$ is an H_0-space. This suggests that the simply connected requirement in Zabrodsky's sequence could probably be weakened to nilpotent.

Grunewald, Pickel, and Segal, [**34**]. They show that the completion genus is finite for any polycyclic-by-finite group.

Notice that the examples N_m that Hilton studied contains nontrivial torsion. If one restricts to torsion-free groups then one must look farther before finding examples of nontrivial genera. Recall that the Hirsch length of a nilpotent group is the total number of infinite cyclic factors occurring in the quotients of its lower central series. Grunewald and Scharlau, [**33**], have shown that $\widehat{G}(N) = *$ for any torsion free nilpotent group N, of class 2 with Hirsch length ≤ 5. However, there is a drastic change at Hirsch length 6.

Let $\mathcal{T}(4,2)$ denote the class of finitely generated, torsion-free, nilpotent groups G of nilpotency class 2 and Hirsch length 6, where G/G' has torsion-free rank 4 and the commutator subgroup G' has rank 2. It follows that the torsion subgroup of G/G' is isomorphic to $\mathbb{Z}/\delta \times \mathbb{Z}/\lambda\delta$ for two uniquely determined positive integers $\delta = \delta(G)$ and $\lambda = \lambda(G)$. On the other hand, given positive integers δ and λ and a binary integral quadratic form

$$\Phi(X,Y) = aX^2 + bXY + cY^2,$$

one can use this data to define a group

$$\begin{aligned} G(\delta,\lambda,\Phi) = \big\langle x_1, x_2, x_3, x_4, u_1, u_2 \colon & u_1 \text{ and } u_2 \text{ are central}, \\ & [x_1,x_2] = [x_3,x_4] = 1, \\ & [x_1,x_3] = u_2^{\delta\lambda}, [x_1,x_4] = u_1^{\delta}, \\ & [x_2,x_3] = u_1^{a\delta}u_2^{b\delta\lambda}, [x_2,x_4] = u_2^{-c\delta\lambda}\big\rangle. \end{aligned}$$

It is not difficult to see that $G \in \mathcal{T}(4,2)$, $\delta(G) = \delta$, and $\lambda(G) = \lambda$ where $G = G(\delta,\lambda,\Phi)$. Grunewald, Segal and Sterling show in [**35**] that *every* group in $\mathcal{T}(4,2)$ is isomorphic to some $G(\delta,\lambda,\Phi)$. To describe the genus of such groups they use the following.

DEFINITION. Let Φ and Ψ be binary quadratic forms over a commutative ring R. These forms are *λ-equivalent over R*, written $\Phi \underset{R}{\overset{\lambda}{\sim}} \Psi$, if there exists a unit $u \in R$ and a matrix $\left(\begin{smallmatrix} \alpha & \beta \\ \lambda\gamma & \delta \end{smallmatrix}\right) \in GL(2,R)$ with $\gamma \in R$ such that

$$\Psi(X,Y) = u\Phi(\alpha X + \beta Y, \lambda\gamma X + \delta Y).$$

The main result in [**35**] is the following.

THEOREM 16. *Let δ and λ be positive integers. The assignment $\Phi \mapsto G(\delta,\lambda,\Phi)$ induces a bijective correspondence between the set of all λ-equivalence classes of binary integral quadratic forms over $\mathbb{Z}$ and the set of all isomorphism classes of groups $G \in \mathcal{T}(4,2)$ with $\delta(G) = \delta$ and $\lambda(G) = \lambda$. Moreover, the profinite completions*

$$\begin{aligned} & \widehat{G}(\delta,\lambda,\Phi) \approx \widehat{G}(\delta',\lambda',\Phi') \\ \iff & \delta = \delta', \lambda = \lambda', \text{ and } \Phi \underset{\mathbb{Z}_p}{\overset{\lambda}{\sim}} \Phi' \text{ for all primes } p \end{aligned}$$

and

$$G(\delta,\lambda,\Phi)_0 \approx G(\delta',\lambda',\Phi')_0 \iff \Phi \underset{\mathbb{Q}}{\overset{1}{\sim}} \Phi'.$$

This theorem shows that the isomorphism types in $\mathcal{T}(4,2)$ cannot be classified by a simple set of numerical invariants, as was the case for groups of Hirsch length

≤ 5. However, if one is willing to settle for a coarser classification in terms of profinite completion or rationalization, then two numerical invariants suffice in many cases. Recall that for $\Phi(X, Y) = aX^2 + bXY + cY^2$ the discriminant of Φ is

$$D(\Phi) = b^2 - 4ac.$$

Define the *content* of Φ to be

$$C(\Phi) = \begin{cases} 0 & \text{if } a = b = c = 0 \\ \gcd(2a, b, 2c) & \text{otherwise .} \end{cases}$$

Both $C(\Phi)$ and $D(\Phi)$ are invariant under λ-equivalence over $\mathbb{Z}$, so each $G \in \mathcal{T}(4,2)$ has well defined invariants $C(G)$ and $D(G)$ given by putting

$$C(G) = C(\Phi) \text{ and } D(G) = D(\Phi)$$

where $G = G(\delta, \lambda, \Phi)$.

COROLLARY 16.1. Let G and H be in $\mathcal{T}(4,2)$. Then

i) $G_0 \approx H_0 \iff$ either $C(G) = C(H) = 0$ or $D(G)/D(H) \in (\mathbb{Q}^*)^2$ and $C(G)C(H) \neq 0$.

ii) If G/G' and H/H' are free Abelian, then the profinite completions

$$\widehat{G} \approx \widehat{H} \iff C(G) = C(H) \text{ and } D(G) = D(H).$$

The authors of [**35**] also show that the complete genus $\widehat{\mathcal{G}}(N)$ can be arbitrarily large[14] for $N \in \mathcal{T}(4,2)$. I assume that the same statement holds for the Mislin genus of groups in this class. I have been told that there exist groups $N \in \mathcal{T}(4,2)$ such that $\mathcal{G}(N)$ is strictly smaller than $\widehat{\mathcal{G}}(N)$ but I cannot cite a specific example at this time. Finally, it seems likely that in Theorem 16, the two groups $G(\delta, \lambda, \Phi)$ and $G(\delta', \lambda', \Phi')$ are in the same Mislin genus if and only if

$$\delta = \delta', \lambda = \lambda', \text{ and } \Phi \underset{\mathbb{Q}}{\overset{1}{\sim}} \text{ for all primes } p.$$

In model theory two models are said to elementarily equivalent if any sentence that is true in one model is also true in the other [**18**]. This notion can be carried over to groups (the proofs often use ultrafilters) and Oger has found some remarkable connections between elementary equivalence and groups in the same genus. He shows in [**79**] that if A and B are groups such that $A \times \mathbb{Z} \approx B \times \mathbb{Z}$, then A and B are elementarily equivalent. The converse is not true in general, but he shows in [**78**] that it is true for finitely generated groups with finite commutator subgroups. Thus it ties in nicely with Warfield's Theorem 15. In [**80**] he shows that if two finitely generated nilpotent groups are elementarily equivalent then they must lie in the same Mislin genus. For further developments along these lines see [**81**].

I will close this section with two examples which I learned from Oger. While I do not know if it is reasonable to seek a characterization of those groups whose genera have just one element, it certainly seems like a good idea, at this point, to look for examples.

EXAMPLE J. *If $F_{r,c}$ denotes the free nilpotent group on r generators of nilpotency class c, then the Mislin genus of $F_{r,c}$ is trivial.*

---

[14] Of course, it must still be finite by Pickel's theorem.

PROOF. Let $F = F_{r,c}$ and assume that N is in the Mislin genus of F. Since F and N are locally isomorphic at each prime, the same is true of the Abelian quotients $\Gamma_k(F)/\Gamma_{k+1}(F)$ and $\Gamma_k(N)/\Gamma_{k+1}(N)$ for each k. It follows that the nilpotency class of N equals that of F; it also follows that the Hirsch lengths of F and N are the same. Since F and N have the same set of finite quotients and N is residually finite, it follows that N is also generated by r elements. This implies, by the freeness of F, that there must be an epimorphism of F onto N. This epimorphism must be an isomorphism; otherwise its kernel, being torsion free, would have Hirsch length at least one which in turn would imply the Hirsch length of F exceeds that of N, a contradiction. Hence the two groups are isomorphic. □

The growth of $\mathcal{G}(\times^k N)$, as a function of k, in the next (and final) example contrasts sharply with that of the examples described at the beginning of this section.

EXAMPLE K. *For each integer $n > 1$, there exists a group N which is finitely generated torsion-free nilpotent of class 2, such that the cardinality of $\mathcal{G}(\times^k N)$ is at least polynomial of degree n in k.*

PROOF. According to Hirschon, [**49**], a finitely generated torsion-free nilpotent group of class 2 can be decomposed in an essentially unique way. In other words, if such a group $G = G_1 \times \cdots \times G_n$ where each G_i is indecomposable, then up to an isomorphism of terms, this is the only way of expressing G as a product of indecomposable groups. Now for any positive integer n, one can find $n+1$ non-isomorphic groups $G_1, G_2, \cdots, G_{n+1}$ each of which is finitely generated, torsion-free, and nilpotent of class 2 and all of which are in the same Mislin genus $\mathcal{G}(N)$. This follows using the results of [**35**] or [**77**]. So if $k = k_1 + k_2 + \cdots + k_{n+1}$ where each k_i is a nonnegative integer, then

$$(\times^{k_1} G_1) \times \cdots \times (\times^{k_{n+1}} G_{n+1})$$

is a member $\mathcal{G}(\times^k N)$. By Hirschon's result, different ordered partitions of k give rise to different members of $\mathcal{G}(\times^k N)$. The total number of ordered partitions of k, with $n+1$ parts and with zero as an allowed entry is $(k+n)\cdots(k+2)(k+1)/n!$ and so the claim follows. □

11. Open problems

Zabrodsky's conjecture on the genus of $SU(n)$ would normally go at the top of my list but since that problem has been discussed at length in Section 6, it will not be repeated here. However, note that an unqualified yes answer to the following question would be enough to prove his conjecture.

1. *Let $W_{n,k}$ denote the Stiefel manifold of complex k-frames in $\mathbb{C}^n$ and let $\pi\colon U(n) \to W_{n,k}$ be the quotient map. Is there a map $j\colon W_{n,k} \to U(n)$ such that the composition has degree $(n-k)!$ on the bottom cell?*

2. *Is having the homotopy type of a 1-connected closed TOP or PL manifold a generic property?*

It was noted earlier that the answer to this question for smooth manifolds is no. However, for $\mathcal{F} = TOP$ or PL, Kahn has shown that the Zabrodsky mixing of $\mathcal{F}$-manifolds is again an $\mathcal{F}$-manifold, [**54**]. For any finite H_0-space X, it is known that every member of $\mathcal{G}(X)$ is a mixing of X and conversely, ([**105**], page 154) and so it follows that the answer to 2 is yes in the special case of manifolds with the homotopy type of an H_0-space. See also [**83**] and [**13**].

3. *Is the obvious map* $\mathcal{G}(X) \longrightarrow \varprojlim \mathcal{G}(X^{(n)})$ *always surjective?*

It was noted in [**63**] that this map is surjective when $SNT(X_{(p)}) = *$ for each p, but it seems doubtful that this condition is necessary as well. If one replaces the Mislin genus by the complete genus $\widehat{G}()$ then the answer is yes; the reason being $SNT(\widehat{X}) = *$ for all finite type X, [**98**]. So the answer to this question would shed some light on how the Mislin genus is situated inside the complete genus.

4. *Suppose that* X *is a* 1*-connected finite CW-complex such that* $\pi_n X \otimes \mathbb{Q} = 0$ *for* $n >> 0$. *Does it follow that the genus of* ΩX *is a finite set?*

An answer of yes here seems reasonable and (if correct) would represent a considerable improvement over Theorem 13. Wilkerson and I showed that, away from a finite set of primes, $\Omega X \simeq M \times \Omega N$, where M and N are finite products of odd dimensional spheres, [**64**]; this seems to make a yes answer more credible. I know the answer is yes for all generalized flag manifolds but I do not know the answer for the homogeneous spaces $\mathrm{Sp}(n)/U(n)$ or $SO(2n)/U(n)$.

5. *If two spaces have the same Mislin genus does it follow that they have the same Lusternik-Schnirelmann category?*

Some special cases of this are known. If one of the spaces is a finite 1-connected co-H-space, then the answer is yes by the Eckmann-Hilton dual of Theorem 2. Recently Cornea has shown that having finite category is a generic property in certain cases. He can prove the following result: Let X be a 1-connected complex with finite category. Suppose that Y is another space such that the localizations,

$$Y_{(S_i)} \simeq X_{(S_i)}, \text{ for } i = 1, 2, \cdots, n,$$

where $S_1, S_2, \cdots S_n$ is a partition of the set of all primes into finitely many sets. Then the category of Y is also finite.

6. *If* X *is a finite complex and* $Y \in \mathcal{G}(X)$, *does it follow that the groups of homotopy classes of self homotopy equivalences,* $\mathrm{Aut}(X)$ *and* $\mathrm{Aut}(Y)$, *are isomorphic?*

7. *Let* X *be a finite co-*H*-space and let* $X^{\wedge k}$ *denote its* k*-fold smash product. How, or to what extent, is* $\mathcal{G}(X^{\wedge k})$ *determined by* $\mathcal{G}(X)$ *and* k?

8. *Given a nilpotent space* X, *call a prime* p irrelevant *if* $\mathcal{G}(X) = \mathcal{G}(X_{(\frac{1}{p})})$. *For finite* H_0*-spaces this definition seems to work as expected; that is, all but a finite number of primes are irrelevant. Does it work equally well for finite complexes which are not* H_0*-spaces?*

If X is a 1-connected finite complex then the genus of X is finite by Wilkerson's theorem. It seems reasonable to suspect that it only depends on a finite number of primes. For H_0-spaces this is clear; the relevant primes are those that divide the number t. Question 8 is an attempt at a general definition of an irrelevant prime.

9. *Let* X *be a simply connected finite CW-complex. The genus of* X *can be regarded as a tree—the nodes at level* n *representing the various* n*-types of the members of* $\mathcal{G}(X)$. *Each* n*-type is connected to its* $(n-1)$*-type by an edge. Does it follow that each node at level* n *has the same degree; that is, the same number of edges having that node as an endpoint?*

10. *Is* X *an* H*-space if each of its Postnikov approximations* $X^{(n)}$ *is?*

If the answer is yes it would follow that being an H-space is generic. (See Section 4, 2.3) This question seems like some sort of $\varprojlim^1$ problem since no compatibility assumptions are being placed on the H-structures for each stage.

11. *Does cancellation hold in general for* p*-local* H*-spaces of finite type?*

12. *Suppose that X is a finite loop space which is not of the homotopy type of a Lie group. Among all the spaces B such that $\Omega B \simeq X$ is there one which somehow stands out from the rest?*

A good example to keep in mind here is the Hilton-Roitberg criminal. It was shown back in Section 4 that among all the loop structures on E_{5_ω} there is none which contains a maximal torus. That fact together with Rector's Theorem 7 and its extension by Notbohm and Smith suggest the following problem.

13. *If ΩB is a finite loop space with a maximal torus does it follow that $B = BG$ for some Lie group G?*

References

1. J. F. Adams, *The sphere considered as an H-space* mod p, Quart. J. Math. Oxford **12** (1961) 52–60.
2. ——, *On the groups $J(X)$—IV*, Topology **5** (1966) 21–71.
3. ——, *Infinite loop spaces*, Annals of Math. Studies, vol. 90, Princeton Univ. Press, 1978.
4. M. Arkowitz, *Localization and H-spaces*, Lecture Note Series, vol. 44, Aarhus Universitet (1976).
5. P. F. Baum, *Local isomorphisms of compact Lie groups*, Pacific J. Math **22** (1967) 197–204.
6. V. Belfi and C. Wilkerson, *Some examples in the theory of p-completions*, Indiana J. Math. **25** (1976) 565–576.
7. I. Bokor, *On the connection between the topological genus of certain polyhedra and the algebraic genus of their Hilton-Hopf quadratic forms*, Publicacions Matematiques, **34** (1990) 323–333.
8. ——, *On genus and cancellation in homotopy*, Israel J. Math., **73** (1991) 361–379.
9. I. Bokor and I. Llerena, *More examples of noncancellation in homotopy*, Springer Lecture Notes in Math. **1509** (1992) 30–34.
10. A. Borel, *Some finiteness properties of adele groups over number fields*, Publ. Math. IHES. **16** (1963) 5–30.
11. A. K. Bousfield and D. M. Kan, *Homotopy limits, completions, and localizations*, Lecture Notes in Math., vol. 304, Springer-Verlag, Berlin-Heidelberg-New York, 1972.
12. R. R. Bruner, F. R. Cohen, and C. A. McGibbon, *On stable homotopy equivalences*, Quart. J. Math. Oxford (to appear).
13. S. Cappell and S. Weinberger, *Which H-spaces are manifolds?* I, Topology **27** (1988) 377–386.
14. C. Casacuberta and P. Hilton, *On the extended genus of finitely generated Abelian groups*, Bull. Soc. Math. Belg. **41** (1989) 51–72.
15. ——, *Calculating the Mislin genus for a certain family of nilpotent groups*, Communications in Algebra **19** (1991) 2051–69.
16. J. W. S. Cassells, *Rational quadratic forms*, Academic Press, New York - London, 1978.
17. C. Cassidy, *Le genre d'un groupe nilpotent avec opérateurs*, Comment. Math. Helv. **53** (1978), 364–384.
18. C. Chang and H. Keisler, *Model Theory*, Studies in Logic, vol. 73, North Holland, Amsterdam, 1973.
19. A. Clark and J. Ewing, *The realization of polynomial algebras as cohomology rings*, Pacific J. Math. **50** (1974) 425–434.
20. D. Davis, *BP-operations and mappings of stunted complex projective spaces*, Springer Lecture Notes in Math. **741** (1979) 373–393.
21. E. Dror, W. G. Dwyer, and D. M. Kan, *Self homotopy equivalences of Postnikov conjugates* Proc. AMS **74** (1979) 183–186.
22. W. G. Dwyer and D. M. Kan, *A classification theorem for diagrams of simplicial sets*, Topology **23** (1984) 139–155.
23. W. G. Dwyer, H. R. Miller, and C. W. Wilkerson, *Homotopical uniqueness of BS^3*, Springer Lecture Notes in Math. **1298** (1987) 90–105.
24. ——, *Homotopical uniqueness of classifying spaces*, Topology **31** (1992) 29–45.
25. W. G. Dwyer and C. W. Wilkerson, *Homotopy fixed point methods for Lie groups and finite loop spaces*, Annals of Math. (to appear).

26. J. Eells and N. H. Kuiper, *Manifolds which are like projective planes*, Publ. Math. IHES **14** (1962) 181–222.
27. J. Ewing, *The non-splitting of Lie groups as spaces*, Topology **14** (1975) 37–47.
28. S. Feder and S. Gitler, *Stable homotopy types of stunted complex projective spaces*, Proc. Camb. Phil. Soc. **73** (1973) 431–438.
29. P. Freyd, *Stable Homotopy* II, AMS Proc. Symp. Pure Math. **18** (1970) 161-183.
30. E. M. Friedlander, *Exceptional isogenies and the classifying spaces of simple Lie groups*, Annals of Math. **101** (1975) 510–520.
31. ——, *Maps between localized homogeneous spaces*, Topology **16** (1977) 205–216.
32. H. H. Glover and G. Mislin, *On the genus of generalized flag manifolds*, Ensign. Math. **27** (1981) 211–219.
33. F. J. Grunewald and R. Scharlau, *A note on finitely generated torsion-free nilpotent groups of class* 2, J. Algebra **58** (1979) 162–175.
34. F. J. Grunewald, P. F. Pickel, and D. Segal, *Polycyclic groups with isomorphic finite quotients*, Annals of Math. **111** (1980) 155–195.
35. F. J. Grunewald, D. Segal, and L. S. Sterling, *Nilpotent groups of Hirsch length six*, Math. Zeit. **179** (1982) 219–235.
36. P. Hilton, *On the Grothendieck group of compact polyhedra*, Fund. Math. **61** (1967) 199–214.
37. ——, *On the genus of nilpotent groups and spaces*, Israel J. Math. **54** (1986) 1–13.
38. ——, *Non-cancellation properties for certain finitely presented groups*, Quaestiones Math. **9** (1986) 281–293.
39. ——, *On the genus of groups with operators*, Topology and its applications, **25** (1987) 113–120.
40. ——, *Groups with isomorphic homology groups and non-cancellation in homotopy theory*, J. Pure Appl. Algebra **44** (1987) 221–226.
41. ——, *On groups of pseudo-integers*, Acta Math. Sinica **4** (1988) 189–192.
42. ——, *On the extended genus*, Acta Math. Sinica, **4** (1988) 372–382.
43. ——, *On induced morphisms of Mislin genera* (to appear).
44. P. Hilton and G. Mislin, *On the genus of a nilpotent group with finite commutator subgroup*, Math. Zeit. **146** (1976) 201–211.
45. P. Hilton, G. Mislin, and J. Roitberg, *Sphere bundles over spheres and non-cancellation phenomena*, Springer Lecture Notes in Math. **249** (1971) 34–46.
46. ——, *H-spaces of rank* 2 *and non-cancellation phenomena*, Inventiones Math. **16** (1972) 325–334.
47. ——, *Localization of nilpotent groups and spaces*, Math. Studies **15**, North-Holland, Amsterdam, 1975.
48. P. Hilton and J. Roitberg, *On principal S^3-bundles over spheres*, Annals of Math. **90** (1969) 91–107.
49. R. Hirshon, *Some cancellation theorems with applications to nilpotent groups*, J. Austral. Math. Soc. **23** (1977) 147–165.
50. J. R. Hubbuck, *Homotopy homomorphisms of Lie groups*, London Math. Soc. Lecture Note Series **11** Cambridge Univ. Press (1974) 33–41.
51. K. Ishiguro, *Unstable Adams operations on classifying spaces* Math. Proc. Camb. Phil. Soc. **102** (1987) 71–75.
52. ——, *Classifying spaces and p-local irreducibility*, J. Pure Appl. Algebra **49** (1987) 254–258.
53. K. Ishiguro, J. Møller, and D. Notbohm, *On the genus of classifying spaces* (in preparation).
54. P. J. Kahn, *Mixing homotopy types of manifolds*, Topology **14** (1975) 203–216.
55. R. Kane, *The homology of Hopf spaces*, Math. Studies, vol. 40, North-Holland, Amsterdam, 1988.
56. I. Llerena, *Localization of fibrations with nilpotent fibre*, Math. Zeit. **188** (1985) 397–410.
57. ——, *Wedge cancellation of certain mapping cones*, Compositio Math. **81** (1992) 1-17.
58. C. A. McGibbon, *Which group structures on S^3 have a maximal torus?*, Springer Lecture Notes in Math. **657** (1978) 353–360.
59. ——, *Stable properties of rank* 1 *loop structures*, Topology **20** (1981) 109–118.
60. ——, *Self maps of projective spaces*, Trans. Amer. Math. Soc. **271** (1982) 325–346.
61. ——, *Spaces which look like quaternionic projective n-space*, Trans. Amer. Math. Soc. **272** (1982) 569–587.
62. ——, *Clones of spaces and maps in homotopy theory*, Comment. Math. Helv. **68** (1993)

263–277.

63. C. A. McGibbon and J. M. Møller, *On spaces with the same n-type for all n*, Topology **31** (1992) 177–201.
64. C. A. McGibbon and C. W. Wilkerson, *Loop spaces of finite complexes at large primes*, Proc. Amer. Math. Soc. **96** (1986) 698–702.
65. G. Mislin, *The genus of an H-space*, Springer Lecture Notes in Math. **249** (1971) 75–83.
66. ——, *Cancellation properties of H-spaces*, Comment. Math. Helv. **49** (1974) 195–200.
67. ——, *Nilpotent groups with finite commutator subgroups*, Springer Lecture Notes in Math. **418** (1974) 103–120.
68. ——, *Wall's obstruction for nilpotent spaces*, Topology **14** (1975) 311–317.
69. ——, *Finitely dominated nilpotent spaces*, Annals of Math. **103** (1976) 547–556.
70. ——, *Groups with cyclic Sylow subgroups and finiteness conditions for certain complexes*, Comment. Math. Helv. **52** (1977) 373–391.
71. J. M. Møller, *The normalizer of the Weyl group*, Math. Ann. **294** (1992) 59–80.
72. ——, *Rational equivalences between classifying spaces*, Preprint 1993.
73. E. Molnar, *Relations between the wedge cancellation and localization for complexes with two cells*, J. Pure Appl. Algebra **3** (1972) 77–81.
74. D. Notbohm, *Fake Lie groups and maximal tori* IV, Math. Ann. **294** (1992) 109–116.
75. D. Notbohm and L. Smith, *Fake Lie groups and maximal tori*, Math. Ann. **288** (1990) 637–661.
76. ——, *Fake Lie groups and maximal tori* III, Math. Ann. **290** (1991) 629–642.
77. F. Oger, *Des groupes nilpotents de classe 2 sans torsion de type fini ayant les mêmes images finies peuvent ne pas être élémentairement équivalents*, Comptes rendus, Acad. Sci. Paris **294** serie I (1982), 1–4.
78. ——, *Equivalence élémentaire entre groupes finis-par-abéliens de type fini*, Comment. Math. Helv. **57** (1982) 460–480.
79. ——, *Cancellation and elementary equivalence of groups*, J. Pure Appl. Algebra **30** (1983) 293–299.
80. ——, *Elementary equivalence and genus of finitely generated nilpotent groups*, Bull. Austral. Math. Soc. **37** (1988) 61–68.
81. ——, *Elementary equivalence and profinite completions: a characterization of finitely generated Abelian-by-finite groups* Proc. Amer. Math. Soc. **103** (1988) 1041–48.
82. S. Papadima, *Rigidity properties of compact Lie groups modulo maximal tori*, Math. Ann. **275** (1986) 637–652.
83. E. K. Pedersen, *Smoothing H-spaces* II, Math. Scand. **46** (1980), 216–222.
84. G. Peschke, *Localization and genus in group theory*, Preprint 1993.
85. G. Peschke and P. Symonds, *Pro-torsion completions of Abelian groups*, J. Pure Appl. Algebra (to appear).
86. P. F. Pickel, *Finitely generated nilpotent groups with isomorphic finite quotients*, Trans. Amer. Math. Soc. **160** (1971) 327–341.
87. D. Rector, *Loop structures on the homotopy type of S^3*, Springer Lecture Notes in Math. **249** (1971) 99–105.
88. C. Schuck, *Some contributions to the study of a class of nilpotent groups*, Ph.D. Dissertation, SUNY Binghamton, 1992.
89. D. Segal, *Polycyclic groups*, Cambridge University Press, 1983.
90. W. Singhof, *On the Lusternik-Schnirelmann category of Lie Groups* II, Math. Zeit. **151** (1976) 143–148.
91. J. D. Stasheff, *Manifolds of the homotopy type of (non-Lie) groups*, Bull. Amer. Math. Soc. **75** (1969) 998–1000.
92. R. J. Steiner, *Localization, completion and infinite complexes*, Mathematika **24** (1977) 1–15.
93. D. Sullivan, *Genetics of homotopy theory and the Adams conjecture*, Annals of math. **100** (1974) 1-79.
94. C. T. C. Wall, *Finiteness obstructions for CW complexes*, Annals of Math. **81** (1965) 56–69.
95. R. B. Warfield, *Nilpotent groups*, Lecture Notes in Math., vol. 513, Springer-Verlag, Berlin-Heidelberg-New York, 1976.
96. ——, *Genus and cancellation for groups with finite commutator subgroup*, J. Pure Appl. Algebra **6** (1975) 125–132.
97. C. W. Wilkerson, *Self-maps of classifying spaces*, Springer Lecture Notes in Math., **418** (1974)

150–157.
98. ———, *Classification of spaces of the same n-type for all n*, Proc. Amer. Math. Soc. **60** (1976) 279–285.
99. ———, *Genus and cancellation*, Topology **14** (1975) 29–36.
100. ———, *Applications of minimal simplicial groups*, Topology **15** (1976) 111–130.
101. A. Zabrodsky, *On the construction of new finite H-spaces*, Inventiones Math. **16** (1972) 260–266.
102. ———, *On the homotopy type of principal classical group bundles over spheres*, Israel J. Math. **11** (1972) 315–325.
103. ———, *On the genus of finite CW H-spaces*, Comment. Math. Helv. **49** (1974) 48–64.
104. ———, *p-equivalences and homotopy type*, Springer Lecture Notes in Math. **418** (1974) 160–171.
105. ———, *Hopf spaces*, Math. Studies, vol. 22, North-Holland, Amsterdam, 1976.

Mathematics Department, Wayne State University, Detroit, MI 48202, USA
E-mail address: mcgibbon@math.wayne.edu

Centre de Recherches Mathématiques
CRM Proceedings and Lecture Notes
Volume 6, 1994

Mapping Class Groups, Characteristic Classes, and Bernoulli Numbers

Guido Mislin

Introduction

The mapping class group Γ_g of a closed, connected and oriented surface S_g of genus g is defined as the group of connected components of the group of orientation preserving diffeomorphisms of S_g. This group has been the object of many recent studies. Of particular interest are its finite subgroups; these are precisely the finite groups which occur as groups of symmetries of the surface S_g equipped with a complex structure (a Riemann surface). The interplay of algebra, topology and analysis in the study of Γ_g make it one of the most fascinating groups. As it is the case for classical arithmetic groups, the finite subgroups of Γ_g are related to certain concepts in number theory. We shall discuss in this essay invariants of Γ_g which are related to number theory via Bernoulli numbers. The invariants we have in mind are firstly certain characteristic classes, associated with a natural flat vector bundle over $\mathrm{B}\Gamma_g$, secondly, the orbifold Euler characteristic of the group Γ_g and thirdly its Yagita invariant. The characteristic classes are related to the denominators of Bernoulli numbers, the Euler characteristic involves the whole Bernoulli numbers, and our theorems concerning the Yagita invariant have to do with the notion of regular primes, which is expressible in terms of numerators of Bernoulli numbers. Although the three concepts which we study seem rather unrelated from the point of view of their definitions, the fact that they all are tightly linked to properties of finite subgroups and their normalizers and centralizers in Γ_g renders it plausible, that the resulting invariants must be somehow linked. The precise relationship, however, remains for the time being a mystery.

We have tried to make these notes easy to read for non-experts. We therefore recall in Sections 1 through 3 many facts and definitions, and state the relevant properties of the mapping class group without proofs, in form of a survey. We also have included a substantial bibliography, helping the reader to find the proofs of the basic theorems of the subject, which are scattered through the literature and which are crucial for analyzing homological properties of the mapping class group. For the classical part, not dealing with the cohomology of the mapping class group, the reader should consult Birman's book [**Bi**]. Section 4 contains a short introduction to the theory of characteristic classes for group representations, and in Section 5 we compute the order of the Euler class $e_{2g}(\Gamma_g)$, associated with the flat bundle over $B\Gamma_g$ induced by the action of Γ_g on the homology group $H_1(S_g;\mathbb{R})$. In Section 6

1991 *Mathematics Subject Classification.* Primary: 55-02; Secondary: 20J05.

This is the final form of the paper.

this Euler class is related to the Euler characteristic $\chi(\Gamma_g)$ of the group Γ_g, and in Section 7 we discuss periodicity phenomena of Γ_g as well as the Yagita invariant.

1. The definition of the mapping class group

Let S_g denote a closed, connected and oriented topological surface of genus g. It is well-known that S_g admits a unique smooth structure; we shall also write S_g for the corresponding smooth (oriented) manifold. There are four basic ways of viewing the mapping class group Γ_g of the surface S_g, one being purely topological, the second more geometric in nature, the third homotopical and the fourth algebraic, involving the fundamental group of the surface in question only. The definitions we have in mind have the following form:

(I) $\Gamma_g = \operatorname{Homeo}_+(S_g)/\operatorname{Homeo}^0(S_g)$
(II) $\Gamma_g = \operatorname{Diffeo}_+(S_g)/\operatorname{Diffeo}^0(S_g)$
(III) $\Gamma_g = \operatorname{Hoequ}_+(S_g)/\operatorname{Hoequ}^0(S_g)$
(IV) $\Gamma_g = \operatorname{Out}_+\big(\pi_1(S_g, s_0)\big)$.

We shall first give some background information and comments concerning these equivalent definitions. Let $s_0 \in S_g$ denote a basepoint. The fundamental group of S_g has a presentation

$$\pi_1(S_g, s_0) = \Big\langle a_1, b_1, \ldots, a_g, b_g \;\Big|\; \prod [a_i, b_i] \Big\rangle$$

and thus $\pi_1(S_g, s_0)_{ab} \cong H_1(S_g; \mathbb{Z}) \cong \mathbb{Z}^{2g}$. Since S_g is assumed to be orientable one has $H_2(S_g; \mathbb{Z}) \cong \mathbb{Z}$. A map $f \colon S_g \to S_g$ is said to be orientation preserving, if the induced map

$$H_2(f) \colon H_2(S_g; \mathbb{Z}) \to H_2(S_g; \mathbb{Z})$$

is the identity map. It is useful to notice that this is equivalent to the requirement that the determinant of

$$H_1(f) \colon H_1(S_g; \mathbb{Z}) \to H_1(S_g; \mathbb{Z})$$

equals one (the multiplicative structure of the cohomology ring $H^*(S_g; \mathbb{Z})$ reveals that $H^2(f)$ is multiplication by $\det H^1(f)$). Let $\operatorname{Homeo}(S_g)$ denote the topological group of homeomorphisms of S_g, with the compact-open topology. We shall write Homeo_+ for the subgroup of orientation preserving homeomorphisms, and Homeo^0 for the connected component of the identity (we use the "+" here as a subscript rather than as a superscript, to avoid confusion with Quillen's *plus*-construction). The mapping class group Γ_g of the surface S_g is then defined as the discrete group of connected components

$$\text{(I)} \qquad \Gamma_g = \operatorname{Homeo}_+(S_g)/\operatorname{Homeo}^0(S_g).$$

We will consider (I) as our basic definition for Γ_g, and want to compare it with (II), (III) and (IV). Consider now S_g as a smooth oriented manifold. In accordance to the notation used above, we write $\operatorname{Diffeo}_+(S_g)$ for the group of orientation preserving diffeomorphisms of S_g with the C^∞-topology, and $\operatorname{Diffeo}^0(S_g)$ for the connected component of the identity. It was proved by Dehn [**De**] that $\operatorname{Homeo}_+(S_g)/\operatorname{Homeo}^0(S_g)$ is generated by "*Dehn twists*", which are diffeomorphisms obtained by splitting S_g along a simple closed smooth curve, rotating one part by 2π, and gluing the surface back together. It follows that the natural map

$$\operatorname{Diffeo}_+(S_g) \to \operatorname{Homeo}_+(S_g)/\operatorname{Homeo}^0(S_g)$$

is surjective. The kernel is precisely $\text{Diffeo}^0(S_g)$; namely, if $f\colon S_g \to S_g$ is a diffeomorphism isotopic to the identity $\big(\text{i.e. } f \in \text{Homeo}^0(S_g)\big)$, then f is a fortiori homotopic to the identity, and therefore, according to Earle and Eells [**Ea-Ee**], the map f can be connected by a path in $\text{Diffeo}_+(S_g)$ to the identity map. We have thus established that

$$\text{(II)} \qquad \Gamma_g = \text{Diffeo}_+(S_g)/\,\text{Diffeo}^0(S_g).$$

In case $g = 0$, that is $S_0 = S^2$ the 2-sphere, $\text{Diffeo}_+(S^2)$ is connected; the inclusion of $SO(3)$ in $\text{Diffeo}_+(S^2)$ is actually a homotopy equivalence by Smale's result [**Sm**]. Thus $\Gamma_0 = \{e\}$. For $g > 0$ however, the mapping class groups Γ_g turn out to be all non-trivial. The group Γ_1 can be most easily understood using the definitions (III) and (IV) respectively, which we shall discuss now. Let $\text{Hoequ}_+(S_g)$ be the topological group of orientation preserving homotopy equivalences of S_g with the compact-open topology, and $\text{Hoequ}^0(S_g)$ the connected component of the identity. By a result due to Nielsen ([**Ni1**]), the natural map

$$\text{Homeo}_+(S_g) \to \text{Hoequ}_+(S_g)/\,\text{Hoequ}^0(S_g)$$

is surjective, and Baer proved [**Ba**] that any homeomorphism which is homotopic to the identity, is actually isotopic to the identity, showing that the kernel is precisely $\text{Homeo}^0(S_g)$, (compare also Mangler [**Ma**]). Therefore, we conclude:

$$\text{(III)} \qquad \Gamma_g = \text{Hoequ}_+(S_g)/\,\text{Hoequ}^0(S_g).$$

Denote the set of free homotopy classes of maps between the spaces X and Y by $[X, Y]$. We have then a natural map

$$\Lambda\colon\ \text{Hoequ}_+(S_g) \to [S_g, S_g].$$

The homotopy set $[S_g, S_g]$ may be identified with the set of orbits of the usual $\pi_1(S_g, s_0)$-action on the pointed homotopy set $\big[(S_g, s_0), (S_g, s_0)\big]_*$ of pointed homotopy classes of pointed maps. Since for $g > 0$ the surface S_g has a contractible universal covering space, there is a natural bijection

$$\big[(S_g, s_0), (S_g, s_0)\big]_* \cong \text{Hom}\big(\pi_1(S_g, s_0), \pi_1(S_g, s_0)\big), \quad (g > 0).$$

Passing to orbit spaces with respect to the $\pi_1(S_g, s_0)$-action, we obtain

$$[S_g, S_g] \cong \text{Rep}(\pi, \pi),$$

where $\text{Rep}(\pi, \pi)$ stands for the set of conjugacy class of homomorphisms $\pi \to \pi$, with $\pi = \pi_1(S_g, s_0)$. Homotopy equivalences correspond then to automorphisms modulo inner automorphisms of π. If we denote by $\text{Out}(\pi)$ the group of outer automorphisms of π, we can view this group as a subset of $\text{Rep}(\pi, \pi)$, and the map Λ defined above yields a surjective homomorphism

$$\text{Hoequ}(S_g) \to \text{Out}\big(\pi_1(S_g, s_0)\big) \subset [S_g, S_g].$$

The kernel consists of course of all homotopy equivalences homotopic to the identity. If we write Out_+ for the "orientation-preserving" outer automorphisms, that is, the subgroup of $\text{Out}\big(\pi_1(S_g, s_0)\big)$ consisting of those elements which act on the abelianized fundamental group $\pi_1(S_g, s_0)_{ab}$ by a homomorphism of determinant one, then we infer that

$$\text{Hoequ}_+(S_g)/\,\text{Hoequ}^0(S_g) \cong \text{Out}_+(S_g).$$

Note that the formula is also correct in case $g = 0$. From (III) we conclude then that

$$\text{(IV)} \qquad \Gamma_g \cong \mathrm{Out}_+\big(\pi_1(S_g, s_0)\big).$$

For the case of $g = 1$ one has $S_1 = S^1 \times S^1$ a torus, and therefore Γ_1 is isomorphic to $\mathrm{Out}_+(\mathbb{Z} \oplus \mathbb{Z}) \cong Sl_2(\mathbb{Z})$.

2. Some algebraic properties of the mapping class group

As mentioned earlier, Γ_g is generated by Dehn twists, associated with (isotopy classes) of simple closed curves on S_g and it follows that Γ_g is finitely generated, see Dehn [**De**]. An explicit finite set of generators is described in Lickorish [**Li**]. Actually, Γ_g is finitely presented. This is obvious for Γ_1. Indeed, $\Gamma_1 \cong Sl_2(\mathbb{Z})$, so that one obtains a finite presentation for Γ_1 from the well-known decomposition of $Sl_2(\mathbb{Z})$ as an amalgamated free product:

$$Sl_2(\mathbb{Z}) \cong \mathbb{Z}/4\mathbb{Z} \underset{\mathbb{Z}/2\mathbb{Z}}{*} \mathbb{Z}/6\mathbb{Z}.$$

Note also that this implies

$$(\Gamma_1)_{ab} \cong \mathbb{Z}/12\mathbb{Z}.$$

Birman and Hilden described in [**Bi-Hi1**] an explicit finite presentation for Γ_2, which can be used to show that

$$(\Gamma_2)_{ab} \cong \mathbb{Z}/10\mathbb{Z}.$$

It was proved by Powell [**Po**] that, for higher genus the mapping class group is always perfect:

$$(\Gamma_g)_{ab} = 0 \quad \text{for} \quad g > 2.$$

That Γ_g is finitely presented for general g was first proved by McCool in [**McCo**], (see also Hatcher-Thurston [**Ha-Th**], as well as Wajnryb [**Wa**] for explicit presentations).

A lot is known about finite subgroups of Γ_g. It follows from Harvey [**Harv2**] that the number of conjugacy classes of finite subgroups of Γ_g is finite. The individual finite subgroups of Γ_g can be described as follows. Let τ denote a complex structure on S_g compatible with the smooth structure. Then the group of holomorphic automorphisms $\mathrm{Aut}(S_g, \tau)$ is a subgroup of $\mathrm{Diffeo}(S_g)$. It is a classical result that for $g > 1$ the group $\mathrm{Aut}(S_g, \tau)$ is finite and the induced map

$$\theta_\tau \colon \mathrm{Aut}(S_g, \tau) \to \Gamma_g \quad (g > 1),$$

is injective (cf. Farkas-Kra [**Fa-Kr**]). According to Kerckhoff [**Ke**], the finite subgroups of Γ_g are precisely the subgroups of the form $\theta_\tau(F)$ for F a finite group of holomorphic automorphisms of (S_g, τ) for some τ, amounting to a positive solution of the *Nielsen realization problem*; for an account on the long history of this problem as well as the partial results proved earlier, the reader should consult Zieschang's book [**Zi**]. Another classical result, due to Hurwitz [**Hu**], states that

$$\big|\mathrm{Aut}(S_g, \tau)\big| \le 84(g - 1), \quad (g > 1).$$

As a consequence, all finite subgroups $F \subset \Gamma_g$ satisfy the *Hurwitz bound*

$$|F| \le 84(g - 1), \quad (g > 1).$$

For a finite cyclic subgroup $F \subset \Gamma_g$ this bound can be improved to

$$|F| \leq 4g + 2,$$

and this bound is sharp, that is, Γ_g always contains a cyclic subgroup of order $4g+2$, see Wiman [**Wi**]. Note also that every finite group F admits an embedding into some Γ_g since, as is well-known, every finite group occurs as a group of symmetries of some Riemann surface (see Broughton [**Br**] for an explicit condition, in terms of a presentation of F, for the existence of an embedding of F in Γ_g). It is even the case that every finite group is isomorphic to the full automorphism group $\mathrm{Aut}(S_g, \tau)$ of some closed Riemann surface, a result due to Greenberg [**Gre**]. The minimal $g = g(F)$ such that F admits an embedding into Γ_g is called the *genus* of F. It is an interesting problem to determine $g(F)$ for certain families of finite groups F. For the case of a finite cyclic group F of prime power order $|F| > 2$ the problem is quite elementary and one finds $g(F) = \frac{1}{2}\phi(|F|)$, where ϕ denotes the Euler-function, see Glover-Mislin [**Gl-Mi2**]; the case of general finite cyclic groups was settled by Harvey [**Harv1**]. For a more involved example, the reader should consult Glover-Sjerve [**Gl-Sj**], where the genus of the group $PSl_2(\mathbb{F}_q)$ is computed for $\mathbb{F}_q$ an arbitrary finite field.

Maps from Γ_g to finite groups were studied by Grossman in [**Gro**]. She proved that Γ_g is residually finite (which means that Γ_g admits an embedding into a product of finite groups). In particular, the elements of Γ_g may therefore be separated by means of finite dimensional linear representations. However, it is still an open question whether Γ_g admits a *faithful* finite-dimensional linear representation. But it is known that for $g > 1$, Γ_g is not isomorphic to an arithmetic group; for a discussion of this fact see Ivanov [**Iv2**] or Harer [**Ha5**].

Another important result is that Γ_g contains a torsion-free subgroup of finite index. This can be seen in an elementary way as follows. From our definition (IV) for Γ_g, there is a natural map $\rho\colon \Gamma_g \subset \mathrm{Out}(\pi) \to \mathrm{Aut}\big((\pi)_{ab}\big)$, where π denotes the fundamental group of S_g, and thus $\pi_{ab} \cong \mathbb{Z}^{2g}$ the abelianized fundamental group. Hence there is a left exact sequence

$$\mathrm{T}\Gamma_g = \ker \rho \to \Gamma_g \to Gl_{2g}(\mathbb{Z}),$$

where $\mathrm{T}\Gamma_g$ denotes the *Torelli* group, which is easily seen to be torsion-free (every element of finite order in Γ_g can be realized as a holomorphic automorphism on some Riemann surface (S_g, τ), and these act non-trivially on $H_1(S_g;\mathbb{Z})$). Since $Gl_{2g}(\mathbb{Z})$ contains a torsion-free subgroup of finite index, it follows that Γ_g too possesses one.

To conclude this section, we want to mention two additional basic results, concerning infinite subgroups of Γ_g. The first one is an analogue of a Theorem of Tits on linear groups. According to McCarthy [**McC**] and Ivanov [**Iv1**] the following *Tits-alternative* holds for the mapping class group: every subgroup of Γ_g either contains a free subgroup on two generators, or a solvable subgroup of finite index. Solvable subgroups of Γ_g were analyzed in Birman-Lubotzky-McCarthy [**B-L-M**]. They proved that every solvable subgroup of Γ_g is virtually abelian, and that the maximal rank of a free abelian subgroup in Γ_g is, for $g > 1$, equal to $3g - 3$.

3. Some cohomological results on Γ_g

Since $\mathrm{Diffeo}^0(S_g)$ is a contractible space for $g > 1$ (see Earle-Eells [**Ea-Ee**]), the natural map of classifying spaces $B\,\mathrm{Diffeo}_+(S_g) \to B\Gamma_g$ is a homotopy equivalence, and thus

$$H^*(\Gamma_g;\mathbb{Z}) \cong H^*\big(B\,\mathrm{Diffeo}_+(S_g);\mathbb{Z}\big), \quad g > 1.$$

One can therefore think of the cohomology elements of Γ_g as universal characteristic classes for smooth orientable S_g-bundles. The most important tool for studying the cohomology of Γ_g is its action on *Teichmüller space* T_g, which is for $g > 1$ a smooth manifold homeomorphic to $\mathbb{R}^{6g-6}$. Teichmüller space is a parameter space for complex structures on the oriented closed smooth surface S_g, where two complex structures on S_g are considered as equivalent if and only if there exists a diffeomorphism $f\colon S_g \to S_g$ diffeotopic to the identity (i.e. $f \in \mathrm{Diffeo}^0(S_g)$), carrying one complex structure to the other. According to Earle-Eells [**Ea-Ee**], one can describe T_g as follows. Consider the space $CS(S_g)$ of complex structures on S_g compatible with the smooth structure and orientation; it carries a natural topology and an obvious action of $\mathrm{Diffeo}_+(S_g)$. Then one has

$$T_g = CS(S_g)/\,\mathrm{Diffeo}^0(S_g).$$

There remains a natural action of $\Gamma_g = \mathrm{Diffeo}_+(S_g)/\,\mathrm{Diffeo}^0(S_g)$ on T_g, which is known to be properly discontinuous. The orbit space

$$M_g = T_g/\Gamma_g = CS(S_g)/\,\mathrm{Diffeo}_+(S_g)$$

is called the *moduli space* of S_g. It has the structure of a complex variety and its points correspond to conformal equivalence classes of complex structures on S_g. Because the action of Γ_g on T_g is properly discontinuous, the stabilizers of points of T_g are finite. The natural projection

$$T_g \to M_g$$

is a branched covering space, and it can be thought of as a resolution of the singularities for the variety M_g. In particular, T_g inherits a natural complex structure such that Γ_g acts by complex automorphisms. By a result due to Royden [**Roy**], Γ_g is actually the full automorphism group of Teichmüller space with this complex structure. If $t \in T_g$ is fixed by the (finite) subgroup $F \subset \Gamma_g$, then we can think of F as a group of symmetries for a complex structure on S_g. On the other hand, by the positive solution of the *Nielsen realization problem* mentioned earlier, every finite subgroup $F \subset \Gamma_g$ is a group of symmetries of some complex structure on S_g and thus has a fixed point when acting on T_g. Harvey proved in [**Harv2**] that there is a contractible simplicial complex $\overline{T}_g$ containing T_g such that the Γ_g-action extends to a proper simplicial action on $\overline{T}_g$, with compact quotient $\overline{T}_g/\Gamma_g$. From general principles (see Brown's book [**Brow1**]), this immediately implies a wealth of finiteness properties for Γ_g, the first of which has also been discussed in the previous section:

1. Γ_g is finitely presented.
2. Γ_g is of finite virtual cohomological dimension (vcd); actually it follows that $\mathrm{vcd}(\Gamma_g) \le 6g-6 = \dim T_g$, $(g > 1)$. Note also that $\mathrm{vcd}(\Gamma_g) \ge 3g-3$, since Γ_g possesses a free abelian subgroup of rank $3g-3$. It has been proved by Harer in [**Ha3**] that, more precisely, one has

$$\mathrm{vcd}(\Gamma_g) = 4g-5, \quad g > 1.$$

Of course, $\mathrm{vcd}(\Gamma_1) = 1$, since $\Gamma_1 \cong Sl_2(\mathbb{Z})$ is an amalgamated free product of finite groups.

3. Γ_g is of type WFL (i.e. all torsion-free subgroups of finite index admit finitely generated free resolutions of finite length).
4. For every subgroup $H \subset \Gamma_g$ of finite index, the homology groups $H_i(H;\mathbb{Z})$ are finitely generated, and they are finite if $i > \mathrm{vcd}(\Gamma_g)$. In particular, the *naive Euler characteristic*

$$\widetilde{\chi}(H) = \sum(-1)^i \mathrm{rk}\, H_i(H;\mathbb{Z})$$

is well defined.

REMARK. There are of course many other ways to define Teichmüller space T_g. We just want to mention one different definitions, which is discussed in Goldman's paper [**Go**], and which is particularly attractive. By choosing a complex structure on S_g one obtains an embedding θ of $\pi_1(S_g)$ into $PSl_2(\mathbb{R})$, which is the group of isometries of the upper half-plane. Teichmüller space can then be identified with the component of θ in $\mathrm{Rep}\big(\pi_1(S_g), PSl_2(\mathbb{R})\big)$, the space of conjugacy classes of homomorphisms from $\pi_1(S_g)$ to $PSl_2(\mathbb{R})$. Since $\pi_1(S_g)$ is finitely presented, this space admits a natural embedding as a real algebraic subvariety in a quotient of a product of copies of $PSl_2(\mathbb{R})$. The action of Γ_g on T_g is just the one induced by the action of the group of orientation preserving automorphisms of $\pi_1(S_g)$ on the space $\mathrm{Hom}\big(\pi_1(S_g), PSl_2(\mathbb{R})\big)$.

3.1 Rational cohomology. Next, we shall describe some results concerning the rational cohomology of Γ_g. We first recall that, because the stabilizers of the Γ_g-action on the contractible space T_g are all finite, the rational cohomology of the moduli space M_g satisfies

$$H^*(M_g;\mathbb{Q}) \cong H^*(\Gamma_g;\mathbb{Q}),$$

which explains the great interest in the rational cohomology of the mapping class group. Obviously, Γ_1 is $\mathbb{Q}$-acyclic, being an amalgamated free product of finite groups. By a result due to Igusa [**Ig**], Γ_2 is $\mathbb{Q}$-acyclic too, but this is not the case for $g \geq 3$. Of course, for $g \geq 3$, $H^1(\Gamma_g;\mathbb{Q})$ is trivial, since Γ_g is perfect. However, Harer proved in [**Ha4**] that

$$H^2(\Gamma_g;\mathbb{Q}) \cong \mathbb{Q} \quad \text{for} \quad g \geq 3,$$

and he also showed that

$$H^3(\Gamma_g;\mathbb{Q}) = 0 \quad \text{for} \quad g \geq 6.$$

For all values of g one can describe as follows a part of the rational cohomology of Γ_g whose dimension increases rapidly with g. According to Miller [**Mil**] and Morita [**Mo**] there are classes y_j in $H^{2j}(\Gamma_g;\mathbb{Q})$, $j \geq 1$, such that the induced map of the polynomial ring

$$\Lambda\colon \mathbb{Q}[y_1, y_2, y_3 \ldots] \to H^*(\Gamma_g;\mathbb{Q})$$

is injective in dimensions less than $g/3$. The classes y_j for odd j can be described as symplectic characteristic classes in the following way, an interpretation which is useful in many contexts. Consider the natural map

$$B\Gamma_g \to BSp_{2g}(\mathbb{R})$$

induced by the action of Γ_g on $H^1(S_g;\mathbb{R})$, viewed as a symplectic space using the cup product. Since a maximal compact subgroup in $Sp_{2g}(\mathbb{R})$ is isomorphic to $U(g)$, one has

$$H^*(BSp_{2g}(\mathbb{R});\mathbb{Z}) = \mathbb{Z}[d_1, d_2, \ldots, d_g],$$

with $d_j \in H^{2j}(BSp_{2g}(\mathbb{R});\mathbb{Z})$ the *universal symplectic characteristic classes*, characterized by the property that d_j restricts to the universal Chern class c_j of the maximal compact subgroup $U(g) \subset Sp_{2g}(\mathbb{R})$. Each even indexed class d_{2j} restrict in $H^{4j}(\Gamma_g;\mathbb{Q})$ to the image of a polynomial in d_i's involving only odd i's; this can easily be seen using the fact that the rational Pontrjagin classes of a flat real vector bundle vanish, and that the Pontrjagin class $p_i \in H^{4i}(BGl_{2g}(\mathbb{R});\mathbb{Z})$ restricts with respect to the inclusion of $Sp_{2g}(\mathbb{R})$ in $Gl_{2g}(\mathbb{R})$ according to the formula

$$\operatorname{im}(1 - p_1 + p_2 - \ldots) = (1 + d_1 + d_2 + \ldots)(1 - d_1 + d_2 - \ldots) \in H^*(BSp_{2g}(\mathbb{R});\mathbb{Z}).$$

From a rational point of view, the only part of the map

$$H^*(BSp_{2g}(\mathbb{R});\mathbb{Z}) \to H^*(\Gamma_g;\mathbb{Z})$$

which is of any relevance, is therefore the induced map

$$H^*(BSp_{2g}(\mathbb{R});\mathbb{Q}) \supset \mathbb{Q}[d_1, d_3, d_5, \ldots] \xrightarrow{\Phi} H^*(\Gamma_g;\mathbb{Q}).$$

It is shown in Miller [**Mi**] and Morita [**Mo**] that the image of this map Φ agrees with the image of the restriction of the map Λ to $\mathbb{Q}[y_1, y_3, y_5, \ldots]$, and in particular Φ is therefore injective in dimensions less than g/3 too.

3.2 Mod-p cohomology in the stable range. Again, we can look at the representation of Γ_g obtained by letting Γ_g act on $H_1(S_g;\mathbb{R})$, yielding a homomorphisms

$$\Gamma_g \to Gl_{2g}(\mathbb{R}) \to Gl_{2g}(\mathbb{C}).$$

Recall that

$$H^*(BGl_{2g}(\mathbb{R});\mathbb{F}_2) = \mathbb{F}_2[w_1, w_2, \ldots, w_{2g}],$$

a polynomial algebra in the universal Stiefel-Whitney classes. Kaufmann showed in [**Kau**] that the odd Stiefel-Whitney classes w_{2j+1} restrict to zero in $H^*(\Gamma_g;\mathbb{F}_2)$, and that the induced map

$$H^*(BGl_{2g}(\mathbb{R});\mathbb{F}_2) \supset \mathbb{F}_2[w_2, w_4, w_6, \ldots] \to H^*(\Gamma_g;\mathbb{F}_2)$$

is injective in dimensions less than $g/3$. He also proved (loc. cit.) a corresponding result for the case of an odd prime p, by considering

$$H^*(BGl_{2g}(\mathbb{C});\mathbb{F}_p) = \mathbb{F}_p[\bar{c}_1, \bar{c}_2, \ldots, \bar{c}_{2g}],$$

where $\bar{c}_i$ stands for the mod-p reduction of the universal Chern class c_i. The result is that the mod-p Chern classes $\bar{c}_i$ restrict to zero in $H^*(\Gamma_g;\mathbb{F}_p)$ in case $i \not\equiv 0$ mod $(p-1)$, whereas the induced map

$$\mathbb{F}_p[\bar{c}_{p-1}, \bar{c}_{2(p-1)}, \ldots] \to H^*(\Gamma_g;\mathbb{F}_p)$$

is injective in dimensions less than $g/3$ (the dimension condition should be thought of as a stability condition, cf. (STAB) in Section 3.4.) Since [**Kau**] is not so readily available, we give a short outline of his proof. First, the vanishing conditions follow from standard results on characteristic classes of classical groups. Namely, the natural inclusion of $Sp_{2g}(\mathbb{R})$ in $Gl_{2g}(\mathbb{R})$ induces a map

$$H^*\big(BGl_{2g}(\mathbb{R}); \mathbb{F}_2\big) \to H^*\big(BSp_{2g}(\mathbb{R}); \mathbb{F}_2\big)$$

for which the odd universal Stiefel-Whitney classes w_{2i+1} restrict to zero; the corresponding result for Γ_g then follows, since the $Gl_{2g}(\mathbb{R})$-representation of Γ_g considered above factors through $Sp_{2g}(\mathbb{R})$. For the case of the Chern classes one considers the restriction map

$$H^*\big(BGl_{2g}(\mathbb{C}); \mathbb{Z}\big) \to H^*\big(BGl_{2g}(\mathbb{Z}); \mathbb{Z}\big)$$

which maps the universal Chern class c_j to $c_j(\mathbb{Z}) \in H^{2j}\big(BGl_{2g}(\mathbb{Z}); \mathbb{Z}\big)$, a torsion class of order prime to p if $j \not\equiv 0 \mod (p-1)$ (see [**Eck-Mi2**]). The mod$p$ Chern classes $\bar{c}_j$ restrict therefore already in $H^*\big(BGl_{2g}(\mathbb{Z}); \mathbb{F}_p\big)$ to zero, if j is not divisible by $(p-1)$; thus $\bar{c}_j$ restricts to zero in $H^{2j}(\Gamma_g; \mathbb{F}_p)$ too, if $j \not\equiv 0 \mod (p-1)$. For the part of Kaufmann's result which deals with injectivity, one proceeds as follows. One makes use of the mapping class groups $\Gamma_{g,1}$ of "oriented surfaces of genus g, with one boundary component", which can be arranged to form a natural increasing sequence

$$\Gamma_{g,1} \hookrightarrow \Gamma_{g+1,1}$$

and one defines the *stable* mapping class group by putting

$$\Gamma_\infty := \cup_g \Gamma_{g,1}.$$

The point is that both natural maps

$$\Gamma_g \leftarrow \Gamma_{g,1} \to \Gamma_\infty$$

induce integral cohomology isomorphisms in dimensions less than $g/3$, see Harer [**Ha2**], so that Γ_∞ is suitable for computing the cohomology of Γ_g in that range. Note also that $\Gamma_{g,1}$ acts on the homology of the surface $S_{g,1} = S_g \setminus D$, with D the interior of a closed disk. Because the inclusion $S_{g,1} \subset S_g$ induces an isomorphism in H^1, one obtains natural representations $\Gamma_{g,1} \to Sl_{2g}(\mathbb{Z})$, which are compatible with the corresponding representations of Γ_g. Moreover, there are natural pairings

$$\Gamma_{g,1} \times \Gamma_{h,1} \to \Gamma_{g+h,1}$$

compatible with the usual pairings

$$Sl_{2g}(\mathbb{Z}) \times Sl_{2h}(\mathbb{Z}) \to Sl_{2g+2h}(\mathbb{Z}),$$

inducing H-space structures on $B\Gamma_\infty^+$ and $BSl(\mathbb{Z})^+$ respectively, where the plus stands for Quillen's *plus*-construction. By naturality, the induced map

$$B\Gamma_\infty^+ \to BSl(\mathbb{Z})^+$$

is an H-map, thus inducing morphisms of Hopf-algebras

$$H^*\big(BGl(\mathbb{C}); \mathbb{F}_p\big) \to H^*\big(BGl(\mathbb{R}); \mathbb{F}_p\big) \to H^*(B\Gamma_\infty; \mathbb{F}_p),$$

with the Hopf-algebra structure on $H^*\big(BGl(\mathbb{C}); \mathbb{F}_p\big)$ and $H^*\big(BGl(\mathbb{R}); \mathbb{F}_p\big)$ induced by the Whitney sum construction. We assume now that p is an odd prime; in case $p = 2$ one argues similarly. The Hopf-algebra structure on

$$A^* = H^*\big(BGl(\mathbb{C}); \mathbb{F}_p\big)$$

is given by

$$\Delta(\bar{c}_k) = \sum_{i=0}^{k} \bar{c}_{k-i} \otimes \bar{c}_i.$$

Now consider $B^* = \mathbb{F}_p[\bar{c}_{p-1}, \bar{c}_{2(p-1)}, \dots]$, with a Hopf-algebra structure defined by

$$\Delta\bar{c}_{k(p-1)} = \sum_{i=0}^{k} \bar{c}_{(k-i)(p-1)} \otimes \bar{c}_{i(p-1)}.$$

Although the inclusion $B^* \subset A^*$ is not a morphism of Hopf-algebras, any Hopf-algebra map $A^* \to H^*(\Gamma_\infty; \mathbb{F}_p)$ which maps $\bar{c}_j$ to zero for j not divisible by $(p-1)$, will restrict to a morphism of Hopf-algebras on B^*. Note also that a morphism of (graded) Hopf-algebras with domain B^* is injective, if it is injective when restricted to the subspace PB^* of primitive elements of B^*. One checks that the graded vector space PB^* has a basis consisting of the Newton polynomials $N_k = N_k(\bar{c}_{p-1}, \bar{c}_{2(p-1)}, \dots)$, given by the usual recursion formula:

$$N_1 = \bar{c}_{p-1}$$

and, for $k > 1$

$$N_k = \bar{c}_{p-1}N_{k-1} - \bar{c}_{2(p-1)}N_{k-2} + \cdots + (-1)^{k-2}\bar{c}_{(k-1)(p-1)}N_1 + (-1)^{k-1}k\bar{c}_{k(p-1)}.$$

Kaufmann proves then that these classes $N_k \in H^{2k(p-1)}\big(BGl(\mathbb{C}); \mathbb{F}_p\big)$ do not restrict to zero in $H^{2k(p-1)}(\Gamma_\infty; \mathbb{F}_p)$, by evaluating them on a suitable subgroup of order p in $\Gamma_{(p^n-1)(p-1)/2}$, with $n \gg k$. His result then follows.

3.3 The case of genus less than three. Since Γ_1 is isomorphic to an amalgamated free product of a cyclic group of order four and a cyclic group of order six over a cyclic group of order two, the Mayer-Vietoris sequence reveals that

$$H^*(\Gamma_1; \mathbb{Z}) = \mathbb{Z}[x]/(12x),$$

with $x \in H^2(\Gamma_1; \mathbb{Z})$. An explicit description of x can be given as follows. Consider the flat real vector bundle over $B\Gamma_1$, induced by the inclusion of $\Gamma_1 = Sl_2(\mathbb{Z})$ into $Sl_2(\mathbb{R})$. One checks that the universal Euler class e_2 in $H^2\big(BSl_2(\mathbb{R}); \mathbb{Z}\big)$ restricts to a generator of $H^2(C; \mathbb{Z})$ for any cyclic subgroup $C \subset Sl_2(\mathbb{R})$. It therefore restricts to a generator $e_2(\Gamma_1) = x$ in $H^2(\Gamma_1; \mathbb{Z})$. Also, it follows that the projection $\Gamma_1 \to (\Gamma_1)_{ab}$, a cyclic group of order 12, induces in integral cohomology an isomorphism.

As mentioned earlier, Γ_2 is $\mathbb{Q}$-acyclic and therefore, because the integral cohomology groups of Γ_g are finitely generated, $H^k(\Gamma_2; \mathbb{Z})$ is a finite group for $k > 0$. Moreover, Lee-Weintraub proved in [**Le-We**] that Γ_2 is $\mathbb{F}_p$-acyclic for all primes $p > 5$. For the remaining primes $p = 2, 3$ and 5 it is useful to study the short exact sequence

$$0 \to \mathbb{Z}/2\mathbb{Z} \to \Gamma_2 \to \Gamma_0^6 \to 0,$$

where Γ_0^6 is the mapping class group of the 2-sphere with 6 punctures, that is

$$\Gamma_0^6 = \pi_0\Big(\mathrm{Diffeo}_+\big(S^2 \setminus \{x_1, \dots, x_6\}\big)\Big).$$

This short exact sequence is discussed in Birman-Hilden [**Bi-Hi1**]. It is obtained by considering the genus two surface S_2 as a branched covering space of the 2-sphere S^2 with six branch points, by forming the quotient surface $S_2/\langle\tau\rangle$, τ the hyperelliptic involution, which is known to generate the center of Γ_2. There is an obvious map $\Gamma_0^6 \to \Sigma_6$, the symmetric group on six letters, with kernel denoted by K_6. An analysis of K_6 led Cohen and Benson [**Co4, Be, Be-Co**] to the following results concerning the mod-p cohomology of Γ_2.

(Γ_2 MOD p).

a) There is a subgroup $\mathbb{Z}/5\mathbb{Z} \subset \Gamma_2$ such that the restriction map induces an isomorphism
$$H^*(\Gamma_2; \mathbb{F}_5) \cong H^*(\mathbb{Z}/5\mathbb{Z}; \mathbb{F}_5).$$

b) There are elements x, y, z and w of degree 3, 4, 4 and 5, such that
$$H^*(\Gamma_2; \mathbb{F}_3) \cong \mathbb{F}_3[x, y, z, w]/\langle x^2, xz, z^2, zw, w^2, yz - xw\rangle.$$

c) The Poincaré series of $H^*(\Gamma_2; \mathbb{F}_2)$ is given by
$$\frac{1 + t^2 + 2t^3 + t^4 + t^5}{(1-t)(1-t^4)} = 1 + t + 2t^2 + 4t^3 + 6t^4 + \dots.$$

A discussion of the integral cohomology of Γ_2 is presented in Cohen [**Co3**], see also [**Co1-2**]; we want to mention in particular the following result, which we will use later on:
$$120 \cdot H^i(\Gamma_2; \mathbb{Z}) = 0 \quad \text{for} \quad i > 0.$$
Away from the prime 2 the cohomology is particularly easy to describe, because it is "periodic with period 4 from dimension four on". It can be expressed as follows:
$$H^*\big(\Gamma_2; \mathbb{Z}[1/2]\big) = \mathbb{Z}[1/2][x, y, z]/\langle 5x, 3y, 3z, z^2\rangle,$$
with $x \in H^2$, $y \in H^4$ and $z \in H^5$. The groups in low dimension are
$$H^i(\Gamma_2; \mathbb{Z}) = \begin{cases} 0, & \text{for } i = 1 \\ \mathbb{Z}/10\mathbb{Z}, & \text{for } i = 2 \\ \mathbb{Z}/2\mathbb{Z}, & \text{for } i = 3 \\ \mathbb{Z}/120\mathbb{Z} \oplus (\mathbb{Z}/2\mathbb{Z})^2, & \text{for } i = 4 \\ \mathbb{Z}/6\mathbb{Z} \oplus (\mathbb{Z}/2\mathbb{Z})^2, & \text{for } i = 5. \end{cases}$$

3.4 Torsion in the cohomology of the mapping class group. The Bernoulli numbers B_n are rational numbers defined recursively by the formula
$$(B+1)^{\downarrow n} - B_n = 0, \quad n \geq 2,$$
where the exponent "$\downarrow n$" means that after evaluating the n'th power of the monomial, one replaces the power B^k by B_k. For $n = 2$ this yields $B_2 + 2B_1 + 1 - B_2 = 0$, hence $B_1 = -1/2$. It turns out that for odd $n > 1$ one always has $B_n = 0$ and, as already observed by Euler, the B_{2k}'s are related to the Taylor series of $\tan(x)$, which is given by
$$x + \frac{1}{3}x^3 + \frac{2}{15}x^5 + \frac{17}{315}x^7 \dots = \sum_{k=1}^{\infty} (-1)^{k-1} \frac{2^{2k}(2^{2k}-1)B_{2k}}{(2k)!} x^{2k-1}.$$
Thus $B_2 = \frac{1}{6}$, $B_4 = -\frac{1}{30}$, $B_6 = \frac{1}{42}$, $\dots$, $B_{12} = -\frac{691}{2 \cdot 3 \cdot 5 \cdot 7 \cdot 13}$, $\dots$ and so on. There are several conflicting notations in use concerning Bernoulli numbers. The one we use here differs by a sign from the one used in [**Gl-Mi1**] and [**Gl-Mi2**], but agrees with the one used in [**Brow**] and [**Ha-Za**]. The Bernoulli numbers turn up in number

theory in several places. Our convention is such that for any integer $k > 0$, the Riemann zeta function satisfies the equation

$$\zeta(1-2k) = -B_{2k}/2k.$$

In Glover-Mislin [**Gl-Mi2**] it is proved that for $g > (8m)^2$ the cohomology group $H^{4m}(\Gamma_g;\mathbb{Z})$ contains an element of order

$$E_{2m} = \operatorname{den}(B_{2m}/2m),$$

the denominator of $B_{2m}/2m$, expressed as a fraction in lowest terms ($E_2 = 12$, $E_4 = 120$, $E_6 = 252, \dots$). On the other hand, Harer proved in [**Ha2**] a *stability result* for Γ_g:

(STAB) $\qquad H^k(\Gamma_g;\mathbb{Z})$ is for $g > 3k$ independent of g.

It follows then that:

(TOR) $\qquad H^{4m}(\Gamma_g;\mathbb{Z})$ contains for $g > 12m$ an element of order E_{2m}.

This result will be improved in Section 5. In [**Gl-Mi2**] the torsion result (TOR) was used to construct *strange* torsion in the cohomology of Γ_g (torsion, which is not present in the group Γ_g itself). It is shown that for p a prime larger than 13 and $g = g(p) = (p^2 - 4p + 1)/2$, the mapping class group $\Gamma_{g(p)}$ is p-torsion-free, but $H^{2(p-1)}(\Gamma_{g(p)};\mathbb{Z})$ contains an element of order p.

REMARK. An alternative way to construct torsion in $H^*(\Gamma_g;\mathbb{Z})$ is to use a result due to Charney-Cohen [**Ch-Co**], which states that, in the stable range, the cohomology of Γ_g contains a direct summand isomorphic to the cohomology of $\operatorname{Im} J_{1/2}$, where $\operatorname{Im} J_{1/2}$ is a space which is a factor of $BGl(\mathbb{Z})^+$, usually referred to as "*the image of the J-homomorphism localized away from* 2".

3.5 Periodicity and Krull dimension. It is well-known that for a finite group F the cohomology ring $H^*(F;\mathbb{F}_p)$ is noetherian. More generally, Quillen's Proposition 14.5 of [**Qu**] implies that if Γ denotes a group of finite virtual cohomological dimension acting simplicially on a finite dimensional contractible simplicial complex, with compact quotient and finite stabilizers, then $H^*(\Gamma;\mathbb{F}_p)$ is noetherian. In particular, this implies that, by considering the action of Γ_g on $\overline{T}_g$:

(NOETH) $\qquad H^*(\Gamma_g;\mathbb{F}_p)$ is a noetherian ring.

To get an idea of the growth rate of H^n as $n \to \infty$, one considers the Krull dimension. Recall that the Krull dimension of a commutative ring R with 1 is defined as the supremum of the lengths n of chains of distinct prime ideals

$$\mathfrak{p}_0 \subset \mathfrak{p}_1 \subset \cdots \subset \mathfrak{p}_n.$$

For an arbitrary (discrete) group Γ and prime p, the Krull dimension of Γ at p, $\kappa(\Gamma,p)$, is per definition the Krull dimension of the commutative ring $H^{ev}(\Gamma;\mathbb{F}_p)$ of even dimensional cohomology classes. If $H^*(\Gamma;\mathbb{F}_p)$ is noetherian, standard results from commutative algebra imply that $\kappa(\Gamma,p)$ is the smallest integer $\kappa \geq 0$ such that there is a constant $C > 0$ satisfying for all $n \geq 0$

$$\sum_{i\leq n} \dim_{F_p} H^i(\Gamma;\mathbb{F}_p) \leq C\cdot n^{\kappa}.$$

Note that in this last formula we did not restrict to the even cohomology; indeed one easily checks that for a finitely generated graded-commutative $\mathbb{F}_p$-algebra H^* satisfying for all $n \geq 0$ the condition $\sum_{2i \leq 2n} \dim H^{2i} \leq C(2n)^\kappa$ with $C > 0$, one can find a constant $D > 0$ such that $\sum_{i \leq n} \dim H^i \leq Dn^\kappa$, and conversely. In particular, if $H^*(\Gamma; \mathbb{F}_p)$ is noetherian then

$$\kappa(\Gamma, p) = 0 \Longleftrightarrow \dim_{\mathbb{F}_p} H^*(\Gamma; \mathbb{F}_p) < \infty.$$

For groups Γ of finite virtual cohomological dimension, the prime ideals of the ring $H^{ev}(\Gamma; \mathbb{F}_p)$ are intimately linked to the elementary abelian subgroups of Γ; in case Γ has only finitely many conjugacy classes of elementary abelian p-subgroups, the minimal prime ideals are in one-one correspondence with the conjugacy classes of maximal elementary abelian p-subgroups (see Quillen [**Qu**]). It was moreover proved in [**Qu**] that for a finite group F the Krull dimension $\kappa(F, p)$ equals the maximal rank of an elementary abelian p-subgroup of F. This result still holds for certain infinite groups, in particular for Γ_g, as was proved by Broughton in [**Br**]. Note that from our description of the cohomology of Γ_1 and Γ_2 one readily sees that:

- $\kappa(\Gamma_1, 2) = \kappa(\Gamma_1, 3) = 1$, and $\kappa(\Gamma_1, p) = 0$ for $p > 3$,
- $\kappa(\Gamma_2, 2) = 2$, $\kappa(\Gamma_2, 3) = \kappa(\Gamma_2, 5) = 1$, and $\kappa(\Gamma_2, p) = 0$ for $p > 5$.

In [**Br**] Broughton established an explicit general formula for the Krull dimension of Γ_g, by determining the maximal rank of an elementary abelian p-subgroup contained in the mapping class group:

(KRULL-DIM I). Let $g > 1$. Then the Krull dimension $\kappa(\Gamma_g, p)$ is the largest integer κ such that there are nonnegative integers $k \neq 1$ and h satisfying:

1) $2g - 2 = p^\kappa(2h - 2) + p^{\kappa-1}(p-1)k$, and
2) $\kappa \leq 2h$ if $k = 0$, and
3) $\kappa < 2h + k$ if $k > 1$.

It will be useful to record the following immediate consequences.

(KRULL-DIM II). Let $g > 1$. Then the Krull dimension of Γ_g satisfies the following:

1) Case $p = 2$: $\kappa(\Gamma_g, 2) \geq 2$.
   1.1) if g is even, $\kappa(\Gamma_g, 2) = 2$
   1.2) if g is odd, $\kappa(\Gamma_g, 2) \geq 3$
2) Case p odd:
   2.1) if $g \not\equiv 1 \mod p$, then $\kappa(\Gamma_g, p) \leq 1$
   2.2) if $g \equiv 1 \mod p$, then $\kappa(\Gamma_g, p) \geq 1$ and if we write g in the form $l\ p^\alpha + 1$ with l prime to p and $\alpha > 0$, then $\kappa(\Gamma_g, p) \leq \alpha + 1$; moreover, if $l \gg p$ one has $\kappa(\Gamma_g, p) = \alpha + 1$.

A particularly interesting case arises when $\kappa(\Gamma_g, p) = 1$. It was observed by Venkov [**Ve**] that if Γ is a group of finite virtual cohomological dimension with $\kappa(\Gamma, p) \leq 1$, then $H^*(\Gamma, \mathbb{F}_p)$ is "periodic in sufficiently high dimensions", meaning that

$$\exists k > 0, \forall i \gg 0\colon H^i(\Gamma; \mathbb{F}_p) \cong H^{i+k}(\Gamma; \mathbb{F}_p).$$

Periodicity phenomena are much easier to handle using *Farrell* cohomology instead of ordinary cohomology. We will write

$$\widehat{H}^i(\Gamma; M), \quad i \in \mathbb{Z}$$

for the i-th Farrell cohomology group [**Fa**] of the group Γ of finite virtual cohomological dimension, and M a Γ-module. For an in depth discussion of the properties of these Farrell cohomology groups, see Brown's book [**Brow1**]. The interested reader might also consult [**Mis2**], where "Tate cohomology groups" $\widehat{H}^i(\Gamma; M)$ are defined for *arbitrary* groups Γ, in a way that for groups of finite virtual cohomological dimension one obtains the Farrell cohomology groups, and thus for finite groups the classical Tate cohomology groups. Here are some basic facts concerning these generalized Tate groups.

(TATE-GROUPS).

a) For an arbitrary group Γ and projective Γ-module P one has $\widehat{H}^*(\Gamma; P) = 0$

b) $\widehat{H}^0(\Gamma; \mathbb{Z}) = 0 \iff cd(\Gamma) < \infty$

c) Suppose $\mathrm{vcd}(\Gamma) < \infty$. Then

  c1) $\widehat{H}^i(\Gamma; M)$ is a torsion group for all i and all Γ-modules M

  c2) if $i > \mathrm{vcd}(\Gamma)$, the natural map $H^i(\Gamma; M) \to \widehat{H}^i(\Gamma; M)$ is an isomorphism for all Γ-modules M

  c3) the following conditions are equivalent:

    c3.1) every abelian p-subgroup of Γ has rank ≤ 1

    c3.2) there exists a $d > 0$ such that $\widehat{H}^i(\Gamma; M)_{(p)} \cong \widehat{H}^{i+d}(\Gamma; M)_{(p)}$ for all Γ-modules M and all $i \in \mathbb{Z}$ (the smallest such d is called the p-period of Γ an is denoted by $p(\Gamma)$; for an abelian torsion group A and prime p we write $A_{(p)}$ for its p-torsion subgroup)

    c3.3) the ring $\widehat{H}^*(\Gamma; \mathbb{Z})_{(p)}$ contains an invertible element of positive degree.

Property a) results from the definition of these general Tate groups, which we will not recall here; the interesting fact b) was proved by Kropholler [**Kr1-2**], and the list c) just recalls some of the basic results on Farrell cohomology. A group satisfying one of the equivalent conditions listed under (c3) above will be called p-periodic (see Xia [**Xi1**] for a discussion of that concept). For a group like Γ_g, being p-periodic is therefore equivalent to the condition that its Krull dimension be less than or equal to one. A basic example of an infinite p-periodic group is the semidirect product

$$\mathrm{S(n,p)} := \mathbb{Z}/p^n\mathbb{Z} \rtimes \mathbb{Z},$$

with p-period equal $2(p-1)p^{n-1}$ in case p is an odd prime, $n \geq 1$, and $\mathbb{Z}$ is acting by means of a surjective map $\mathbb{Z} \to \mathrm{Aut}(\mathbb{Z}/p^n\mathbb{Z})$. It was proved in [**G-M-X1**] that in general, the p-period of a p-periodic group Γ divides $2(p-1)p^n$ for some $n \geq 0$. For Γ_g one has a stronger result. It was shown in [**G-M-X1**] that the p-period of a p-periodic Γ_g actually always divides $2(p-1)$, and the exact value of the p-period as a function of p and g was computed in an explicit way. Moreover, the values of g for which Γ_g is p-periodic were determined by Xia in [**Xi2**]. The results on the p-period of Γ_g may be summarized as follows:

(p-PERIOD). Let $g > 1$. Then

a) Γ_g is never 2-periodic

b) for an odd prime p, Γ_g is p-periodic if and only if one of the following two condition holds:

  b1) $g \not\equiv 1 \mod p$

  b2) g is of the form $kp+1$ with $k \not\equiv 0, -1 \mod p$ and the interval $\big[(2k+3)/p, (2k+2)/(p-1)\big]$ does not contain any integer

c) The p-period of p-periodic Γ_g, denoted by $p(\Gamma_g)$, is given by

$$p(\Gamma_g) = \operatorname{lcm}\{2[N(\pi)\colon C(\pi)] | \pi \in P\},$$

where π ranges over the set P of subgroups of order p in Γ_g, and $N(\pi)$ (respectively $C(\pi)$) denotes the normalizer (respectively centralizer) of π in Γ_g. In particular, the p-period $p(\Gamma_g)$ divides $2(p-1)$.

We use the convention that $\operatorname{lcm}\{2[N(\pi)\colon C(\pi)] \mid \pi \in P\} = 1$ in case P is the empty set; in that case $p(\Gamma_g) = 1$ too, according to our definition. It is possible to convert the general formula for $p(\Gamma_g)$ in an explicit formula in terms of g and p as follows. If $\pi \subset \Gamma_g$ is a subgroup of order p then, as discussed earlier, one can lift π to a subgroup of $\mathrm{Diffeo}_+(S_g)$, and it is a classical result, that the number of fixed points $n(\pi)$ of this π-action on S_g does not depend on the lift chosen. It was proved in [**G-M-X1**] that for $g > 1$ one has

$$\operatorname{lcm}\Big\{2\big[N(\pi)\colon C(\pi)\big] \mid \pi \in P\Big\} = \operatorname{lcm}\Big\{\gcd\big(2(p-1), 2n(\pi)\big) \mid \pi \in P\Big\}.$$

According to Xia [**Xi2**], the numbers $n(\pi)$, which occur as cardinalities of such fixed-point sets, form a set denoted by $B_{g,p}$, which is given by the following formula. We may restrict to the case p odd, since Γ_g is never 2-periodic for $g > 1$. Write $2g-2$ in the form $mp - i$ with $0 \le i < p$. Then

$$B_{g,p} = \begin{cases} \Big\{i, i+p, \dots, i + \big[2g/(p-1) - m\big]p\Big\}, & \text{if } i \not\equiv 1 \mod p \\ \Big\{1+p, \dots, 1 + \big[2g/(p-1) - m\big]p\Big\}, & \text{if } i \equiv 1 \mod p. \end{cases}$$

As usual, the notation $[x]$ stands for the integral part of the rational number x. We also use the natural convention that $\gcd\big(2(p-1), 0\big) = 2(p-1)$ and, as before, the lcm of an empty set of numbers is understood to equal 1. For example, one easily checks that for $p = 3$ the sets $B_{g,p}$ contain always an even number. We also note, by using (KRULL-DIM I) or by checking the condition (b2) of (p-PERIOD), that for $g > 1$ and $g \equiv 1 \mod 3$, Γ_g is never 3-periodic. This shows that the following holds.

(3-PERIOD). Suppose $g > 1$. Then Γ_g is 3-periodic if and only if $g \not\equiv 1 \mod 3$ and the 3-period of Γ_g is always 4.

It is easy to see from the definition of the p-period that a subgroup of a p-periodic group Γ is also p-periodic, with p-period dividing the p-period of Γ. As a result, one concludes for example:

The groups $S(2,3) = \mathbb{Z}/9\mathbb{Z} \rtimes \mathbb{Z}$ *cannot be embedded into any mapping class group* Γ_g *with* $g \not\equiv 1 \mod 3$.

Indeed, using the same notation as earlier, $S(2,3)$ is 3-periodic with 3-period 12, which is larger than 4, the 3-period of a 3-periodic Γ_g. Returning to the case of an arbitrary prime p, we like to mention one more result, which follows from (p-PERIOD). Consider Γ_g with $g \equiv 1 \mod p$. As mentioned earlier, there can only be a finite number of such Γ_g's which are p-periodic (p a fixed prime). On the other hand, because S_g can be considered as an unramified covering space of S_h with h given by $p(2-2h) = 2-2g$, it follows that Γ_g contains a subgroup π of order p satisfying $n(\pi) = 0$, corresponding to a fixed-point-free action. From our formula for the p-period we infer thus:

Suppose Γ_g *is p-periodic and* $g \equiv 1 \mod p$. *Then its p-period is* $2(p-1)$.

We would like to conclude this section by mentioning computations, which involve the mapping class groups $\Gamma_{(p-1)/2}$, respectively $\Gamma_{(p-1)}$, and which demonstrate the power of using p-periodicity and the p-period in the course of computing Farrell cohomology. For an odd prime p, the smallest value $g \geq 1$ such that Γ_g has p-torsion, is $g = (p-1)/2$, and $g = p-1$ is the second smallest such value. The Krull dimension at $p > 2$ is 1 for $\Gamma_{(p-1)/2}$ and Γ_{p-1}, so that these groups are p-periodic. The p-primary part of the Farrell cohomology has been completely computed for these two families of examples by Xia in his papers [**Xi3-4**].

4. Characteristic classes for group representations

In Eckmann-Mislin [**Eck-Mi1-5**] characteristic classes of group representations where discussed in relationship with their field of definition. In our applications here we will mainly be concerned with representations defined over $\mathbb{Q}$. But first we shall recall some basic facts; a general reference on characteristic classes of representations is Thomas' book [**Th1**].

4.1 Chern classes. If $\rho\colon G \to Gl_n(\mathbb{C})$ denotes a complex representation of the (discrete) group G, then the induced map of classifying spaces

$$(B\rho)^*\colon H^*\big(BGl_n(\mathbb{C});\mathbb{Z}\big) \to H^*(G;\mathbb{Z})$$

maps the universal Chern classes $c_i \in H^{2i}\big(BGl_n(\mathbb{C});\mathbb{Z}\big)$ to the Chern classes of ρ, defined by

$$c_i(\rho) := (B\rho)^* c_i \in H^{2i}(G;\mathbb{Z}).$$

To understand the Chern classes of representations of finite groups, it is useful to analyze first the case of cyclic groups, which we will quickly review. The boundary homomorphism associated to the short exact sequence

$$0 \to \mathbb{Z} \to \mathbb{C} \xrightarrow{\exp} Gl_1(\mathbb{C}) \to 1$$

induces an isomorphism

$$c_1\colon \operatorname{Hom}\big(\mathbb{Z}/n\mathbb{Z}, Gl_1(\mathbb{C})\big) \cong H^2(\mathbb{Z}/n\mathbb{Z};\mathbb{Z}),$$

mapping a one dimensional representation to its first Chern class. Since every complex representation ρ of $\mathbb{Z}/n\mathbb{Z}$ decomposes as a sum $\oplus\rho_i$ of one dimensional representations, we can express the total Chern class

$$c(\rho) = 1 + c_1(\rho) + c_2(\rho) + \dots$$

as

$$c(\rho) = \prod c(\rho_i) = \prod \big(1 + c_1(\rho_i)\big).$$

Let $\phi(n)$ denote the number of generators of the cyclic group $\mathbb{Z}/n\mathbb{Z}$ (the Euler ϕ-function). Note that there are $\phi(n)$ faithful irreducible $\mathbb{C}$-representations of $\mathbb{Z}/n\mathbb{Z}$, and their sum, which we denote by σ_n, is a representation of degree $\phi(n)$, which is defined over $\mathbb{Z}$; the representation σ_n is sometimes called the *cyclotomic* representation, since as a $\mathbb{Q}$-representation it is equivalent to the one obtained from the Galois action of $\operatorname{Gal}\big(\mathbb{Q}(\zeta_n)/\mathbb{Q}\big)$ on the cyclotomic extension of $\mathbb{Q}$ gotten by adjoining a primitive n-th root of unity ζ_n. It is obvious that any faithful irreducible representation of $\mathbb{Z}/n\mathbb{Z}$ over $\mathbb{Q}$ must involve σ_n and thus σ_n is characterized as being the smallest faithful irreducible $\mathbb{Q}$-representation of $\mathbb{Z}/n\mathbb{Z}$. Note that its top Chern class

$$c_{\phi(n)}(\sigma_n) \in H^{2\phi(n)}(\mathbb{Z}/n\mathbb{Z};\mathbb{Z})$$

has (maximal) order n, because the total Chern class $c(\sigma_n)$ is given by

$$\prod_{j,(j,n)=1} (1 + jc_1(\sigma_n)) = 1 + \cdots + \prod_{j,(j,n)=1} j \cdot c_1(\sigma)^{\phi(n)}.$$

Recall that E_{2m} denotes the denominator of $B_{2m}/2m$. From the well-known divisibility properties of Bernoulli numbers one can infer that a prime power $p^k > 1$ divides E_{2m} if and only if $(p-1)p^{k-1}$ divides $2m$ (von Staudt's Theorem). Thus

$$E_{2m} = \operatorname{lcm}\{n \mid 2m \equiv 0 \mod \phi(n)\}.$$

Therefore, if $p^s > 1$ denotes the highest power of a prime p dividing E_{2m}, we can write $2m$ in the form $(p-1)p^{s-1} \cdot t$ with t prime to p. If σ denotes the representation given by taking a t-fold sum of the cyclotomic representation of $\mathbb{Z}/p^s\mathbb{Z}$ composed with the projection of $\mathbb{Z}/E_{2m}\mathbb{Z}$ onto $\mathbb{Z}/p^s\mathbb{Z}$, then $c_{2m}(\sigma)$ will have order p^s. Taking a sum of representations of this type, one for each prime divisor of E_{2m}, one ends up with a representation

$$\theta_m \colon \mathbb{Z}/E_{2m}\mathbb{Z} \to Gl(\mathbb{Z})$$

with $c_{2m}(\theta_m) \in H^{4m}(\mathbb{Z}/E_{2m}\mathbb{Z};\mathbb{Z})$ of (maximal) order E_{2m}. The main result of [**Eck-Mi1**] says that this order is optimal for $\mathbb{Q}$-representations of *arbitrary* finite groups, in the following sense:

If ρ: $F \to Gl(\mathbb{Q})$ denotes an arbitrary representation of a finite group F over $\mathbb{Q}$, then the order of $c_i(\rho)$ is at most two for i odd, and it divides E_i for $i > 0$ even.

The bound is actually also best possible for $\mathbb{Q}$-representations of infinite groups, up to possibly a factor 2. This follows from [**Eck-Mi4**] in conjunction with Arlettaz's result [**Ar**] who proved that the universal Chern classes restrict to torsion classes in the cohomology of the group $Gl(\mathbb{Q})$ considered as a discrete group (the complication to overcome is the fact that the integral homology of $Gl_n(\mathbb{Q})$ is not finitely generated); see also [**Mis1**] for a general torsion result on characteristic classes.

4.2 The Euler class. It is well-known that the integral cohomology ring of $BSl_n(\mathbb{R})$ is generated by elements of order two, the Euler class

$$e_n \in H^n\big(BSl_n(\mathbb{R});\mathbb{Z}\big),$$

and the Pontrjagin classes

$$p_j \in H^{4j}\big(BSl_n(\mathbb{R});\mathbb{Z}\big), \quad 2j \leq n.$$

For n odd, the Euler class has order 2. If n is even, say $n = 2m$, one has a relation of the form

$$e_{2m}^2 = p_m = (-1)^m \operatorname{res}(c_{2m})$$

with $c_{2m} \in H^{4m}\big(BGl(\mathbb{C});\mathbb{Z}\big)$ the universal Chern class, and $\operatorname{res}(c_{2m})$ the image under the restriction map induced by the inclusion of $Sl_{2m}(\mathbb{R})$ in $Gl(\mathbb{C})$. For any subring R of $\mathbb{R}$, we will write $e_n(R)$ for the restriction $\operatorname{res}(e_n) \in H^n\big(BSl_n(R)^\delta;\mathbb{Z}\big)$, where $BSl_n(R)^\delta$ stands for the classifying space of the group $Sl_n(R)$, considered as a discrete group. In case that R is a discrete subring of $\mathbb{R}$, we will omit the superscript δ; a similar convention is used for the case of the Chern classes. The Euler class

$$e_{2m}(\mathbb{Q}) \in H^{2m}(BSl_{2m}(\mathbb{Q})^{\delta};\mathbb{Z})$$

has infinite order (see Milnor [**Miln**]), but, according to Sullivan [**Su**]

$e_{2m}(\mathbb{Z}) \in H^{2m}\big(BSl_{2m}(\mathbb{Z});\mathbb{Z}\big)$ *is a torsion class.*

The order of $e_{2m}(\mathbb{Z})$ is — the same number is coming up again — E_{2m} or $2E_{2m}$. For the restriction of $e_{2m}(\mathbb{Q})$ to the cohomology of a finite subgroup of $Sl_{2m}(\mathbb{Q})$ it was proved in [**Eck-Mi1**] that the "universal bound" for the order is precisely E_{2m}:

If ρ denotes a representation $F \to Sl_{2m}(\mathbb{Q})$ of a finite group F, then the order of $e_{2m}(\rho) := (B\rho)^* e_{2m}(\mathbb{Q})$ divides E_{2m}, and E_{2m} is the best universal bound for the order of the Euler class of such representations.

One should note that a finite subgroup of $Sl_n(\mathbb{Q})$ is not necessarily conjugate to a subgroup of $Sl_n(\mathbb{Z})$!

5. Torsion in Γ_g and the homology representation

Our goal in this section is to improve the (TOR)-result concerning torsion in the integral cohomology of the mapping class group, stated in Section 3.3. The techniques we use are essentially the ones used in Glover-Mislin [**Gl-Mi2**], just a little bit refined.

5.1 Fixed point data. A basic invariant of an orientation preserving diffeomorphism of finite order $n > 1$, $f \in \text{Diffeo}_+(S_g)$, is its *fixed point data*. It is defined as follows. Because f preserves orientation, the singular set of f, that is, the points $x \in S_g$ for which the orbit $\{f^k(x) \mid k \in \mathbb{Z}\}$ has fewer than n elements, is necessarily a finite set. Let $\{x_i\}$ be a set of representatives of the singular orbits of f and write n_i for the order of $\text{stab}_f(x_i)$, the stabilizer of $\langle f\rangle$ at $x_i \in S_g$, where $< f >$ denotes the subgroup generated by f. Then f^{n/n_i} generates $\text{stab}_f(x_i)$ and, with respect to a fixed Riemannian structure, the differential of f^{n/n_i} acts by rotation on the tangent space at x_i. Let k_i be an integer such that $f^{k_i n/n_i}$ acts by rotation through $2\pi/n_i$. The number k_i is well defined modulo n_i, and k_i is prime to n_i. The fixed point data of f, denoted by $\delta(f)$, is then the collection

$$\delta(f) = \langle g, n \mid k_1/n_1, \ldots, k_q/n_q\rangle$$

where g is the genus of the surface S_g, n the order of f, and q the number of singular orbits of the f-action; the numbers $k_1/n_1, \ldots, k_q/n_q$ are unique up to order, if we choose k_i so that $1 \le k_i < n_i$.

A classical theorem of Nielsen [**Ni2**] states that two diffeomorphisms of finite order are conjugate in $\text{Diffeo}_+(S_g)$ if and only if they have the same fixed point data. Symonds [**Sy**] proved that the fixed point data of a diffeomorphism of finite order depends only upon its isotopy class, that is, its image in Γ_g. By the classical case of the "Nielsen Realization Theorem" (cf. Fenchel [**Fe**]), every element of finite order of Γ_g can be represented by a diffeomorphism of the same order. It follows that one can define the fixed point data $\delta(x)$ for a torsion element $x \in \Gamma_g$ by putting $\delta(x) = \delta(f)$, where f denotes any lift of x to $\text{Diffeo}_+(S_g)$ of the same order. As a consequence we infer:

Two elements of finite order in Γ_g are conjugate if and only if they have the same fixed point data.

If $x \in \Gamma_g$ has finite order and has fixed point data

$$\delta(x) = \langle g, n \mid k_1/n_1, \dots, k_q/n_q \rangle$$

and if f denotes a diffeomorphism of order n representing x, then the orbit space $S_g/ < f >$ is a surface S_h such that the natural projection $\pi \colon S_g \to S_h$ is an n-sheeted branched covering, with q branch points in S_h corresponding to the singular orbits of the action of the group $\langle f \rangle$ generated by f. The order of a branch point $P \in S_h$ is defined as $n/|\pi^{-1}(P)|$, and these orders correspond therefore to the orders of stabilizers, denoted by n_i above. From covering space theory one sees that the genus h is determined by g and the fixed point data $\delta(x)$ via the *Riemann-Hurwitz Relation*:

$$\text{(R-H)} \qquad 2g - 2 = n\left((2h-2) + \sum_{i=1}^{q}(1 - 1/n_i)\right).$$

REMARK. The branching orders can be computed by counting the numbers of fixed points of the different powers of the map f. The number of such fixed points can be computed from the representation of the group $\langle f \rangle \cong \mathbb{Z}/n\mathbb{Z}$ on $H_1(S_g;\mathbb{R})$, by the Lefschetz-Hopf trace formula. From character theory it is then plain that the conjugacy class of the image of f in $Gl_{2g}(\mathbb{R})$ with respect to this representation, is determined by the branching numbers $\{n_i\}$. We could also look at the conjugacy class of f in $Sp_{2g}(\mathbb{Z})$, by considering the action of f on the symplectic space $H_1(S_g;\mathbb{Z})$, the symplectic structure being given by intersection product. In case of $n = p$ a prime, this conjugacy question was analyzed by Edmonds and Ewing [**E-E**]. The fixed point data of an $x \in \Gamma_g$ of order p has the form

$$\delta(x) = \langle g, p \mid k_1/p, \dots, k_q/p \rangle$$

where $0 < k_i < p$; the number q here equals the cardinality of the fixed point set of any diffeomorphism of order p representing x, and k_i is called the *type* of the i'th fixed point. Note that q can be computed as the *Lefschetz number* $\Lambda(x)$ of x, that is

$$\Lambda(x) = \sum(-1)^i \operatorname{trace}\bigl(x_* \colon H_i(S_g;\mathbb{Q}) \to H_i(S_g;\mathbb{Q})\bigr) = q.$$

The fixed point types come up in the formula for the *signature* of x, given by

$$\operatorname{sign}(x) = \sum_{0<k<p} N_k(\zeta^k + 1)/(\zeta^k - 1)$$

with N_k denoting the number of fixed points of type k, and $\zeta = \exp(2\pi\sqrt{-1}/p)$. This signature invariant corresponds to the *equivariant signature* of the linear transformation induced by x on the hermitian space $H_1(S_g;\mathbb{C})$; see also Ewing's paper [**Ew**] for the relationship with the *Eichler Trace Formula*. The main result of [**E-E**] states that:

Two elements of prime order in Γ_g have conjugate images in $Sp_{2g}(\mathbb{Z})$ if and only if they have the same Lefschetz number and signature.

Closely related results were also proved by Symonds [**Sy**].

5.2 The homology representation of Γ_g. We consider again the natural action of Γ_g on the symplectic space $H_1(S_g;\mathbb{Z})$, yielding the canonical representation

$$\rho_g \colon \Gamma_g \to Sp_{2g}(\mathbb{Z}).$$

We will write

$$c_i(\Gamma_g) \in H^{2i}(\Gamma_g;\mathbb{Z}),$$

respectively

$$e_{2g}(\Gamma_g) \in H^{2g}(\Gamma_g;\mathbb{Z})$$

for the images of the universal Chern classes, respectively the universal Euler class, under the restriction maps induced via

$$\Gamma_g \to Sp_{2g}(\mathbb{Z}) \subset Sl_{2g}(\mathbb{R}) \subset Gl(\mathbb{C}).$$

From our earlier discussions it is plain that the Chern classes $c_i(\Gamma_g)$ for $i > 0$, and the Euler class $e_{2g}(\Gamma_g)$, are torsion classes. We want to establish the following result concerning their order.

TOR-THEOREM. *Denote as before the denominator of $B_{2m}/2m$ by E_{2m}, so that $E_2 = 12$, $E_4 = 120$, $E_6 = 252$ and so on. Then the following holds.*

1. $c_2(\Gamma_1) \in H^4(\Gamma_1;\mathbb{Z})$ *and* $e_2(\Gamma_1) \in H^2(\Gamma_1;\mathbb{Z})$ *both have order* 12.
2. $c_4(\Gamma_2) \in H^8(\Gamma_2;\mathbb{Z})$ *and* $e_4(\Gamma_2) \in H^4(\Gamma_2;\mathbb{Z})$ *both have order* 120.
3. *For $g > 2$, the order of $c_{2g}(\Gamma_g) \in H^{4g}(\Gamma_g;\mathbb{Z})$ and $e_{2g} \in H^{2g}(\Gamma_g;\mathbb{Z})$ is either E_{2g} or $2E_{2g}$.*

PROOF. Because of the results on the order of $c_i(\mathbb{Z})$ and $e_n(\mathbb{Z})$ mentioned earlier, it suffices to show that $c_{2g}(\Gamma_g)$ has order at least E_{2g}. It follows then that $e_{2g}(\Gamma_g)$ has order at least E_{2g}, as

$$e_{2g}(\Gamma_g)^2 = (-1)^g c_{2g}(\Gamma_g).$$

Also, for $g \leq 2$ the computations concerning $H^*(\Gamma_g;\mathbb{Z})$ rule out the existence of elements of order $2E_{2g}$ in $H^*(\Gamma_g;\mathbb{Z})$. To get the general lower bound on the order of $c_{2g}(\Gamma_g)$, we proceed as follows. Suppose that p^β (with $\beta \geq 1$) is the largest power of a prime p, which divides E_{2g}. Then, by von Staudt's theorem, $p^{\beta-1}(p-1)$ divides $2g$, say

$$2g = l \cdot p^{\beta-1}(p-1).$$

Next, we construct a subgroup

$$\pi \subset \Gamma_g, \quad \pi \cong \mathbb{Z}/p^\beta\mathbb{Z}$$

such that the restriction $c_{2g}(\pi) \in H^{2g}(\pi;\mathbb{Z})$ of $c_{2g}(\Gamma_g)$ has order p^β. To this end, we consider the branched covering space

$$S_g \to S^2$$

with two branch points of order p^β and l of order p. We recall the classical construction of such a covering space. One begins by deleting $2 + l$ points from the 2-sphere to get

$$X = S^2 \setminus \{x_1, x_2, y_1, \dots y_l\}$$

and chooses a suitable surjective homomorphism

$$\partial \colon \pi_1(X) \to \mathbb{Z}/p^n\mathbb{Z}.$$

To describe ∂ more explicitly, we choose a presentation of $\pi_1(X)$ of the form

$$\langle u_1, u_2, v_1, \ldots, v_l \mid u_1 u_2 v_1 \ldots v_l = 1\rangle$$

and a generator $x \in \mathbb{Z}/p^\beta\mathbb{Z}$. Let's restrict to the case p odd and $\beta > 1$; the other cases are similar. Put $\partial(v_i) = y$ for $1 \leq i \leq l$, where y denotes a fixed element of $\mathbb{Z}/p^\beta\mathbb{Z}$ of order p, and define $\partial(u_1) = x$ respectively $\partial(u_2) = -(x + ly)$ so that ∂ is well defined and surjective. By compactifying the regular covering space associated to the kernel of ∂, one obtains a branched covering $S_g \to S^2$, with genus g determined by the Riemann-Hurwitz relation (H-R),

$$2g - 2 = p^\beta\big(-2 + (2 - 2/p^\beta) + (l - l/p)\big),$$

yielding $2g = l(p-1)p^{\beta-1}$, as desired. The associated covering transformation group is cyclic of order p^β and acts with $2 + l$ branch points, two of order p^β and l of order p. It defines therefore a subgroup $\pi \subset \Gamma_g$, generated by an element z with fixed point data

$$\delta(z) = \langle l(p-1)p^{\beta-1}/2,\, p^\beta \mid k_1/p^\beta, k_2/p^\beta, k_3/p, \ldots, k_{2+l}/p\rangle.$$

Consider now the representation of π gotten by the composite

$$\rho\colon \pi \subset \Gamma_g \to Sp_{2g}(\mathbb{Z}) \subset Gl_{2g}(\mathbb{C}).$$

We claim that ρ is equivalent the sum of l copies of the cyclotomic representation σ_{p^β}. This is established by comparing characters. It is easy to see that σ_{p^β} is induced from the reduced regular representation of a subgroup of order p (just use the fact that the cyclotomic representation is the unique faithful irreducible representation over $\mathbb{Q}$). Thus, by the well-known formula for the character of an induced representation, one infers that the character χ of $\sigma_{p\beta}$ is given by

$$\chi(w) = \begin{cases} 0, & \text{if } pw \neq 0 \\ -p^{\beta-1}, & \text{if } pw = 0 \text{ but } w \neq 0 \\ p^{\beta-1}(p-1), & \text{if } w = 0. \end{cases}$$

Here w denotes a general element in $\mathbb{Z}/p^\beta\mathbb{Z}$. On the other hand, if $\tilde{z}$ denotes a lift of order p^β in $\text{Diffeo}_+(S_g)$ of the generator $z \in \pi \subset \Gamma_g$, then by construction $\tilde{z}$ has precisely 2 fixed points (we are still assuming that $\beta > 1$). More generally we can say that if $\tilde{z}^j$ has order greater then p, then it has precisely two fixed points, and when its order is p, it has $2 + lp^{\beta-1}$ fixed points. The Lefschetz number computes the number of fixed points $FP(\tilde{z}^j)$ of the map $\tilde{z}^j$, in case $\tilde{z}^j$ is not the identity map, by

$$\Lambda(z^j) = 2 - \text{trace}\big(z_*^j \mid H_1(S_g;\mathbb{Q})\big) = FP(\tilde{z}^j).$$

We infer readily that

$$\text{trace}\big(z_*^j \mid H_1(S_g;\mathbb{Q})\big) = \begin{cases} 0, & \text{if } z^{pj} \neq 1 \\ -lp^{\beta-1}, & \text{if } z^{pj} = 1 \text{ but } z^j \neq 1 \\ 2g, & \text{if } z^j = 1. \end{cases}$$

Comparing with the computation for the trace of the cyclotomic representation, we see that ρ is equivalent to the sum of l copies of σ_{p^β}. Since the top Chern class of σ_{p^β} has order p^β (compare Section 4.1), we conclude that the top Chern class

$$c_{2g}(\rho) \in H^{4g}(\pi;\mathbb{Z})$$

is the l'th power of an element of order p^β and has therefore order p^β. This implies that $c_{2g}(\Gamma_g)$ has order at least p^β, and we are done.

6. The Euler characteristic

A group Γ is said to be of *finite homological type* if it has finite virtual cohomological dimension and if for every Γ-module M which is finitely generated as an abelian group, $H_i(\Gamma; M)$ is finitely generated for every i. For instance, if Γ has finite vcd and acts properly and simplicially on a finite dimensional simplicial complex, with compact quotient, then Γ is homologically of finite type. Thus, Γ_g is of finite homological type, as we see from its action on the extended Teichmüller space $\overline{T}_g$.

If Γ is an arbitrary group of finite homological type, its Euler characteristic is defined by

$$\chi(\Gamma) = \frac{\chi(\Delta)}{[\Gamma\colon \Delta]},$$

where $\Delta \subset \Gamma$ denotes a torsion-free subgroup of finite index $[\Gamma\colon \Delta]$ in Γ, and where

$$\chi(\Delta) = \sum (-1)^i \dim_{\mathbb{Q}} H_i(\Delta; \mathbb{Q})$$

is the topological Euler characteristic of the classifying space of Δ. One checks that $\chi(\Gamma)$ is well-defined (cf. Brown's book [**Brow1**]); for background and various results concerning the Euler characteristic, see also Brown [**Brow2-4**]). Notice that for a finite group F one has obviously

$$\chi(F) = \frac{1}{|F|},$$

and for a free group L of rank n, whose classifying space is homotopy equivalent to a wedge of n circles, one has

$$\chi(L) = 1 - n.$$

It follows then that for the case of $\Gamma_1 \cong Sl_2(\mathbb{Z})$, whose commutator subgroup is free of rank two and whose abelianized group is of order 12, the Euler characteristic is given by

$$\chi\big(Sl_2(\mathbb{Z})\big) = -\frac{1}{12}.$$

The denominator of the Euler characteristic is also in the general case closely linked to the torsion of the group. Indeed, a basic result [**Brow2**] is the following:

If a prime power p^n divides the denominator of $\chi(\Gamma)$, then Γ possesses a subgroup of order p^n.

One also defines the *naive* Euler characteristic of a group Γ of finite homological type by

$$\widetilde{\chi}(\Gamma) = \sum (-1)^i \dim_{\mathbb{Q}} H_i(\Gamma; \mathbb{Q}).$$

Note that for the mapping class group $\widetilde{\chi}(\Gamma_g)$ is just the topological Euler characteristic of the moduli space M_g. Brown proved in [**Brow4**] the following general relationship between χ and $\widetilde{\chi}$:

Let Γ be a group such that for all elements of finite order $x \in \Gamma$ the centralizers $C(x)$ have finite homological type. Then

$$\chi(\Gamma) = \widetilde{\chi}(\Gamma) - \sum_{x \in T} \chi\big(C(x)\big),$$

where T denotes a set of representatives for the conjugacy classes of nontrivial torsion elements of Γ.

Using this result, and the fact that for an $x \in \Gamma_g$ of finite order the centralizer $C(x)$ can be identified with a mapping class group, Harer-Zagier [**Ha-Za**] were able to prove the following amazing result

$$\chi(\Gamma_g) = \frac{1}{2-2g}\zeta(1-2g), \quad g > 1. \tag{Ha-Za 1}$$

It is convenient, and natural from the point of view of the formula (Ha-Za 1), to work with the *pointed* mapping class group $\widetilde{\Gamma}_g$, which is defined to be the group of connected components of $\mathrm{Diffeo}_+(S_g, s_0)$, the group of *pointed* orientation preserving diffeomorphisms of S_g. The obvious map

$$\widetilde{\Gamma}_g \to \mathrm{Aut}_+\big(\pi_1(S_g, s_0)\big)$$

is an isomorphism, compare with (IV) of Section 1. Thus

$$\widetilde{\Gamma}_1 \cong \Gamma_1 \cong Sl_2(\mathbb{Z})$$

and for $g > 1$ one has a natural short exact sequence

$$1 \to \pi_1(S_g, s_0) \to \widetilde{\Gamma}_g \to \Gamma_g \to 1,$$

where we have identified the group of inner automorphisms of $\pi_1(S_g, s_0)$ with $\pi_1(S_g, s_0)$, a group with trivial center if $g > 1$. It follows then readily, by taking in account that $\chi\big(\pi_1(S_g)\big) = 2 - 2g$, that

$$\chi(\widetilde{\Gamma}_g) = (2-2g)\chi(\Gamma_g), \quad g > 1.$$

Thus we can rewrite (Ha-Za 1) to get

$$\chi(\widetilde{\Gamma}_g) = \zeta(1-2g), \quad g \geq 1. \tag{Ha-Za 2}$$

Note that the formula is indeed also correct for $g = 1$, as both sides of the equation equal then $-\frac{1}{12}$. It is a classical result that the values of the ζ-function at negative integers can be expressed in terms of Bernoulli numbers as follows:

$$\zeta(1-2g) = -\frac{B_{2g}}{2g}, \quad g \geq 1.$$

Our computation of the order of $e_{2g}(\Gamma_g)$ shows now that the denominator

$$\mathrm{den}\big(\chi(\widetilde{\Gamma}_g)\big) = \mathrm{den}\left(-\frac{B_{2g}}{2g}\right) = E_{2g}$$

corresponds up to possibly a factor two to the order of $e_{2g}(\Gamma_g)$:

For $g \geq 1$ the denominator of $\chi(\widetilde{\Gamma}_g)$ equals the order of $e_{2g}(\Gamma_g)$ or half that order.

In this statement, we could have used equally well the Euler class $e_{2g}(\widetilde{\Gamma}_g)$ in place of $e_{2g}(\Gamma_g)$, i.e., the Euler class of the flat bundle induced by the composite map

$$\widetilde{\Gamma}_g \to \Gamma_g \to Sp_{2g}(\mathbb{Z}) \subset Sl_{2g}(\mathbb{R}).$$

Indeed, the cyclic subgroups $\pi \subset \Gamma_g$ of order p^β, which were used to detect the order of $e_{2g}(\Gamma_g)$, are generated by an element which lifts to a periodic diffeomorphism of

S_g with a fixed point, as we see by looking at its fixed point data. It follows that π lifts to $\widetilde{\Gamma}_g$ and detects the order of $e_{2g}(\widetilde{\Gamma}_g)$ as well. It would be good to have a more direct way of understanding the relationship between the Euler class and the Euler characteristic. Is the order of $e_{2g}(\Gamma_g)$ precisely the denominator of $\chi(\widetilde{\Gamma}_g)$? For small values of g, the explicit computations show that this is indeed the case: $e_2(\Gamma_1)$ has order 12 and $e_4(\Gamma_2)$ has order 120; on the other hand,

$$\chi(\widetilde{\Gamma}_1) = -\frac{1}{12}, \quad \text{and} \quad \chi(\widetilde{\Gamma}_2) = \frac{1}{120}.$$

7. The Yagita invariant of the mapping class group

The invariant, which we call the *Yagita invariant*, was first introduced by Yagita [**Ya**] in the case of finite groups, and in Thomas [**Th2**] for more general groups. It is related to group actions on products of spheres in a similar way as finite groups with periodic cohomology are related to actions on spheres (a more precise statement is given below). We recall the definition of the Yagita invariant and some background.

Let Γ be a group of finite virtual cohomological dimension and $\pi \subset \Gamma$ any subgroup of prime order p. Because π injects into any finite quotient of the form Γ/Δ, where Δ is a torsion-free normal subgroup of finite index in Γ, the image $\operatorname{Im}(H^k(\Gamma;\mathbb{Z}) \to H^k(\pi;\mathbb{Z}))$ of the restriction map in cohomology is non-zero for some degree $k > 0$. Reduction mod-p maps $H^*(\pi;\mathbb{Z})$ onto $\mathbb{F}_p[u] \subset H^*(\pi;\mathbb{F}_p)$ with u a generator in $H^2(\pi;\mathbb{F}_p)$. Thus, there exists a maximum value $m = m(\pi,\Gamma)$ such that

$$\operatorname{Im}\big((H^*(\Gamma;\mathbb{Z}) \to H^*(\pi;\mathbb{F}_p)\big) \subset \mathbb{F}_p[u^m] \subset H^*(\pi;\mathbb{F}_p).$$

Note that $m(\pi,\Gamma)$ is bounded by $m(\pi,\Gamma/\Delta)$, where Δ denotes as before a torsion-free normal subgroup of finite index. Since Γ/Δ is finite, we conclude that $m(\pi,\Gamma)$ is bounded by a bound depending on Γ only. The Yagita invariant $p(\Gamma)$ of Γ with respect to the prime p is then defined to be the least common multiple of values $2m(\pi)$, where π ranges over all subgroups of order p of Γ. We use the convention that $p(\Gamma) = 1$ if Γ is p-torsion-free. The invariant $p(\Gamma)$ agrees with the p-period of a p-periodic group and one can show [**G-M-X2**] that $p(\Gamma)$ has in general the form $l \cdot p^k$ with l dividing $2(p-1)$; in particular, for the prime 2 the Yagita invariant is a power of 2.

The interest in $p(\Gamma)$ stems from the fact that it provides a lower bound for the dimension of a complex, which admits a certain type of action of Γ. For instance, using the same reasoning as in [**Ya**], where only finite groups were considered, one finds that if Γ acts properly discontinuously on $\mathbb{R}^n \times (S^m)^k$ and trivially on $H^*(\mathbb{R}^n \times (S^m)^k;\mathbb{Z})$, in a way that the stabilizer of any point $x \in \mathbb{R}^n \times (S^m)^k$ is a p-torsion-free group, then $m+1$ is a multiple of the Yagita invariant $p(\Gamma)$.

The mapping class group Γ_g is never 2-periodic for $g > 1$, since the Krull dimension $\kappa(\Gamma_g, 2)$ is at least two. For an odd prime p and p-periodic Γ_g we have described $p(\Gamma_g)$ earlier. We recall that for an odd prime p and genus $g \not\equiv 1 \mod p$, Γ_g is always p-periodic; thus for the discussion of the Yagita invariant of Γ_g we will only need to be concerned with the case $g \equiv 1 \mod p$. In [**G-M-X2**] a complete result is given in case p is an odd regular prime, and partial results for general primes. Recall that a prime p is called *regular* if it does not divide the class number of the cyclotomic field $\mathbb{Q}(\exp(2\pi\sqrt{-1}/p))$. A famous criterion of Kummer states that a prime p is regular if and only if p does not divide the numerator of any Bernoulli number B_{2i} with $2 \le 2i < p-1$. Thus 691, the numerator of B_{12}, is an example of an *irregular* prime; the first three irregular primes are 37, 59 and 67. The following terminology was introduced in [**G-M-X2**].

Let p be a prime. We say that an integer g satisfies the (p)-condition if and only if g is of the form $lp^\alpha+1$ with l prime to p, $\alpha > 0$, and $2l = p(2h-2)+k(p-1)$ for some integers $h > 0$, $k \geq 0$ with $k \neq 1$.

We can now state the main results concerning the Yagita invariant of the mapping class group, as proved in [**G-M-X2**].

(Y-INVARIANT 1). Let p be an odd regular prime and assume that $g = lp^\alpha+1$ with l prime to p and $\alpha > 0$. Then the Yagita invariant $p(\Gamma_g)$ is determined as follows.

(i) If g does not satisfy the (p)-condition, then $p(\Gamma_g)$ equals $2(p-1)p^{\alpha-1}$.
(ii) If g satisfies the (p)-condition, then $p(\Gamma_g)$ equals $2(p-1)p^\alpha$.

For the case of a general odd prime, only a partial result is available, which however underlines the role of the (p)-condition.

(Y-INVARIANT 2). Let p be an odd prime and $g = lp^\alpha+1$ with l prime to p and $\alpha > 0$. Then the following holds.

(i) $p(\Gamma_g)$ has the form $2(p-1)p^\alpha$ or $2(p-1)p^{\alpha-1}$.
(ii) If g satisfies the (p)-condition, then $p(\Gamma_g) = 2(p-1)p^\alpha$.
(iii) If $1 < 2l < p-1$ then $p(\Gamma_g) = 2(p-1)p^{\alpha-1}$.

For the prime 2, the following result on the Yagita invariant is due to Xia [**Xi5**].

For even genus g the Yagita invariant $p(\Gamma_g)$ at the prime 2 equals 4.

We want to sketch the strategy involved in the proofs concerning the Yagita invariant of Γ_g. As explained, we can assume that p is an odd prime and $g = lp^\alpha+1$, $\alpha > 0$ and l prime to p.

FIRST STEP. Show that $p(\Gamma_g)$ is of the form $2(p-1)p^\beta$ with $\beta \geq \alpha - 1$.

For this, one observes that Γ_g contains a subgroup π isomorphic to $\mathbb{Z}/p^\alpha\mathbb{Z}$ suitable to provide a lower bound for $p(\Gamma_g)$. Indeed, the the covering transformation group in the cyclic p^α-sheeted unramified covering space

$$S_{lp^\alpha+1} \to S_{l+1},$$

which one can construct by mapping $\pi_1(S_{l+1})$ onto $\mathbb{Z}/p^\alpha\mathbb{Z}$, is suitable. All generators of that group $\pi \subset \Gamma_g$ have the same fixed-point data, and they are therefore conjugate in Γ_g. This implies that the restriction map

$$H^*(\Gamma_g;\mathbb{Z}) \to H^*(\pi;\mathbb{F}_p)$$

is zero in dimensions not divisible by $2(p-1)p^{\alpha-1}$, by looking at the action of $\mathrm{Aut}(\pi)$ on the cohomology of π. Therefore, $p(\Gamma_g)$ is a multiple of $2(p-1)p^{\alpha-1}$. On the other hand, as mentioned earlier, the Yagita invariant of any group of finite vcd is always a factor of $2(p-1)p^n$ for some $n \geq 0$, and the result follows.

SECOND STEP. Show that $p(\Gamma_g)$ divides $2(p-1)p^\alpha$.

It suffices to show that for every subgroup $\pi \subset \Gamma_g$ of order p one can find a number $j(\pi)$ prime to p such that the restriction map

$$H^{2j(\pi)p^\alpha}(\Gamma_g;\mathbb{Z}) \to H^{2j(\pi)p^\alpha}(\pi;\mathbb{Z})$$

is non-trivial. This can be achieved, for details consult [**G-M-X2**]. Besides of studying the restriction of various characteristic classes, one also makes use of the action of Γ_g on the complement of the union of the singular sets in Teichmüller space of the actions of all subgroups of Γ_g which are conjugate to π. This complement is a smooth non-compact manifold, which is used to construct a cohomology element in $H^*(\Gamma_g)$, whose restriction to the cohomology of π is non-zero.

THIRD STEP. Settle the case when g satisfies the p-condition.

The p-condition is just the condition needed to be able to construct a subgroup $\mathbb{Z}/p^{\alpha+1}\mathbb{Z}$ in Γ_g, with a fixed point data for a generator x to be of the form

$$\delta(x) = \langle g, p^{\alpha+1} \mid k_1/p, \dots, k_l/p \rangle,$$

a fixed point data which can be used in a manner similar to the argument in step one to show that $p(\Gamma_g)$ is a multiple of p^α. It follows then that $p(\Gamma_g)$ must be equal to $2(p-1)p^\alpha$.

FOURTH STEP. Show that if p is a regular prime and g does not satisfy the p-condition, then $p(\Gamma_g) < 2(p-1)p^\alpha$.

At this point we know that $p(\Gamma_g)$ is either equal to $2(p-1)p^\alpha$ or $2(p-1)p^{\alpha-1}$. It was proved in [**G-M-X2**] that for every subgroup π of order p in Γ_g, p a regular prime, one can find a symplectic characteristic class $d_i(\Gamma_g)$ which restricts non-trivially to $H^*(\pi;\mathbb{Z})$ and which satisfies $0 < i < p^\alpha$. This implies that $p(\Gamma_g) \neq 2(p-1)p^\alpha$ and thus $p(\Gamma_g) = 2(p-1)p^{\alpha-1}$. The regularity condition on p is used as follows. The case of $p = 3$ can be checked directly, so we can assume $p \geq 5$. To get the non-vanishing of the symplectic characteristic class one is led to consider the $(p-1)/2 \times (p-1)/2$ matrix

$$A = \begin{pmatrix} 1 & 1 & 1 & \dots & 1 \\ 2 & 4 & 6 & \dots & p-1 \\ 3 & 6 & 9 & \dots & \\ \vdots & \vdots & & & \\ \frac{p-1}{2} & p-1 & \dots & & \end{pmatrix}$$

with the (i,j) entry in the rows below the first row being the the smallest positive residue of the product ij mod p. This matrix, which is very closely related to the matrix considered by Carlitz and Olson in [**Ca-Ol**] turns out to have

$$|\det(A)| = 2p^{(p-5)/2} \cdot h_1, \quad p \geq 5,$$

where h_1 denotes a well understood divisor of the class number of the maximal real subfield of the cyclotomic field $\mathbb{Q}\big(\exp(2\pi\sqrt{-1}/p)\big)$. It is well-known that p is irregular if and only if p divides h_1, which is usually called the first factor of the class number of $\mathbb{Q}\big(\exp(2\pi\sqrt{-1}/p)\big)$. But the argument concerning the symplectic characteristic classes requires that

$$p^{(p-3)/2} \nmid \det A,$$

which is equivalent with the condition that p be regular.

References

[Ar] D. Arlettaz, *On the homology of the special linear group over a number field*, Comment. Math. Helv. **61** (1986), 556–564.

[Ba] R. Baer, *Isotopie von Kurven auf orientierbaren, geschlossenen Flächen und ihr Zusammenhang mit topologischen Deformationen der Flächen*, J. reine angew. Math. **159** (1928), 101–116.

[Be] D. J. Benson, *Specht modules and the cohomology of mapping class groups*, Mem. Amer. Math. Soc. **443** (1991), 29–92.

[Be-Co] D. Benson and F. R. Cohen, *The mod 2 cohomology of the mapping class group for a surface of genus 2*, Mem. Amer. Math. Soc. **443** (1991), 93–104.

[Bi] J. S. Birman, *Braids, Links and Mapping Class Groups*, Ann. of Math. Studies, vol. 82, Princeton University Press, 1974.

[Bi-Hi1] J. Birman and H. Hilden, *On mapping class groups of closed surfaces as covering spaces*, Ann. of Math. Studies **66** (1971), 81–115.

[Bi-Hi2] ——, *On isotopies of homeomorphisms of Riemann surfaces*, Ann. of Mathematics **97** (1973), no. 3, 424–439.

[B-L-M] J. Birman, A. Lubotzky and J. McCarthy, *Abelian and solvable subgroups of the mapping class group*, Duke Math. J. **50** (1983), no. 4, 1107–1120.

[Br] S. A. Broughton, *The equisymmetric stratification of the moduli space and the Krull dimension of mapping class groups*, Topology and its Applications **73** (1990), 101–113.

[Brow1] K. S. Brown, *Cohomology of Groups*, Graduate Texts in Math., vol. 87, Springer Verlag, 1982.

[Brow2] ——, *Euler characteristics of discrete groups and G-spaces*, Inventiones math. **27** (1974), 229–264.

[Brow3] ——, *Groups of virtually finite dimension*, London Math. Soc. Lecture Notes Series **36** (1977), 27–70.

[Brow4] ——, *Complete Euler characteristics and fixed point theory*, J. Pure Appl. Algebra **24** (1982), 103–121.

[Ca-Ol] L. Carlitz and F. R. Olson, *Maillet's determinant*, Proc. Amer. Math. Soc **6** (1955), 265–269.

[Ch-Co] R. M. Charney and F. R. Cohen, *A stable splitting for the mapping class group*, Michigan Math. J. **35** (1988), 269–284.

[Co1] F. R. Cohen, *Homology of mapping class groups for surfaces of low genus*, Contemp. Math. **58**, Part II (1987), 21–30.

[Co2] ——, *Mapping class groups and classical homotopy theory*, Contemp. Math. **150** (1993), 63–74.

[Co3] ——, *On the mapping class groups for punctured spheres, the hyperelliptic mapping class groups, $SO(3)$, and $Spin^c(3)$*, Amer. Jour. of Math. **115** (1993), 389–434.

[Co4] ——, *Artin's braid group and the homology of certain subgroups of the mapping class group*, Mem. Amer. Math. Soc. **443** (1991), 6–28.

[De] M. Dehn, *Die Gruppe der Abbildungsklassen*, Acta Math. **69** (1938), 135–206.

[Ea-Ee] C. J. Earle and J. Eells, *The diffeomorphism group of a compact Riemann surface*, Bull. Amer. Math. Soc. **73** (1967), 557–559.

[Eck-Mi1] B. Eckmann and G. Mislin, *Rational representations of finite groups and their Euler class*, Math. Ann. **245** (1979), 45–54.

[Eck-Mi2] ——, *Chern classes of group representations over a number field*, Math. Ann. **271** (1985), 349–358.

[Eck-Mi3] ——, *On the Euler class of representations of finite groups over real fields*, Comment. Math. Helv. **55** (1980), 319–329.

[Eck–Mi4] ——, *Profinite Chern classes for group representations*, London Math. Soc. Lecture Notes Series **86** (1983), 103–119.

[Eck–Mi5] ——, *Galois action on algebraic matrix groups, Chern classes, and the Euler class*, Math. Ann **271** (1985), 349–358.

[E-E] A. L. Edmonds and J. H. Ewing, *Surface symmetry and homology*, Math. Proc. Camb. Phil. Soc. **99** (1986), 73–77.

[Ew] J. Ewing, *Automorphisms of surfaces and class numbers: An illustration of the G-index theorem*, London Math. Soc. Lecture Notes Series **86** (1983), 120–127.

[Fa-Kr] H. Farkas and I. Kra, *Riemann surfaces*, Graduate Texts in Mathematics, vol. 71, Springer Verlag, 1982.

[Fa] F. T. Farrell, *An extension of Tate cohomology to a class of infinite groups*, J. Pure Appl. Algebra **10** (1977), 153–161.

[Fe] W. Fenchel, *Estensioni gruppi descontinui e transformazioni periodiche delle surficie*, Rend. Acc. Naz. Linzei (Se. fis., mat. e nat.) **5** (1948), 326–329.

[Gl-Mi1] H. Glover and G. Mislin, *On the stable cohomology of the mapping class group*, Springer Lecture Notes in Math. **1172** (1985), 80–84.

[Gl-Mi2] ——, *Torsion in the mapping class group and its cohomology*, J. Pure Appl. Algebra (1987), 177–189.

[G-M-X1] H. Glover, G. Mislin and Y. Xia, *On the Farrell cohomology of mapping class groups*, Invent. Math. **109** (1992), 535–545.

[G-M-X2] ——, *On the Yagita invariant of mapping class groups*, Topology (to appear).

[Gl-Sj] H. Glover and D. Sjerve, *The genus of $PSl_2(q)$*, J. reine angew. Math. **380** (1987), 59–86.

[Go] W. M. Goldman, *Representations of fundamental groups of surfaces*, Springer Lecture Notes in Math. **1167** (1980), 95–117.

[Gre] L. Greenberg, *Maximal groups and signatures*, Discontinuous Groups and Riemann Surfaces (L. Greenberg, ed.), vol. 79, Annals of Math. Studies, Princeton University Press, 1974, pp. 207–226.

[Gro] E. Grossman, *On the residual finiteness of certain mapping class groups*, J. London Math. Soc. **9** (1974), 160–164.

[Ha1] J. Harer, *The second homology group of the mapping class group of an orientable surface*, Invent. Math. **72** (1983), no. 2, 221–239.

[Ha2] ——, *Stability of the homology of the mapping class groups of orientable surfaces*, Ann. of Math. **121** (1985), 215–249.

[Ha3] ——, *The virtual cohomological dimension of the mapping class group of an orientable surface*, Invent. Math. **84** (1986), 157–176.

[Ha4] ——, *The third homology group of the moduli space of curves*, Duke Math. Journal **63** (1991), no. 1, 25–55.

[Ha5] ——, Lecture Notes in Math. **1337** (1985), Springer Verlag, 138–221.

[Ha-Za] J. Harer and D. Zagier, *The Euler characteristic of the moduli space of curves*, Invent. Math. **85** (1986), 457–485.

[Harv1] W. J. Harvey, *On cyclic groups of automorphisms of a compact Riemann surface*, Quart. J. Math. Oxford **17** (1966), no. 2, 86–97.

[Harv2] ——, *Branch loci in Teichmüller space*, Trans. Amer. Math. Soc. **153** (1971), 387–399.

[Ha-Th] A. Hatcher and W. Thurston, *A presentation for the mapping class group of a closed orientable surface*, Topology **19** (1980), 221–237.

[Hu] A. Hurwitz, *Über algebraische Gebilde mit eindeutigen Transformationen in sich*, Math. Ann. **41** (1893), 403–442.

[Ig] J.-I. Igusa, *Arithmetic variety of moduli for genus two*, Ann. of Math. **72** (1960), 612–649.

[Iv1] N. V. Ivanov, *Algebraic properties of the mapping class groups of surfaces*, Soviet Math. Dokl. **29** (1984), 288–291.

[Iv2] ——, *Algebraic properties of the mapping class groups of surfaces*, Preprint (1985), Steklov Mathematical Institute, Leningrad.

[Kau] M. Kaufmann, *Über die Cohomologie der Abbildungsklassengruppe*, Thesis (1988), Eidgenössische Technische Hochschule, Zürich (Switzerland).

[Ke] S. Kerckhoff, *The Nielsen realization problem*, Ann. of Math. **117** (1983), 235–265.

[Kr1] P. H. Kropholler, *On groups of type $(FP)_\infty$*, J. Pure Appl. Algebra (to appear).

[Kr2] ——, *Hierarchical decompositions, Tate cohomology, and groups of type $(FP)_\infty$*, Proc. Edin. Math. Soc. Geometric Group Theory Conference 1993 (to appear).

[Le-We] R. Lee and S. Weintraub, *The cohomology of $Sp(4,\mathbb{Z})$ and related groups*, Topology **24** (1985), 391–410.

[Li] W. B. R. Lickorish, *A finite set of generators for the homeotopy group of a 2-manifold*, Proc. Cam. Phil. Soc. **60** (1964), 769–778.

[Ma] W. Mangler, *Die Klassen von topologischen Abbildungen einer geschlossenen Fläche auf sich*, Math. Zeitsch. **44** (1939), 541–554.

[McC] J. McCarthy, *A Tits alternative for subgroups of surface mapping class groups*, Trans. Amer. Math. Soc. **291** (1985), 583–612.

[McCo] J. McCool, *Some finitely generated subgroups of the automorphism group of a free group*, J. of Algebra **35** (1975), 205–213.

[Mil] E. Y. Miller, *The homology of the mapping class group*, J. Differ. Geom **24** (1986), 1–14.

[Miln] J. Milnor, *On the existence of a connection with curvature zero*, Comment. Math. Helv. **58** (1983), 215–223.

[Mis1] G. Mislin, *On the characteristic ring of flat bundles defined over a number field*, Contemporary Mathematics **96** (1989), 291–296.

[Mis2] ——, *Tate cohomology for arbitrary groups via satellites*, Topology and its Applications (to appear).

[Mo] S. Morita, *Characteristic classes of surface bundles*, Invent. math. **90** (1987), 551–577.

[Ni1] J. Nielsen, *Untersuchungen zur Topologie der geschlossenen zweiseitigen Flächen*, Acta Mathematica **50** (1927), 189–358.

[Ni2] ——, *Die Struktur periodischer Transformationen von Flächen*, Danske Vid. Selsk. Mat.-Fys. Medd. **15** (1937), 1–77.

[Po] J. Powell, *Two theorems on the mapping class group of a surface*, Proc. Amer. Math. Soc. **68** (1978), no. 3, 347–350.

[Qu] D. Quillen, *The spectrum of an equivariant cohomology ring*: I, II, Annals of Math. **94** (1971), 549–572, 573–602.

[Roy] H. L. Royden, *Isometries of Teichmüller space*, Annals of Math. Studies **66** (1971), 369–383.

[Sm] S. Smale, *Diffeomorphisms of the 2-sphere*, Proc. Amer. Math. Soc. **10** (1959), 621–626.

[Su] D. Sullivan, *La classe d'Euler réelle d'un fibré vectoriel à groupe structural $Sl_n(\mathbb{Z})$ est nulle*, C. R. Acad. Sci. Ser. A **281** (1975), 17–18.

[Sy] P. Symonds, *The cohomology representation of an action of C_p on a surface*, Trans. Amer. Math. Soc. **306** (1988), no. 2, 389–400.

[Th1] C. B. Thomas, *Characteristic classes and the cohomology of finite groups*, Cambridge University Press, 1986.

[Th2] ——, *Free actions by p-groups on products of spheres and Yagita's invariant $po(G)$*, Lecture Notes in Math. **1375** (1989), Springer, 326–338.

[Ve] B. B. Venkov, *On homologies of groups of units in division algebras*, Trudy Mat. Inst. Steklov **80** (1965), 66–89; English transl. in Proc. Steklov Inst. Math. **80** (1965), 73–100.

[Wa] B. Wajnryb, *A simple presentation for the mapping class group of an orientable surface*, Israel Journal of Math **45** (1983), 157–174.

[Wi] A. Wiman, *Ueber die hyperelliptischen Curven und diejenigen vom Geschlecht $p = 3$, welche eindeutige Transformationen in sich zulassen*, Bihang Kongl. Svenska Vetenskaps Akademiens Handlingar (1895–6), Stockholm.

[Xi1] Y. Xia, *The p-period of an infinite group*, Publications Matemàtiques **36** (1992), 241–250.

[Xi2] ——, *The p-periodicity of the mapping class group and the estimate of its p-period*, Proc. Amer. Math. Soc. **116** (1992), 1161–1169.

[Xi3] ——, *The p-torsion of Farrell-Tate cohomology of the mapping class group $\Gamma_{(p-1)/2}$*, Topology'90 (B. Apanasov, W. D. Neumann, A. W. Reid and L. Siebenmann, eds.), Walter de Gruyter, 1992, pp. 391–398.

[Xi4] ——, *The p-torsion of the Farrell-Tate cohomology of the mapping class group Γ_{p-1}*, J. Pure Appl. Algebra **78** (1992), 319–334.

[Xi5] ——, *An obstruction in the mapping class group*, Bulletin of the London Math. Soc. (to appear).

[Ya] N. Yagita, *On the dimension of spheres whose product admits a free action by a nonabelian group*, Quart. J. Math. Oxford **36** (1985), 117–127.

[Zi] H. Zieschang, *Finite groups of mapping classes of surfaces*, Lecture Notes in Math., vol. 875, Springer Verlag, 1981.

Mathematik, ETH Zürich, 8092 Zurich, Switzerland and Department of Mathematics, Ohio State University, Columbus, Ohio 43210, USA

E-mail address: mislin@math.ethz.ch

Centre de Recherches Mathématiques
CRM Proceedings and Lecture Notes
Volume 6, 1994

$H^*(\mathrm{RP}^\infty \times \cdots \times \mathrm{RP}^\infty)$ as a Module Over the Steenrod Algebra

V. Giambalvo, Nguyên H. V. Hung, and F. P. Peterson

Introduction and Background

Let $P^k = H^*(\mathrm{RP}^\infty \times \cdots \times \mathrm{RP}^\infty; \mathbb{Z}/2)$, k-times, i.e. $P^k = \mathbb{Z}/2[x_1, \ldots, x_k]$, where $|x_i| = 1$. P^k has been studied for many years as a module over the mod 2 Steenrod algebra $\mathcal{A}$. The authors have been interested in three aspects of this structure and these will be described in detail in the following three sections.

PROBLEM 1. What is a minimal set of generators for P^k as a module over $\mathcal{A}$? That is, describe $\mathbb{Z}/2 \otimes_{\mathcal{A}} P^k$.

PROBLEM 2. The general linear group $G = \mathrm{GL}(k, \mathbb{Z}/2)$ acts on P^k, and the ring of invariants $D_k = (P^k)^G$ was described by Dickson [**3**]. What is a minimal set of generators for D_k as a module over $\mathcal{A}$? That is, describe $\mathbb{Z}/2 \otimes_{\mathcal{A}} D_k$. The results on this problem are proved in Hung-Peterson [**13**].

PROBLEM 3. Let X be a space. Then an element of $\mathcal{A}$ acts as an endomorphism on $H^*(X; \mathbb{Z}/2)$. Let $\Phi_X : \mathcal{A} \to \mathrm{End}\, H^*(X)$ be the algebra homomorphism induced by this. What is $\mathrm{Ker}\, \Phi_X$ when $X = \mathrm{RP}^\infty$ or $X = \mathrm{RP}^\infty \times \cdots \times \mathrm{RP}^\infty$, k-times? The results on this problem are proved in Giambalvo-Peterson [**4**].

We first give a brief description of the Steenrod algebra and how it acts on P^k. The Steenrod algebra $\mathcal{A}$ is defined as the free associative algebra generated by the Steenrod operations $\{Sq^i; i > 0\}$ modulo the so-called Adem relations that are true for every element of $H^*(X)$ for all spaces X. Here

$$Sq^i : H^q(X) = H^q(X; \mathbb{Z}/2) \to H^{q+i}(X)$$

are natural homomorphisms defined for every space satisfying the following properties:

(a) $Sq^0 =$ identity
(b) $Sq^i(x) = 0$ if $i > |x|$
(c) $Sq^i(x) = x^2$ if $i = |x|$
(d) $Sq^i(x \cdot y) = \sum_{j=0}^{i} Sq^j(x) \cdot Sq^{i-j}(y)$.

One can easily see that these properties determine how Sq^i acts on P^k. One easy formula that we will need is:

(e) $Sq^i(x^q) = \binom{q}{i} x^{i+q}$ if $|x| = 1$.

1991 *Mathematics Subject Classification.* Primary 55S10; Secondary 08A35.

Key words and phrases. Steenrod Algebra, Dickson Invariants.

The research of the second author was supported in part by NSF Grant DMS-8505550 and a JSPS fellowship.

It is an exposition of two papers which will be published elsewhere.

1065-8580/94 $1.00 + $.25 per page

We note that P^k is an algebra over $\mathcal{A}$, but we will be studying P^k only as a module over $\mathcal{A}$.

Next we describe the Dickson algebra, D_k. Dickson [3] showed that

$$D_k \cong \mathbb{Z}/2[Q_{k-1}, \dots, Q_0],$$

where $|Q_s| = 2^k - 2^s$. Note that Q_s depends on k, and when necessary, will be denoted $Q_{k,s}$. An inductive definition of $Q_{k,s}$ is given by

$$Q_{k,s} = (Q_{k-1,s-1})^2 + V_k.Q_{k-1,s},$$

where, by convention, $Q_{k-1,k-1} = 1$, $Q_{k-1,s} = 0$ for $s < 0$ and

$$V_k = \prod_{\lambda_i \in \mathbb{Z}/2} (\lambda_1 x_1 + \cdots + \lambda_{k-1} x_{k-1} + x_k).$$

Since the operations of $\mathrm{GL}(k, \mathbb{Z}/2)$ on P^k commute with the action of the Steenrod algebra, D_k is also a module (in fact an algebra) over $\mathcal{A}$. The importance of D_k to topologists was first shown by Madsen [10] who calculated the dual to the Dyer-Lashof operations of length k, and by Mùi [5] who related this to D_k. Furthermore, the cohomology operations derived from the invariants in D_k are exactly the Steenrod-Milnor ones of length k (see Madsen-Milgram [11], Mùi [6]).

1. $\mathbb{Z}/2 \otimes_{\mathcal{A}} \mathbf{P^k}$

The motivation for this problem is provided by R. Wood [21] who shows its relationship to modular representation theory, by W. Singer [17] who shows its relationship to the E^2-term of the Adams spectral sequence converging to the stable homotopy groups of spheres, and by S. Priddy [15] and others who have studied the stable homotopy type of BG, where G is a finite group.

There are two approaches to this problem, the quantitative and the qualitative. We first describe the quantitative.

Let $k = 1$. Then $P^1 = \mathbb{Z}/2[x]$. Using (e) above, it is easy to check that $\mathbb{Z}/2 \otimes_{\mathcal{A}} P^1 = \{x^{2^r - 1}\}, r \geq 0$. This was known over forty years ago.

Let $k = 2$. This case is not very difficult, and the answer is

$$\mathbb{Z}/2 \otimes_{\mathcal{A}} P^2 = \{x_1^{2^r-1}, x_2^{2^r-1}, x_1^{2^{r_1}-1} x_2^{2^{r_2}-1}, x_1^{2^t-2^s-1} x_2^{2^s}\},$$

with $r \geq 0, r_1 \geq 1, r_2 \geq 1, t > s \geq 1$.

Let $k = 3$. This case is very complicated and difficult and was solved by M. Kameko [7].

It seems unlikely that a very explicit description of $\mathbb{Z}/2 \otimes_{\mathcal{A}} P^k$ for general k will be appear in the near future.

The second approach is the qualitative. By this we mean giving conditions on elements of P^k to show that they go to zero in $\mathbb{Z}/2 \otimes_{\mathcal{A}} P^k$, i.e. belong to $\bar{\mathcal{A}} P^k$.

The third author made a conjecture which was proved by R. Wood [21]. The key formula used in the proof is

$$\chi(a)(x) \cdot y + x \cdot a(y) \in \bar{\mathcal{A}} P^k$$

for $x, y \in P^k$ and $a \in \mathcal{A}$. Here χ denotes the canonical anti-automorphism in $\mathcal{A}$.

THEOREM 1.1. *$\mathbb{Z}/2 \otimes_{\mathcal{A}} P^k = 0$ in dimension d such that $\alpha(d + k) > k$. (Here $\alpha(n)$ = the number of ones in the dyadic expansion of n.)*

Let $\theta\colon P^k \to (\overline{H}^*(\mathrm{BO}))^k/(\overline{H}^*(\mathrm{BO}))^{k+1}$ be the $\mathcal{A}$-linear epimorphism, of degree k, given by $\theta(x_i^j) = w_{j+1}, j \geq 0$. Using θ, one proves the following (unlikely sounding) corollary [**14**].

COROLLARY 1.2. *Let M^d be a closed C^∞-manifold of dimension d, with $\alpha(d) > k$. Assume all $(k+1)$-fold products of Stiefel-Whitney classes of M^d vanish. Then M^d is cobordant to zero.*

Recently, W. Singer [**18**] and J. Silverman and W. Singer [**16**] have refined the methods of R. Wood to show that many more monomials in P^k are in $\bar{A}P^k$.

2. $\mathbb{Z}/2\otimes_{\mathcal{A}} \mathbf{D_k}$

The motivation for this problem is provided by the fact that a calculation of $\mathbb{Z}/2\otimes_{\mathcal{A}} D_k$ for all k also calculates the $\mathcal{A}$-annihilated elements of $PH_*(Q_0S^0)$ which form the bottom line of the E^2-term of the unstable Adams spectral sequence which converges to $\pi_*(Q_0S^0)(=\pi_*^S(S^0))$ (this has been studied by Wellington [**19**] and others). This problem is also related to the study of the Hurewicz map $\pi_*(Q_0S^0)\to H_*(Q_0S^0;\mathbb{Z}/2)$ (see Lannes-Zarati [**8**], [**9**]).

We will use the quantitative approach and try to get explicit answers.

Our computation is based on the explicit formula for the action of $\mathcal{A}$ on D_k given by N. Hai and N. Hung (see [**12**]):

$$Sq^i(Q_{k,s}) = \begin{cases} Q_{k,r} & i = 2^s - 2^r,\ r\leq s,\\ Q_{k,t}Q_{k,r} & i = 2^k-2^t+2^s-2^r,\ r\leq s<t,\\ Q_{k,s}^2 & i = 2^k-2^s,\\ 0 & \text{otherwise.}\end{cases}$$

Let $k=1$. Then $\mathrm{GL}(1,\mathbb{Z}/2)=\{I\}$, and $D_1 = P^1$, and the answer is given in §1.

Let $k=2$. Then $D_2 \cong H^*(\mathrm{B\,SO}(3)) = \mathbb{Z}/2[w_2,w_3]$, and it is not very difficult to show that

$$\mathbb{Z}/2\otimes_{\mathcal{A}} D_2 = \{w_2^{2^r-1}\} = \{Q_{2,1}^{2^r-1}\}, r\geq 0.$$

We wish to generalize these results.

Let u be a non-negative integer. Define $s(u)=s_1(u)$ to be the non-negative integer with $2^{s(u)}$ being the first missing 2 power in the dyadic expansion of u. Define $s_2(u)$ to be the positive integer with $2^{s_2(u)}$ being the second missing 2 power in the dyadic expansion of u. Thus $u = 2^{s_1(u)}-1+2^{s_1(u)+1}+\cdots+2^{s_2(u)-1}+2^{s_2(u)+1}\cdot t$, where t is a non-negative integer.

DEFINITION 2.1. $u\rhd v$ if $v<2^{s_1(u)}$. Here u,v are non-negative integers.

DEFINITION 2.2. $u \rhd\!\rhd v$ if $2^{s_1(u)}\leq v < 2^{s_2(u)}$ and $s_1(u)=s_1(v)=s_2(v)-1$.

Let $I=(i_{k-1},i_{k-2},\cdots,i_0)$ be a sequence of k non-negative integers, with corresponding monomial $Q^I = Q_{k-1}^{i_{k-1}}\cdots Q_0^{i_0}\in D_k$. Let $h(I)=\sum_{j=0}^{k-1} i_j$, the classical degree of Q^I.

DEFINITION 2.3. (a) I is strongly allowed if $h(I)\rhd i_{k-1}\rhd i_{k-2}\rhd\cdots\rhd i_0$.

(b) I is weakly allowed if one or more $\rhd$ in the above definition is replaced by $\rhd\!\rhd$.

(c) I is allowed if it is either strongly or weakly allowed.

(d) Q^I is (strongly, weakly) allowed if I is.

Let us consider small values of k. If $k=1$, then it is easy to check that $i_0 \rhd i_0$ if and only if $i_0 = 2^s - 1$, so the strongly allowed monomials form a basis for $\mathbb{Z}/2 \otimes_{\mathcal{A}} P^1$.

If $k=2$, then it is easy to check that $i_1 + i_0 \rhd i_1 \rhd i_0$ if and only if $i_1 = 2^s - 1$ and $i_0 = 0$. Thus the strongly allowed monomials form a basis of $\mathbb{Z}/2 \otimes_{\mathcal{A}} D_2$.

Our first theorem is the following one.

THEOREM 2.4. *Let $k=3$. Then the set of strongly allowed monomials forms a basis for $\mathbb{Z}/2 \otimes_{\mathcal{A}} D_3$.*

We note which I are strongly allowed for $k=3$

$$\begin{array}{ll} (2^s-1,0,0), & s \geq 0 \\ (2^r-2^s-1, 2^s-1, 1), & r > s > 0. \end{array}$$

This pattern does not continue, as the set of strongly allowed monomials does not span D_k for $k > 3$. It should also be noted that there is no weakly allowed sequence for $k \leq 3$.

Our second theorem handles the case $k=4$.

THEOREM 2.5. *Let $k=4$. Then the set of allowed monomials is a basis for $\mathbb{Z}/2 \otimes_{\mathcal{A}} D_4$.*

We show that the only allowed sequences for $k=4$ are the following ones:

$$\begin{array}{ll} (2^s-1,0,0,0), & s \geq 0 \\ (2^r-2^s-1,\ 2^s-1,1,0), & r > s > 0 \\ (2^t-2^r-1, 2^r-2^s-1, 2^s-1, 2), & t > r > s > 1 \\ (2^r-2^{s+1}-2^s-1, 2^s-1, 2^s-1, 2), & r > s+1 > 2. \end{array}$$

Note that the last sequence with $r > s+2$ is the only weakly allowed one. It has

$$h(I) \rhd\!\rhd i_3 \rhd i_2 \rhd i_1 \rhd i_0.$$

From the definition of strongly allowed, if I is not strongly allowed, then it is easy to find a sequence K and an integer a so that $Sq^a(Q^K) = Q^I +$ other terms. For example, if $i_m \not\rhd i_{m-1}$, $s = s(i_m)$, then $Sq^{2^{s+m-1}} Q^{(\cdots, i_m+2^s, i_{m-1}-2^s, \cdots)} = Q^I +$ other terms. Also, if $h(I) \not\rhd i_{k-1}$, $s = s(h)$, then $Sq^{2^{s+k-1}} Q^{(i_{k-1}-2^s, i_{k-2}, \cdots)} = Q^I +$ other terms.

The trouble comes from the fact that a given pair (K,a) might be chosen by 2 or more different I's. The pair (K,a) is not uniquely determined by I and the proof shows that there are enough (K,a) for all of the I when $k=3$. When $k=4$, there are not enough pairs (K,a) and we need to add the weakly allowed generators.

More precisely, let $I = (i_{k-1}, \dots, i_0)$, define

$$m(I) = \min\Big\{ s\big(h(I)\big), s(i_{k-2} + \cdots + i_0 - 1), \dots, s(i_1 + i_0 - k + 2) \Big\}.$$

The idea of the proof for $k=3$ and 4 is to prove the following result. Let I be not allowed for $k=3$ or 4. Then we find a pair (K,a) such that $Sq^a(Q^K) \equiv Q^I + \sum Q^J$ mod $Im\overline{\mathcal{A}}$, where either $m(J) < m(I)$ or $m(J) = m(I)$ and $h(J) < h(I)$. These arguments show that the set of allowed sequences for $k=3$ or 4 forms a spanning set for $\mathbb{Z}/2 \otimes_{\mathcal{A}} D_k$.

One step in the proof of linear independence is to prove the following proposition.

PROPOSITION 2.6. *Let $k = 3$ or 4. Suppose given a strongly allowed I and*

$$Sq^a Q^K = Q^I + \text{ other terms.}$$

Then $a = 0$ and $K = I$.

We note that this is not true for a weakly allowed I.

3. The kernel of Φ_X

The motivation for this problem is wholly algebraic and the fact that the answer is interesting even for $X = \mathrm{RP}^\infty$. This problem was suggested by D. Carlisle and R. Wood [**2**]. Note that $\operatorname{Ker}\Phi_X$ is the ideal in $\mathcal{A}$ of cohomolgy operations which vanish on every cohomology class of X.

We begin this section with a result that has little to do with RP^∞, except perhaps through classifying spaces.

THEOREM 3.1. *Let $K(\mathbb{Z}/2, n)$ denote the Eilenberg-Maclane space. If $n > 1$, then $\mathcal{K}(K(\mathbb{Z}/2, n)) = 0$.*

In order to understand $\mathcal{K}(\mathrm{RP}^\infty \times \cdots \times \mathrm{RP}^\infty)$, we use the following product theorem, which is true for any spaces X and Y.

THEOREM 3.2. $\mathcal{K}(X \times Y) = \{a \in \mathcal{A} \mid \psi(a) \in \mathcal{K}(X) \otimes \mathcal{A} + \mathcal{A} \otimes \mathcal{K}(Y)\}$.

We finally turn to $\mathcal{K}(\mathrm{RP}^\infty)$, which we will denote by $\mathcal{K}$. This is a two-sided ideal in $\mathcal{A}$. Since $\binom{n}{2^r}$ is non-zero mod 2 if and only if 2^r appears in the binary expansion of n, it follows from (e) above, that $Sq^{2^r}Sq^{2^r} \in \mathcal{K}$, for $r \geq 0$. Let I_0 be the two sided ideal in $\mathcal{A}$ generated by $\{Sq^{2^r}Sq^{2^r}\}$. It is not hard to determine that in low dimensions $\mathcal{K}$ and I_0 are the same, say up to dimension 8. In particular, the element of least dimension is Sq^2Sq^2, which in the Milnor basis is written as $Sq(1,1)$. Now theorem 3.2 implies that the element of smallest dimension in $\mathcal{K}(P^s)$ is $Sq(1,1,\ldots,1)$, with $s+1$ ones. This gives a rapidly increasing filtration of $\mathcal{A}$ by two sided ideals, which may be useful in verifying identities.

The next theorem says that I_0 is not all of $\mathcal{K}$, and describes the rest of a set of ideal generators.

THEOREM 3.3. *There exists elements $c_{r,m} \in \mathcal{A}$ in dimensions $2^m(2^r+2^{r-1}+5)$, $m \geq 0$, $r \geq 4$, such that $\mathcal{K}$ is the two sided ideal generated by $Sq^{2^r}Sq^{2^r}$, $r \geq 1$, and $c_{r,m}$.*

The elements $c_{r,m}$ are rather mysterious. The lowest dimensional one is $c_{4,0}$ in dimension 29, and it was discovered by computer calculations. $c_{r,m+1}$ is the double of $c_{r,m}$, so the $c_{r,0}$ are the important elements to study. Some of the mystery may be dispelled by explicit formulae, which will be given after a discussion of some of the tools needed to prove this.

A main technical tool is a new additive basis for the Steenrod algebra discovered by D. Arnon [**1**]. This basis consists of monomials in the indecomposable elements in the Steenrod algebra, i.e., Sq^{2^i}, $i \geq 0$. For any two integers i, j, $j \geq i$ we define a "string" s_i^j to be the product $Sq^{2^j}Sq^{2^{j-1}}\ldots Sq^{2^i}$. The basis elements consist of all finite products $s_{i_1}^{j_1}s_{i_2}^{j_2}\ldots s_{i_q}^{j_q}$, where $j_s \leq j_{s+1}$, and if $j_s = j_{s+1}$, then $i_s < i_{s+1}$. We call the sequence of j's the "tops" of the basis element, and the sequence of i's the "bottoms". We also write s_i for $s_i^i = Sq^{2^i}$.

This basis has some interesting properties. Let $\mathcal{A}_r$ be the subalgebra of $\mathcal{A}$ generated by $\{Sq^{2^i}, i \leq r\}$. Then a monomial in the basis is in $\mathcal{A}_r$ if and only if it contains no Sq^{2^k}, with $k > r$. The highest dimensional element in $\mathcal{A}_r$ can be written as the product of all the basis elements in $\mathcal{A}_r$ in order, i.e. $s_0^0 s_0^1 s_1^1 \dots s_{r-1}^r s_r^r$.

The same property of the binomial coefficients that gives $Sq^{2^k} Sq^{2^k} \in \mathcal{K}$ gives that the product $Sq^{2^k} P Sq^{2^k}$, where P is a product of Sq^{2^j} with $j > k$, is also in $\mathcal{K}$ (It is also in I_0). This tell us that any Arnon basis element with non-increasing bottoms must be in $\mathcal{K}$. The converse is not true, in fact not even I_0 is spanned by basis elements with non-increasing bottoms.

Since $I_0 \subset \mathcal{K}$ we can look at the induced map from $\mathcal{A}/I_0$ to End P^1, but $\mathcal{A}/I_0$ is still too complicated to work with effectively. But if we add to it the endomorphism which is multiplication by the generator $x \in H^1(\mathrm{RP}^\infty)$ we get an algebra that is very nice to work with, as expressed in the following result.

THEOREM 3.4. *There is an algebra S such that the map $\mathcal{A} \to \mathcal{E}$ factors*

$$\mathcal{A} \to \mathcal{A}/\mathcal{I}_0 \to S \to \mathcal{E}$$

and so that the last map in the sequence is a monomorphism.

Construct the algebra S as a quotient of the free $Z\mathbb{Z}/2$ algebra with generators s_i, $i \geq 0$, and t, $|s_i| = 2^i$, $|t| = 1$, with the relations:

(1) $s_i^2 = 0$.
(2) $s_i s_{i+1} s_i = 0$
(3) $s_i s_j + s_j s_i + s_{i-1} s_j s_{i-1} = 0$, $j < i - 1$.
(4) $[s_j, s_{j-1}] = t^{2^j} s_{j-1}$.
(5) $[s_j, t] = t^2 s_0 s_1 \dots s_{j-1}$, $j > 0$.
(6) $[s_0, t] = t^2$.

Some simple computations from these give the more familiar relations:

(7) $[t^{2^j}, s_j] = t^{2^{j+1}}$.
(8) $[t^{2^j}, s_k] = 0$, $k < j$.

The map $S \to \mathcal{E}$ just sends s_i to the endomorphism given by Sq^{2^i} and t to the map which is multiplication by $x \in H^1(\mathrm{RP}^\infty)$.

We note for the record formulae for $c_{4,0}$. In the Adem basis it is

$$Sq^{25}Sq^4 + Sq^{22}Sq^7 + Sq^{21}Sq^8 + Sq^{24}Sq^4Sq^1 + Sq^{20}Sq^6Sq^3,$$

while in the Milnor basis it is $Sq(8,2,0,1) + Sq(2,9)$, and in the Arnon basis it is

$$Sq^2Sq^1Sq^2\left[Sq^{16}, Sq^8\right] + Sq^2Sq^1Sq^4Sq^2Sq^8Sq^4Sq^8 + Sq^4Sq^2Sq^1Sq^8Sq^4Sq^2Sq^8.$$

In the Milnor basis $c_{r,0} = Sq(2^{r-1}, 2, 0, \dots, 0, 1) + Sq(2, 2^{r-1} + 1)$, where the 1 is in the $(r+1)^{st}$ place. We also have a formula for $c_{r,0}$ in the Arnon basis, but not in the Adem basis.

Now that you have seen some explicit formulas, let's take another look at where these might come from. The key is not in the standard (Milnor or Adem) bases, but in the Arnon basis. Looking at the above formula for $c_{4,0}$ one sees the commutator of Sq^{16} and Sq^8. In the Steenrod algebra there is no relation involving this commutator, but when acting in $H^*(\mathrm{RP}^\infty)$ we get one. Since $H^*(\mathrm{RP}^\infty)$ is a polynomial algebra on one generator an element is determined by its exponent, and we can regard the $\mathcal{A}$-action as an action on the positive integers. Since the action of each element of $\mathcal{A}$ is eventually periodic, we can extend backwards to an action on the integers. Note that every element of $\mathcal{A}$ has period a power of 2. The period of Sq^{2^r} is 2^{r+1} (from property (5) above). With a small amount of calculation, one can verify that every two power periodic function can be realized as a Steenrod operation. For example the four functions of period 2, which we can describe by their coefficients on x and x^2 by $\{(0,0),(0,1),(1,0),(1,1)\}$ can be realized by the Steenrod operations $\{0,[Sq^2,Sq^1],Sq^1,Sq^0\}$ respectively. Note that these have varying degrees in $\mathcal{A}$. In order to bring back the degree, we need the periodic function of period one, which is just multiplication by x, the generator in $H^*(\mathrm{RP}^\infty)$. This corresponds to the element t in the algebra S in theorem 3.4. One can say a little more. Any $\alpha \in \mathcal{A}_r$ has period at most 2^{r+1}, and any function with period 2^{r+1} can be realized by an element in $\mathcal{A}_r$. Now the commutator $[Sq^{2^r}, Sq^{2^{r-1}}]$ has period 2^r, not 2^{r+1}, so it can be realized by an element in $\mathcal{A}_{r-1}$. In fact this corresponds to relation (4) in theorem 3.4, and it is $t^{2^r} Sq^{2^{r-1}}$. This is not a relation in $\mathcal{A}$ however, and to get one we need to eliminate the power of t. In order to do this, and keep the dimension as small as possible, multiply both sides by $Sq^2Sq^1Sq^2$ and then reduce the right hand side using the relations in theorem 3.4. This gives the element $c_{r,0}$.

References

1. D. Arnon, *Monomial Bases in the Steenrod Algebra*, (to appear).
2. D. Carlisle and R. Wood, *Facts and Fancies about Relations in the Steenrod Algebra*, (to appear).
3. L. E. Dickson, *A Fundamental System of Invariants of the General Modular Linear Group with a Solution of the Form Problem*, Trans. Amer. Math. Soc. **12** (1911), 75–98.
4. V. Giambalvo and F. P. Peterson, *The Annihilator Ideal of the Action of the Steenrod Algebra on $H^*(\mathrm{RP}^\infty)$*, (to appear).
5. Huỳnh Mùi, *Modular Invariant Theory and Cohomology Algebras of Symmetric Groups*, Jour. Fac. Sci. Univ. Tokyo **22** (1975), 310–369.
6. Huỳnh Mùi, *Dickson invariants and Milnor basis of the Steenrod algebra*, Eger Internat. Colloq. Topology (1983), 345–355.
7. M. Kameko, *Products of Projective Spaces as Steenrod Modules*, Ph.D Thesis, The Johns Hopkins University.
8. J. Lannes and S. Zarati, *Invariants de Hopf d'ordre supérieur et suite spectrale d'Adams*, C. R. Acad. Sci. **296** (1983), 695–698.
9. J. Lannes and S. Zarati, *Sur les foncteurs dérivés de la déstabilisation*, Math. Z. **194** (1987), 25–59.
10. I. Madsen, *On the Action of the Dyer-Lashof Algebra in $H_*(G)$*, Pacific Journal of Math. **60** (1975), 235–275.
11. I. Madsen and J. Milgram, *The classifying spaces for surgery and cobordism of manifolds*, Ann. of Math. Studies No. **92**, Princeton Univ. Press, 1979.
12. Nguyên H. V. Hung, *The action of the Steenrod squares on the modular invariants of linear groups*, Proc. Amer. Math. Soc. **113** (1991), 1097–1104.
13. Nguyên H. V. Hung and F. P. Peterson, *$\mathcal{A}$-Generators for the Dickson Algebra*, (to appear).
14. F. P. Peterson, *$\mathcal{A}$-Generators for Certain Polynomial Algebras*, Math. Proc. Camb. Phil. Soc. **105** (1989), 311–312.
15. S. Priddy, *On Characterizing Summands in the Classifying Space of a Group,* I, Amer. Jour. Math. **112** (1990), 737–748.
16. J. H. Silverman and W. Singer, *On the Action of Steenrod Squares on Polynomial Algebras,* II, (to appear).
17. W. Singer, *The Transfer in Homological Algebra*, Math. Zeit. **202** (1989), 493–523.
18. W. Singer, *On the Action of Steenrod Squares on Polynomial Algebras*, Proc. Amer. Math. Soc. **111** (1991), 577–583
19. R. Wellington, *The Unstable Adams Spectral Sequence for Free Iterated Loop Spaces*, Memoirs A.M.S. 258, 1982.
20. R. M. W. Wood, *Steenrod Squares of Polynomials and the Peterson Conjecture*, Math. Proc. Camb. Phil Soc. **105** (1984), 307–309

21. R. M. W. Wood, *Modular Representations of* $\mathrm{GL}(n, F_p)$ *and Homotopy Theory*, Lecture Notes in Math. 1172, Springer Verlag, 1985.

Department of Mathematics, University of Connecticut, Storrs, CT 06268

E-mail address: vince@math.uconn.edu

Department of Mathematics, University of Hanoi, 90 Nguyen Trai Str., Hanoi 10000, Vietnam

Current address: Centre de Recerca Mathematica, Institut d'Estudes Catalans, Apartat 50, E08193 Bellaterra, Spain

Department of Mathematics, Massachusetts Institute of Technology, Cambridge, MA 01239

E-mail address: fpp@math.mit.edu

Centre de Recherches Mathématiques
CRM Proceedings and Lecture Notes
Volume 6, 1994

Computing Homotopy Classes of Phantom Maps

Joseph Roitberg

ABSTRACT. We make an excursion into the slums of algebraic topology to survey an emerging branch of homotopy theory centered around the notion of phantom map. Emphasis is placed on illustrating techniques for computing the set $\mathrm{Ph}(X,Y)$ of homotopy classes of phantom maps from a space X to a space Y, particularly when $\mathrm{Ph}(X,Y)$ possesses a natural group structure, and also for computing the maps $f^*\colon \mathrm{Ph}(X',Y) \to \mathrm{Ph}(X,Y)$ and $g_*\colon \mathrm{Ph}(X,Y) \to \mathrm{Ph}(X,Y')$ induced by maps $f\colon X \to X'$ and $g\colon Y \to Y'$.

1. Introduction

As originally conceived, the notion of phantom map was both a homotopy notion and a CW-complex notion. In the first published paper on phantom maps [**3**], Adams and Walker responded to a question raised by Olum by finding an (essential) map $f\colon X \to Y$, X and Y CW-complexes, with the property that $f \mid X_n$, the restriction of f to the n-skeleton of X, is inessential for all $n \geq 1$. In order to define the notion of phantom map in a homotopy-invariant fashion, we may proceed as follows:

DEFINITION 1.1. Let S be a class of pointed topological spaces. A (pointed) map $\phi\colon X \to Y$ is said to be an S-phantom map provided that, for any $W \in S$ and any $j\colon W \to X$, the composition

$$W \xrightarrow{j} X \xrightarrow{\phi} Y$$

is inessential. (We count the constant map to the basepoint as an S-phantom map.)

The two most important examples of S are $S = \mathcal{FD}$, the class of all spaces supporting the structure of a finite-dimensional CW-complex (or even of all spaces homotopy equivalent to such spaces) and $S = \mathcal{F}$, the class of all spaces supporting the structure of a finite CW-complex (or even of all spaces homotopy equivalent to such spaces). We note immediately that the notions of $\mathcal{FD}$-phantom map and $\mathcal{F}$-phantom map are, in general, different.

1991 *Mathematics Subject Classification.* Primary: 55Q05, 55P60; Secondary: 20K25, 20K40.

Key words and phrases. Phantom map, inverse system of groups, finite type domain, finite type target, skeleton-finite CW-complex, finite Postnikov space, localization, completion, rational homotopy equivalence, n-type, weak identity.

Research supported (in part) by a grant from the City University of New York PSC-CUNY Research Award Program.

This is the final version of the paper.

1065-8580/94 $1.00 + $.25 per page

EXAMPLE 1.1 (SEE ALSO [**26**, footnote 6], [**43**, Ex. 4.3]). Let $X = K(\mathbb{Q}, 3)$, $Y = K(\mathbb{Z}, 4)$ or S^4. Then any non-0 element of $[X, Y]$,[1] the set of all (pointed) homotopy classes of maps from X to Y, is (the homotopy class of) a map $f \colon X \to Y$ which is an $\mathcal{F}$-phantom map but not an $\mathcal{FD}$-phantom map.

To see that $f \colon X \to Y$ is an $\mathcal{F}$-phantom map, note that any map $W \to K(\mathbb{Q}, 3)$, with W a finite CW-complex, factors as

$$W \to S^3 \to K(\mathbb{Q}, 3);$$

on the other hand, $f \colon X \to Y$ is not an $\mathcal{FD}$-phantom map since $K(\mathbb{Q}, 3)$ itself has the structure of a finite-dimensional CW-complex.

The following easy proposition shows, nevertheless, that for a large and natural class of domains, the notions of $\mathcal{FD}$-phantom map and $\mathcal{F}$-phantom map coincide.

PROPOSITION 1.1. *If X is a skeleton-finite CW-complex, then $f \colon X \to Y$ is an $\mathcal{FD}$-phantom map $\iff$ it is an $\mathcal{F}$-phantom map.*

If X now is a CW-space, that is has the homotopy type of a CW-complex, then the notion of $\mathcal{FD}$-phantom map "agrees" with the notion in [**3**]. More precisely, if $h \colon X \to X'$ is a homotopy equivalence from X to a space X' furnished with a particular CW-complex structure, then a map $\phi \colon X \to Y$ is an $\mathcal{FD}$-phantom map $\iff$ $f' = f \circ h^{-1} \colon X' \to Y$ satisfies: $f' \mid X'_n$ is inessential for all $n \geq 1$; here, of course, h^{-1} denotes (any) homotopy inverse of h. In this discussion, it is unnecessary to assume that Y is a CW-space; for, if X is a CW-space and Y' is a CW-space together with a *weak* homotopy equivalence $k \colon Y' \to Y$, then

$$k_* \colon [X, Y'] \to [X, Y]$$

is a bijection. However, if X is no longer assumed to be a CW-space, the notion of $\mathcal{FD}$-phantom map is probably not the proper phantom notion. For example, suppose that X is the Warsaw Circle, described, among other places, in [**42**, Ch. 2, Sec. 4, Ex. 8]. Since X has the singular homotopy type of a point, a map $j \colon W \to X$, with W *any* CW-space, is inessential, hence any map $f \colon X \to Y$ is, a fortiori, an $\mathcal{FD}$-phantom map. In particular, the standard quotient map $q \colon X \to S^1$ to the ordinary circle (a generator of the first integral Čech cohomology group of X) is an $\mathcal{FD}$-phantom map. If one wants to allow spaces such as the Warsaw Circle (and worse!) into the discussion, it would seem preferable to enlarge the class $\mathcal{FD}$ to the class $\widetilde{\mathcal{FD}}$ consisting of all finite-dimensional spaces (in the sense of Lebesgue); of course, q is *not* an $\widetilde{\mathcal{FD}}$-phantom map. We will not pursue such speculations, nor the possibility of developing a theory of phantom morphisms in the shape categories, in this paper.

Up to this point, the domain, X, of the map $f \colon X \to Y$ has played the dominant role. One may, however, dualize the notion of S-phantom map so as to give the target, Y, center stage and move X to the background.

DEFINITION 1.1′. Let S' be a class of spaces. A map $\phi \colon X \to Y$ is said to be an S'-co-phantom map provided that, for any $Z \in S'$ and any $k \colon Y \to Z$, the composition

$$X \xrightarrow{\phi} Y \xrightarrow{k} Z$$

is inessential.

[1]It should be noted that $[X, Y] \neq 0$. In fact, $[X, Y] \cong H^4\big(K(\mathbb{Q}, 3); \mathbb{Z}\big) \cong \operatorname{Ext}(\mathbb{Q}, \mathbb{Z})$, and the latter is a rational vector space of uncountable dimension.

Natural choices for S' are $S' = \mathcal{FD}'$ the class of all finite Postnikov spaces, that is CW-spaces[2] with only finitely many non-0 homotopy groups and $S' = \mathcal{F}'$, the class of all finite Postnikov spaces each of whose homotopy groups is finitely generated. By classical homotopy theory, we readily infer:

PROPOSITION 1.2. *A map $f\colon X \to Y$ of CW-spaces is an $\mathcal{FD}$-phantom map $\iff$ it is an $\mathcal{FD}'$-co-phantom map.*

An equivalent formulation of this proposition, assuming a particular CW-complex structure on X, is the following: the composition

$$X_n \xrightarrow{\iota_n} X \xrightarrow{f} Y,$$

$\iota_n\colon X_n \to X$ the inclusion of the n-skeleton, is inessential for all $n \geq 1 \iff$ the composition

$$X \xrightarrow{f} Y \xrightarrow{\pi^{(m)}} Y^{(m)},$$

$\pi^{(m)}\colon Y \to Y^{(m)}$ the projection to the mth Postnikov approximation, is inessential for all $m \geq 1$.

For the remainder of the paper, we assume all spaces to be path-connected (pointed) CW-spaces. Later, and especially when applying localization/completion theory to our spaces, additional finiteness and connectivity conditions need to be imposed; these conditions will be introduced as needed. We write $\mathrm{Ph}_S(X,Y)$ for the subset of $[X,Y]$ consisting of homotopy classes of S-phantom maps but confer preferred status to the case $S = \mathcal{F}$ and simply write $\mathrm{Ph}(X,Y)$ in that case; moreover, the elements of $\mathrm{Ph}(X,Y)$ will be referred to simply as phantom maps, using the usual abuse of terminology. We note the evident fact that, for any S, $\mathrm{Ph}_S(X,Y)$ is functorial in both variables.

2. Background/Plan of the paper

Before delving into the various technologies for computing $\mathrm{Ph}(X,Y)$ and giving some applications thereof, a brief historical review/guide to the journal literature may be in order. In doing this, we make no pretentions at completeness.

We have already referred to the first published example of a phantom map [**3**]; in this example, $X = \Sigma K(\mathbb{Z},2)$ and Y is a countably infinite bouquet of S^4's. Subsequently, Gray published a short but seminal paper [**9**] based on his thesis, constructing, in particular, phantom maps from $X = K(\mathbb{Z},2)$ to $Y = S^3$. Thus Gray's examples were an "improvement" over that of Adams-Walker in that both domain and target were of finite type. Gray's paper was further notable for highlighting the central role played by the $\varprojlim{}^1$ functor, introduced a few years earlier by several writers (see [**13, 5**] for references); the $\varprojlim{}^1$ description of $\mathrm{Ph}(X,Y)$ provides one of the principal means for computing $\mathrm{Ph}(X,Y)$. Additional examples of phantom maps, with familiar domains and targets (both of finite type), were found by Anderson and

[2]If Y is not required to be a CW-space, then to avoid pathology, one should suitably enlarge $\mathcal{FD}'$, as in the dual situation. For example, if Y is the "mildly schizophrenic interval"—take the ordinary interval $[0,1]$, "split" $1/2$ into two distinct points P, Q and take $\{(1/2-\epsilon,1/2)\cup\{P\}\cup(1/2,1/2+\epsilon)\}$, $\{(1/2-\epsilon,1/2)\cup\{Q\}\cup(1/2,1/2+\epsilon)\}$, $\epsilon>0$, as neighborhood bases for P, Q—then Y is a non-Hausdorff $K(\mathbb{Z},1)$ and a map of Y into any CW-space (indeed, any Hausdorff space) is inessential. We note that Y, while having first singular cohomology group isomorphic to $\mathbb{Z}$, has trivial (reduced) Čech cohomology groups and so may be regarded as a sort of dual to the Warsaw Circle.

Hodgkin [**4**]. Sullivan's theory of localization and completion in homotopy theory [**44**] gave new impetus to the study of phantom maps, particularly through the key work of Meier [**23, 24, 25**]. Meier's work was, in turn, recast by Zabrodsky [**47**], who also brought to bear on the subject a powerful theorem of H. Miller [**25**] affirming the so-called Sullivan conjecture. Following the appearance of the preprint version of [**47**], Friedlander and Mislin [**7**] made a rather definitive study of $\mathrm{Ph}(X,Y)$ for X the classifying space of a Lie group and Y a finite-dimensional CW-complex. Also, Oda and Shitanda published several papers about phantom maps [**27, 28, 29**]; they were expecially concerned to extend Zabrodsky's work to the equivariant case. In the last few years, both the $\varprojlim^1$ and Meier-Zabrodsky approaches to phantom map theory have been vigorously developed by several writers—in alphabetical order: Gray, Harper, McGibbon, Moller, the author, Shitanda and Touhey—see [**10, 11, 14, 15, 18, 19, 20, 31, 32, 33, 34, 36, 37, 38, 39, 40, 41, 45**].

More than forty years ago, J.H.C. Whitehead introduced the notion of n-type; for our purposes, it is convenient to say that spaces X and Y have the same n-type if the nth Postnikov approximation $X^{(n)}$ and $Y^{(n)}$ are homotopy equivalent. An example of non-homotopy equivalent spaces having the same n-type for all n was given in [**1**], in response to a question of Whitehead; the spaces in this example were not of finite type. A new and "improved" such example (this time the spaces were of finite type) was given in [**9**], in response to a problem posed by Adams; in this paper, Gray put in evidence a connection between the phenomenon in question and phantom maps. Wilkerson [**46**] placed the examples of [**1**] and [**9**] into a conceptual framework, basing himself on $\varprojlim^1$ techniques and the aforementioned theory of Sullivan. In recent years, the Wilkerson framework for studying the set $\mathrm{SNT}(X)$ of all homotopy types of CW-spaces having the same n-type as X for all n has been successfully exploited by McGibbon and Moller [**16, 17, 18, 15**]. In addition, phantom map theory was applied in [**11**] to sharpen the results in [**16, 17**] and, independently, Shitanda [**37, 38, 39**] extended the work in [**9, 16, 17**].

The rest of the paper is organized as follows. Sections 3 and 4 are foundational in nature and review the two existing approaches to phantom map theory: Section 3 treats the $\varprojlim^1$ approach and shows, among other things, that under rather general finiteness and "algebraic" conditions on X and $Y, \mathrm{Ph}(X,Y)$ is a divisible, Abelian group of a very special kind; Section 4 treats the localization/completion approach and features a first look at some non-trivial computations of $\mathrm{Ph}(X,Y)$, based on Zabrodsky's extension of H. Miller's theorem on the Sullivan conjecture. Section 5 takes up the notion of *special* phantom map, a particularly elusive creature which, however, appears regularly. Section 6 is devoted mainly to reviewing results of [**14, 19, 20**], which exhibit the critical role in phantom map theory played by maps satisfying certain rational conditions; in particular, a "practical" criterion is given for determining the vanishing of $\mathrm{Ph}(X,-)$ for all finite type targets and the map $f^*\colon \mathrm{Ph}(X',Y) \to \mathrm{Ph}(X,Y)$ induced by a rational homotopy equivalence $f\colon X \to X'$ is studied. The next three sections consist of explicit computations of $\mathrm{Ph}(X,Y)$ for various X and Y and may be regarded as the heart of the paper. Section 7 deals with variants of an example of Meier [**24**]; these may be viewed as more elaborate forms of the examples in Section 4. Sections 8 and 9 reexamine three examples introduced in [**10**] in the light of the theorems in [**19, 20**] expounded in Section 6. One of these examples depends on results of Cohen, Moore and Neisendorfer contained in a series of papers beginning with [**6**] while the other two examples rely on more classical results of homotopy theory. The brief Section 10 considers the question of obtaining Eckmann-Hilton duals of various results in phantom map theory. The final two sections are concerned with "applications" of phantom map theory to related parts of homotopy theory. In Section 11, we consider the set SNT( ) described earlier. Though we refrain from giving a detailed survey of recent activity in SNT-theory—such a survey would increase the length of this paper unduly—we indicate how phantom map theory allows for the construction of many non-trivial elements of SNT(Z) for special spaces Z. Section 12 applies phantom map theory to a study of the set of *weak identities* of a space, a notion introduced by the author in [**30**] in connection with a homotopy-theoretic analogue of a theorem of G. Baumslag. It is proper and fitting to mention that [**30**] is part of a volume consisting of papers dedicated to Peter Hilton on the occasion of his *sixtieth* birthday!

3. The $\varprojlim^1$ approach to phantom maps

We recall that for an inverse system of groups

$$1 \leftarrow G_0 \xleftarrow{j_1} G_1 \leftarrow \dots \xleftarrow{j_n} G_n \leftarrow \dots,$$

$\varprojlim^1 G_n$ is defined as the orbit set of $\prod G_n$ with respect to the left action

$$(g_n)\cdot(x_n) = \big(g_n x_n (j_{n+1} g_{n+1})^{-1}\big);$$

see [**13, 5**], which we use as the basic references for this section. If the groups G_n are non-Abelian, then $\varprojlim^1 G_n$ is merely a pointed set; but if the groups G_n are Abelian, then $\varprojlim^1 G_n$ may be viewed as the cokernel of the endomorphism

$$(g_n) \to \big(g_n - j_{n+1}(g_{n+1})\big)$$

of $\prod G_n$, hence is itself an Abelian group.

We recall also a well-known result [**5**, Cor. 3.3, Cor. 3.2].

PROPOSITION 3.1. (i) *If X is a CW-complex with n-skeleton X_n, there is a short exact sequence of pointed sets*

$$\varprojlim^1[\Sigma X_n, Y] \xrightarrow{\mu} [X,Y] \xrightarrow{\epsilon} \varprojlim[X_n, Y];$$

(ii) *If $Y^{(n)}$ is the nth Postnikov approximation of Y, there is a short exact sequence of pointed sets*

$$\varprojlim^1[\Sigma X, Y^{(n)}] \xrightarrow{\mu'} [X,Y] \xrightarrow{\epsilon'} \varprojlim[X, Y^{(n)}].$$

Since, by definition,

$$\mathrm{Ph}_{\mathcal{FD}}(X,Y) = \ker\epsilon,$$

and since, by Proposition 1.4,

$$\mathrm{Ph}_{\mathcal{FD}}(X,Y) = \ker\epsilon',$$

we infer, from Proposition 3.1:

COROLLARY 3.1. *There are bijections*
(i) $\mathrm{Ph}_{\mathcal{FD}}(X,Y) \cong \varprojlim^1[\Sigma X_n, Y]$;
(ii) $\mathrm{Ph}_{\mathcal{FD}}(X,Y) \cong \varprojlim^1[\Sigma X, Y^{(n)}]$.

We may exploit Corollary 3.1(ii) as follows. If either X is a co-H-space or Y is an H-space, the groups $[\Sigma X, Y^{(n)}]$ are Abelian and independent of the particular co-H-structure on X or H-structure on Y. Thus, in either case, $\mathrm{Ph}_{\mathcal{FD}}(X,Y)$ has a natural, Abelian group structure. Actually, considerably more is true, provided we place suitable finiteness restrictions on X and Y. Namely, let X be a *finite type domain*, that is the integral homology groups of X are finitely generated, and let Y be a *finite type target*, that is the higher homotopy groups of Y are finitely generated [**19, 20**]. Then, still assuming either X a co-H-space or Y an H-space, the groups $[\Sigma X, Y^{(n)}]$ are finitely generated Abelian. But groups of the form $\varprojlim{}^1 G_n$, with each G_n finitely generated Abelian, are completely understood.

Proposition 3.2 ([**13**, Section 2]). *A group G of the form $\varprojlim{}^1 G_n$, with each G_n finitely generated Abelian, is isomorphic to a group of the form* $\mathrm{Ext}(M,\mathbb{Z})$, *with M a countable, torsion-free Abelian group. Such a group is divisible, hence isomorphic to a direct sum of copies of $\mathbb{Q}$ and copies of $\mathbb{Z}_{p^\infty}$ (the p-torsion subgroup of $\mathbb{Q}/\mathbb{Z}$), p running over the set of primes; moreover, if $G \neq 0$, the number of $\mathbb{Q}$-summands is uncountable and, for each p, the number of $\mathbb{Z}_{p^\infty}$-summands is either finite or uncountable.*

Theorem 3.1. (i) *If X is a finite type domain, Y is a finite type target and if either X is a co-H-space or Y is an H-space, then* $\mathrm{Ph}_{\mathcal{FD}}(X,Y)$ *has a natural, Abelian group structure of the type described in Proposition* 3.2, *which is moreover independent of the particular co-H-structure on X or H-structure on Y;*

(ii) *If X and Y are nilpotent spaces of finite type and if either X has the rational homotopy type of a co-H-space or Y has the rational homotopy type of an H-space*[3] *then* $\mathrm{Ph}_{\mathcal{FD}}(X,Y)$ *has a natural, Abelian group structure of the type described in Proposition* 3.2.

Part (i) is an immediate corollary of Proposition 3.2 and the prior discussion. Part (ii) is proved in [**15**]: the hypotheses are utilized to conclude that the finitely generated nilpotent groups $[\Sigma X, Y^{(n)}]$, modulo their torsion subgroups $T([\Sigma X, Y^{(n)}])$, are Abelian; then, a standard $\varprojlim{}^1$ argument is invoked to identify $\varprojlim{}^1[\Sigma X, Y^{(n)}]$ with $\varprojlim{}^1[\Sigma X, Y^{(n)}]/T([\Sigma X, Y^{(n)}])$.

Theorem 3.1 suggests the problem of determining which of the groups described in Proposition 3.2 actually occur as $\mathrm{Ph}_{\mathcal{FD}}(X,Y)$. In Section 7, we show that, in fact, any such group can arise as $\mathrm{Ph}_{\mathcal{FD}}(X,Y)$ for suitable (explicit) 1-connected finite type spaces X, Y with Y a loop space.

We observe that, in the context of Theorem 3.1, maps $f\colon X \to X'$, $g\colon Y \to Y'$ induce *group homomorphisms*

$$f^*\colon [\Sigma X', Y^{(n)}] \to [\Sigma X, Y^{(n)}], \quad g_*\colon [\Sigma X, Y^{(n)}] \to [\Sigma X, Y'^{(n)}],$$

whether or not f is a (rational) co-H-map or g is a (rational) H-map. It is even possible to begin with a map $F\colon \Sigma X \to \Sigma X'$ *not* of the form $\Sigma f, f\colon X \to X'$, and infer that

$$F^*\colon [\Sigma X', Y^{(n)}]/T([\Sigma X', Y^{(n)}]) \to [\Sigma X, Y^{(n)}]/T([\Sigma X, Y^{(n)}])$$

is a group homomorphism provided F is a (rational) co-H-map or Y is a (rational)

[3]While an H-space is always nilpotent, a co-H-space need not be nilpotent.

H-space.[4] It follows that, in the circumstances just described, there are induced group homomorphisms

$$f^*\colon \mathrm{Ph}_{\mathcal{FD}}(X',Y) \to \mathrm{Ph}_{\mathcal{FD}}(X,Y), \quad g_*\colon \mathrm{Ph}_{\mathcal{FD}}(X,Y) \to \mathrm{Ph}_{\mathcal{FD}}(X,Y'),$$
$$F^*\colon \mathrm{Ph}_{\mathcal{FD}}(X',Y) \to \mathrm{Ph}_{\mathcal{FD}}(X,Y),$$

a point overlooked in [**32**]. For a fuller discussion, see [**20**].

We emphasize that all the results of this section are valid with $\mathrm{Ph}(X,Y)$ replacing $\mathrm{Ph}_{\mathcal{FD}}(X,Y)$, provided we insist that X has the homotopy type of a skeleton-finite CW-complex (Proposition 1.1). Beginning with the next section, $\mathrm{Ph}(X,Y)$ will assume the prominent role.

4. The localization/completion approach to phantom maps

Throughout this section, unless explicitly stated to the contrary, the spaces X and Y are assumed to be nilpotent of finite type[5] and with finite fundamental group. The basic references for this section are [**47, 31, 32**].

Whereas in Section 3 we used as point of departure the observation that $\mathrm{Ph}_{\mathcal{FD}}(X,Y)$ may be identified as the kernel of the map

$$\epsilon'\colon [X,Y] \to \varprojlim [X, Y^{(n)}],$$

we begin here with the identification of $\mathrm{Ph}(X,Y)$ as the kernel of the map

$$c_*\colon [X,Y] \to [X,\widehat{Y}]$$

induced by Sullivan's profinite completion $c\colon Y \to \widehat{Y}$ [**44**, Th. B(b)]. Since the homotopy-fiber of c is a rational space [**28**] [**11**, Remark after Th. 2.1], we may argue (ibid.) that if $r\colon X \to X_{(0)}$ is a rationalization map, then

$$\mathrm{Ph}(X,Y) = r^*[X_{(0)}, Y] \subset [X,Y]. \tag{4.1}$$

To effectively exploit (4.1), we consider the homotopy-fiber X_τ of r—a "torsion" space—and make the rather remarkable observation [**47**, 1.2(b)],[**31**, Lem. 2.1] that the fibration sequence

$$X_\tau \xrightarrow{i} X \xrightarrow{r} X_{(0)} \tag{4.2}$$

is also a cofibration sequence. Hence (4.2) may be extended indefinitely to the right as

$$X_\tau \xrightarrow{i} X \xrightarrow{r} X_{(0)} \xrightarrow{q} \Sigma X_\tau \xrightarrow{\Sigma i} \Sigma X \xrightarrow{\Sigma r} \Sigma X_{(0)} \to \dots \tag{4.3}$$

Next we take an *integral approximation* of Y [**47**, Def. 1.4.1], that is, a rational homotopy equivalence $j\colon Y \to L$ with the property that ΩL is a product of $K(\mathbb{Z},n)$'s. If Y is a rational H-space, we can and do take L to be a product of $K(\mathbb{Z},n+1)$'s, and if Y is, moreover, grouplike, $j\colon Y \to L$ may be assumed to be an H-map with

[4] If F is a co-H-map or Y is an H-space, we need not pass to the quotient. We may also dualize this last remark: begin with a map $G\colon \Omega Y \to \Omega Y'$ not of the form $\Omega g, g\colon Y \to Y'$, and infer, under suitable conditions, that

$$G_*\colon \left[X, \Omega Y^{(n)}\right]/T\left([X,\Omega Y^{(n)}]\right) \to \left[X, \Omega Y'^{(n)}\right]/T\left([X,\Omega Y'^{(n)}]\right)$$

is a group homomorphism.

[5] It suffices for X, Y to be nilpotent of finite type over $\mathbb{Z}_A, \mathbb{Z}_B$ respectively, where $\mathbb{Z}_C$ is the integers localized at a set of primes C, and where $B \neq \phi$.

respect to the standard grouplike structure on L. The long cofibration sequence (4.3) gives rise to a commutative diagram

$$\begin{array}{ccccccccccccc}
[X_\tau,Y] & \longleftarrow & [X,Y] & \longleftarrow & [X_{(0)},Y] & \overset{q^*}{\longleftarrow} & [\Sigma X_\tau,Y] & \longleftarrow & [\Sigma X,Y] & \longleftarrow & [\Sigma X_{(0)},Y] & \longleftarrow & \dots \\
\downarrow & & \downarrow & & \downarrow\cong & & \downarrow & & \downarrow & & \downarrow\cong & & \\
[X_\tau,L] & \longleftarrow & [X,L] & \overset{0}{\longleftarrow} & [X_{(0)},L] & \longleftarrow & [\Sigma X_\tau,L] & \longleftarrow & [\Sigma X,L] & \longleftarrow & [\Sigma X_{(0)},L] & \longleftarrow & \dots,
\end{array} \tag{4.4}$$

where, as indicated, $r^*\colon [X_{(0)}, L] \to [X, L]$ is the 0 map and

$$j_*\colon [X_{(0)}, Y] \to [X_{(0)}, L], \quad j_*\colon [\Sigma X_{(0)}, Y] \to [\Sigma X_{(0)}, L]$$

are bijections [**47**, Th. B, 1.2(e)], [**31**, Lem. 2.2, Lem. 2.3], [**32**, Lem. 2.1]. From (4.4), we infer:

THEOREM 4.1 ([**47**, Th. B],[**31**, Th. 2.1],[**32**, Th. 3.1]). *If X is a 1-connected cogroup with $\pi_2 X$ finite, or Y is grouplike, the group* $\mathrm{Ph}(X,Y)$ *is isomorphic to the quotient group $[X_{(0)}, Y]/q^*[\Sigma X_\tau, Y]$ and the group $[X_{(0)}, Y]$ is isomorphic to the product group $\prod_{n\geq 1} \mathrm{Ext}(H_{n-1}X_{(0)}, \pi_n Y)$.*

With some additional effort [**31**, Th. 2.1], it may be seen that Theorem 4.1 holds when Y is an H-space which is not necessarily grouplike. Even if Y is merely a rational H-space, we may use the bijection

$$j_*\colon [X_{(0)}, Y] \xrightarrow{\cong} [X_{(0)}, L]$$

of (4.4) to transport the group structure on $[X_{(0)}, L]$ to a group structure on $[X_{(0)}, Y]$ with respect to which j_* becomes an isomorphism. Inspection of the middle square in (4.4) shows that

$$q^*\colon [\Sigma X_\tau, Y] \to [X_{(0)}, Y]$$

is then a group homomorphism. Hence $\mathrm{Ph}(X,Y)$ acquires a group structure by virtue of identification of the latter with the quotient group $[X_{(0)}, Y]/q^*[\Sigma X_\tau, Y]$.

The effectiveness of Theorem 4.1 is obviously dependent on our ability to analyze the group $q^*[\Sigma X_\tau, Y]$. Zabrodsky's generalization of H. Miller's affirmative solution of the Sullivan conjecture provides a situation where such an analysis can be made.

THEOREM 4.2 ([**47**, Th. D]). *If X is finite Postnikov space, or an iterated suspension of such a space, and if Y has the homotopy type of a finite CW-complex, or an iterated loop space of such a space, then*

$$[X_\tau, Y] = 0 = [\Sigma X_\tau, Y],$$

so that we have bijections of sets

$$[X_{(0)}, Y] \cong [X, Y] = \mathrm{Ph}(X, Y).$$

If X and Y are as in Theorem 4.1, we thereby have a group isomorphism

$$\mathrm{Ph}(X, Y) \cong \prod_{n\geq 1} \mathrm{Ext}(H_{n-1}X_{(0)}, \pi_n Y).$$

As a simple illustration of Theorem 4.2, we present a more complete treatment of a slightly generalized version of [**32**, Ex. 4.1]; see also [**20**, Ex. F]. More complex illustrations of Theorems 4.1 and 4.2 will be given in Sections 7 and 9.

EXAMPLE 4.1. Let $K = K(\mathbb{Z},2)$ $\big($or $K(\mathbb{Z},2n)\big)$ and $Y = \Omega^k S^{2n+1+k}$, $n \geq 1$, $k \geq 0$. Then

(i) $\mathrm{Ph}(K,Y) \cong \mathrm{Ext}(H_{2n}K_{(0)}, \pi_{2n+1}Y) \cong \mathrm{Ext}(\mathbb{Q},\mathbb{Z}) \cong \mathbb{R}$, viewing $\mathbb{R}$ as a $\mathbb{Q}$-vector space of uncountable dimension;

(ii) The maps in the "suspension sequence"

$$\mathrm{Ph}(K, S^{2n+1}) \xrightarrow{\Sigma_1} \mathrm{Ph}(\Sigma K, S^{2n+2}) \xrightarrow{\Sigma_2} \mathrm{Ph}(\Sigma^2 K, S^{2n+3}) \to \dots$$

are group isomorphisms, where the group structure on $\mathrm{Ph}(K, S^{2n+1})$ comes from the rational H-space structure on S^{2n+1}. (The case $n = 1$, $k = 0$ in (i) is a precise version of [9], originally carried out by a $\varprojlim^1$ argument.)

Part (i) is an immediate consequence of Theorem 4.2. For part (ii), we are content to show that the first two maps Σ_1, Σ_2 in the sequence are isomorphisms. We have

$$\mathrm{Ph}(\Sigma K, S^{2n+2}) = [\Sigma K, S^{2n+2}] \cong [K, \Omega S^{2n+2}] = \mathrm{Ph}(K, \Omega S^{2n+2}).$$

Thus Σ_1 may be identified with the homomorphism

$$\iota_*\colon \mathrm{Ph}(K, S^{2n+1}) \to \mathrm{Ph}(K, \Omega S^{2n+2})$$

induced by the canonical map $\iota\colon S^{2n+1} \to \Omega S^{2n+2}$. Similarly, the composition $\Sigma_2 \circ \Sigma_1$ may be identified with the homomorphism

$$\iota'_*\colon \mathrm{Ph}(K, S^{2n+1}) \to \mathrm{Ph}(K, \Omega^2 S^{2n+3})$$

induced by the canonical map $\iota'\colon S^{2n+1} \to \Omega^2 S^{2n+3}$. As ι' is a rational homotopy equivalence, we see immediately that

$$\iota'_*\colon \mathrm{Ph}(K, S^{2n+1}) \cong [K_{(0)}, S^{2n+1}] \to [K_{(0)}, \Omega^2 S^{2n+3}] \cong \mathrm{Ph}(K, \Omega^2 S^{2n+3})$$

is an isomorphism $\big($see (4.4)$\big)$; thus $\Sigma_2 \circ \Sigma_1$ is an isomorphism. However, as ι is not a rational homotopy equivalence, we have to work a bit harder to establish that Σ_1—and therefore also Σ_2—is an isomorphism. To this end, consider the long fibration sequence

$$\dots \to \Omega T \to \Omega S^{2n+2} \xrightarrow{\gamma} S^{2n+1} \to T \to S^{2n+2}, \tag{4.5}$$

where T is the total space of the tangent sphere bundle over S^{2n+2}. Since the composition $\gamma \circ \iota\colon S^{2n+1} \to S^{2n+1}$ is a map of degree 2, hence a rational homotopy equivalence, the induced homomorphism

$$\gamma_* \circ \iota_*\colon \mathrm{Ph}(K, S^{2n+1}) \cong [K_{(0)}, S^{2n+1}] \to [K_{(0)}, S^{2n+1}] \cong \mathrm{Ph}(K, S^{2n+1})$$

is an isomorphism. In particular, γ_* is an epimorphism. But γ_* is also a monomorphism since, in the exact sequence

$$\dots \to [K_{(0)}, \Omega T] \to [K_{(0)}, \Omega S^{2n+2}] \xrightarrow{\gamma_*} [K_{(0)}, S^{2n+1}] \to \dots$$

derived from (4.5), $[K_{(0)}, \Omega T] = 0$. We conclude that ι_* is an isomorphism, as claimed.

We should point out some limitations of the approach in this section as compared to the $\varprojlim^1$ approach in Section 3. In the first place, Theorems 4.1 and 4.2 definitely do not apply to spaces X with infinite fundamental group. For example, if $X = S^1 \times S^1$ and $Y = S^3$, then $\mathrm{Ph}(X,Y) = 0$ although $\Pi_{n \geq 1} \mathrm{Ext}(H_{n-1}X_{(0)}, \pi_n Y) \cong$

$\mathbb{R}$ [**31**, Remark after Th. 2.2]. Secondly, if the finite type assumption on Y is removed, even (4.1) fails. Indeed, in [**10**] (see also [**14, 19, 20**]), it is observed that $\mathrm{Ph}\big(K(\mathbb{Z}/2,1),Y\big)$ is non-0—actually, uncountable—for a certain countable type target Y, even though $K(\mathbb{Z}/2,1)$ has the rational homotopy type of a point, so that—by (4.1)—$\mathrm{Ph}\big(K(\mathbb{Z}/2,1),Y\big) = 0$ for all finite type targets. We refer to [**10, 14**] for a detailed discussion of the theory of phantom maps with non-finite type targets.

5. Not all phantom maps are created equal

Some phantom maps are even more difficult to detect than others.

DEFINITION 5.1. If Y is nilpotent, an S-phantom map $\phi\colon X \to Y$ is said to be *special* if for each prime p, the composition

$$X \xrightarrow{\phi} Y \xrightarrow{e_p} Y_{(p)}, \tag{5.1}$$

with $e_p\colon Y \to Y_{(p)}$ a p-localization map, is inessential. The set of homotopy classes of special S-phantom maps is denoted by $S\,\mathrm{Ph}_S(X,Y)$.

If we set

$$\check{Y} = \prod Y_{(p)},$$

the "local expansion" of Y [**12**], and $\check{e}\colon Y \to \check{Y}$ the map with components e_p, then we may reformulate Definition 5.1 as:

$$S\,\mathrm{Ph}_S(X,Y) = \ker\big\{\mathrm{Ph}_S(X,Y) \xrightarrow{\check{e}_*} \mathrm{Ph}_S(X,\check{Y})\big\}.$$

Special phantom maps were introduced in [**18, 15**] under the more colorful name "clones of the constant map" (in the case $S = \mathcal{FD}$) and in [**11**] (in the case $S = \mathcal{F}$). In the latter case, the inessentiality of the compositions in (5.1) for each p actually implies that ϕ is phantom [**12**, Th. II 5.3].

Here are some simple examples:

EXAMPLE 5.1 (EXAMPLE 1.1 CONTINUED). We have seen that

$$\mathrm{Ph}\big(K(\mathbb{Q},3),K(\mathbb{Z},4)\big) = \big[K(\mathbb{Q},3),K(\mathbb{Z},4)\big] \cong \mathrm{Ext}(\mathbb{Q},\mathbb{Z}).$$

By definition,

$$\begin{aligned} S\,\mathrm{Ph}\big(K(\mathbb{Q},3),K(\mathbb{Z},4)\big) &= \ker\Big\{\mathrm{Ph}\big(K(\mathbb{Q},3),K(\mathbb{Z},4)\big) \to \mathrm{Ph}\big(K(\mathbb{Q},3),K(\check{\mathbb{Z}},4)\big)\Big\} \\ &= \ker\big\{\mathrm{Ext}(\mathbb{Q},\mathbb{Z}) \to \mathrm{Ext}(\mathbb{Q},\check{\mathbb{Z}})\big\}. \end{aligned}$$

According to [**11**, Lem. 2.3], there is a short exact sequence

$$0 \longrightarrow \mathrm{Hom}(\mathbb{Q},\check{\mathbb{Z}}/\mathbb{Z}) \longrightarrow \mathrm{Ext}(\mathbb{Q},\mathbb{Z}) \longrightarrow \mathrm{Ext}(\mathbb{Q},\check{\mathbb{Z}}) \longrightarrow 0,$$

with all three terms being $\mathbb{Q}$-vector spaces of uncountable dimension. The elements of $\mathrm{Hom}(\mathbb{Q},\check{\mathbb{Z}}/\mathbb{Z})$ thus represent all the special phantom maps = special $\mathcal{F}$-phantom maps.

See also [**12**, Prop. II 5.5] for an explicit example of a special $\mathcal{F}$-phantom map which is not an $\mathcal{FD}$-phantom map.

EXAMPLE 5.2 (EXAMPLE 4.1(i) CONTINUED). Arguing much as in Example 5.1, we find that

$$S\,\mathrm{Ph}(K, \Omega^k S^{2n+1+k}) \cong \mathrm{Hom}(\mathbb{Q}, \check{\mathbb{Z}}/\mathbb{Z}).$$

In this case,

$$S\,\mathrm{Ph}(K, \Omega^k S^{2n+1+k}) = S\,\mathrm{Ph}_{\mathcal{FD}}(K, \Omega^k S^{2n+1+k}).$$

The following theorem of McGibbon shows that while special phantom maps may be hard to see, they are there—in abundance.

THEOREM 5.1 ([**15**, Th. 4]). *With respect to Theorem* 3.1(ii), $\check{e}_*\colon \mathrm{Ph}(X,Y) \to \mathrm{Ph}(X,\check{Y})$ *is an epimorphism; moreover, the subgroup* $S\,\mathrm{Ph}(X,Y)$ *of* $\mathrm{Ph}(X,Y)$ *is divisible and is non-*0 *whenever* $\mathrm{Ph}(X,Y)$ *is non-*0.

McGibbon's result is actually stated for X a rational suspension space or Y a rational loop space, but his proof works just as well in the more general situation. Note that because of the divisibility of $S\,\mathrm{Ph}(X,Y)$, the short exact sequence

$$0 \longrightarrow S\,\mathrm{Ph}(X,Y) \longrightarrow \mathrm{Ph}(X,Y) \longrightarrow \mathrm{Ph}(X,\check{Y}) \longrightarrow 0 \tag{5.2}$$

splits. This leads us to ask:

QUESTION 5.1. Under the conditions of Theorem 5.1, is there a *natural* splitting $\mathrm{Ph}(X,Y) \to S\,\mathrm{Ph}(X,Y)$?

It is not generally true that $\mathrm{Ph}(X,Y) \neq 0$ implies $\mathrm{Ph}(X,\check{Y}) \neq 0$. Interesting examples where $\mathrm{Ph}(X,Y) \neq 0$ but $\mathrm{Ph}(X,\check{Y}) = 0$ were described in [**10, 14**] and will be studied in Sections 8 and 9. However, there is one situation where $\mathrm{Ph}(X,Y) \neq 0$ does imply $\mathrm{Ph}(X,\check{Y}) \neq 0$.

THEOREM 5.2 ([**11**, Th. 2.2]). *In the context of Theorems* 4.1 *and* 4.2, *all the groups in* (5.2) *are* $\mathbb{Q}$*-vector spaces of uncountable dimension whenever* $\mathrm{Ph}(X,Y)$ *is non-*0.

The non-vanishing of $\mathrm{Ph}(X,\check{Y})$ under the conditions of Theorem 5.2 turns out to be useful in constructing various examples [**11**]; we return to such matters in Section 11.

For a (weaker) version of Theorem 5.1 in the case that $\mathrm{Ph}(X,Y)$ does not have a group structure, we refer to a very recent paper of McGibbon and Steiner [**21**, Th.1.1.].

6. Phantom maps and rational homotopy equivalences

We have seen that rational homotopy equivalences crop up naturally in the theory of phantom maps. In this section, we discuss some of the results of [**19, 20**] (see also [**14**]) which involve rational homotopy equivalences in their statements. These results, together with Theorems 4.1 and 4.2, form the basis for some interesting computations to be carried out in subsequent sections.

To set the stage, we make a few preliminary remarks of a general character about rational homotopy equivalences. By definition, a rational homotopy equivalence $f\colon X \to X'$ of nilpotent spaces induces a homotopy equivalence $f_{(0)}\colon X_{(0)} \to X'_{(0)}$. However, a homotopy equivalence $X_{(0)} \simeq X'_{(0)}$ need not come from a rational homotopy equivalence $f\colon X \to X'$. For instance, S^3 and $K(\mathbb{Z},3)$ certainly have the same rational homotopy type, that is $S^3_{(0)} \simeq K(\mathbb{Q},3)$, but while there is a rational homotopy equivalence $S^3 \to K(\mathbb{Z},3)$, there is no rational homotopy equivalence $K(\mathbb{Z},3) \to S^3$; indeed, by Theorem 4.2, $\big[K(\mathbb{Z},3), S^3\big] = 0$. It is even possible to have $X_{(0)} \simeq X'_{(0)}$ but with no rational homotopy equivalences $X \to X'$, $X' \to X$ in *either* direction. As an example, take $X = S^3 \vee K(\mathbb{Z}, 5)$, $X' = K(\mathbb{Z}, 3) \vee S^5$ [**20**, footnote 3]; in this example, there are rational homotopy equivalences $S^3 \vee S^5 \to X$, $S^3 \vee S^5 \to X'$ as well as rational homotopy equivalences $X \to K(\mathbb{Z}, 3) \vee K(\mathbb{Z}, 5)$, $X' \to K(\mathbb{Z}, 3) \vee K(\mathbb{Z}, 5)$. As another example (which the reader may verify as an entertaining exercise), take $X = BS^3$, $X' = \Omega S^5$; in this example, there are rational homotopy equivalences $X \to K(\mathbb{Z}, 4)$, $X' \to K(\mathbb{Z}, 4)$. Finally, we recall that a suspension space ΣX has the rational homotopy type of a bouquet of spheres $\mathcal{B}$; in fact, there is always a rational homotopy equivalence $\mathcal{B} \to \Sigma X$. As the following theorem shows, the question of the existence of a rational homotopy equivalence in the opposite direction, $\Sigma X \to \mathcal{B}$, is intimately related to our concerns in this paper.

THEOREM 6.1 ([**19**, Th. 1]). *If X has the homotopy type of a skeleton-finite CW-complex,*[6] *then the following statements are equivalent:*
(i) $\mathrm{Ph}(X, Y) = 0$ *for all finite type targets* Y;
(ii) $\mathrm{Ph}(X, S^n) = 0$ *for all* n;
(iii) *There exists a rational homotopy equivalence from ΣX to a bouquet of spheres $\mathcal{B}$.*

We also mention a variant of Theorem 6.1, in the case that X is a loop space of a special kind.

THEOREM 6.2 ([**14**, Th. 1]). *Let X be a loop space having the rational homotopy type of a product P of odd-dimensional spheres and loop spaces of odd-dimensional spheres. Then* $\mathrm{Ph}(X, S^n) = 0$ *for all n* $\Longleftrightarrow$ *there exists a rational homotopy equivalence $X \to P$.*

(In the light of the equivalence of (i) and (ii) in Theorem 6.1, we may restate the conclusion of Theorem 6.2: $\mathrm{Ph}(X, Y) = 0$ for all finite type targets $\Longleftrightarrow$ there exists a rational homotopy equivalence $X \to P$.)

As a very simple illustration of Theorem 6.1 (or Theorem 6.2), we see that $\mathrm{Ph}(\Omega S^n, Y) = 0$ for all finite type targets $Y, n \geq 2$. However, this example is not too compelling an advertisement for Theorems 6.1 and 6.2 since a more general result follows from first principles. Namely, if W has the homotopy type of a (skeleton-finite) bouquet of finite CW-complexes, then $\mathrm{Ph}(\Omega\Sigma W, Y) = 0$ for *any* (not necessarily finite type) target Y. Indeed, reverting to Corollary 3.1, one need only observe that the inverse system of groups $\{G_n\}$, $G_n = [\Sigma\Omega\Sigma W, Y^{(n)}]$, is *Mittag-Leffler*—that is, for each m, the images of G_{m+k} in G_m stabilize for sufficiently large k—to conclude that $\mathrm{Ph}(\Omega\Sigma W, Y) \cong \varprojlim{}^1 G_n = 0$.

Another way to see that $\mathrm{Ph}(\Omega\Sigma W, Y) = 0$ for all Y is to invoke the following theorem, which, in sharp contrast to Theorem 6.1, does not involve rational conditions.

THEOREM 6.3 ([**10**, Th. 2]). *Let X come with a particular skeleton-finite CW-complex structure. Then* $\mathrm{Ph}(X, Y) = 0$ *for all targets Y* $\Longleftrightarrow$ *ΣX is a retract of the bouquet $V\Sigma X_n$.*

Several non-trivial applications of Theorem 6.1 will be given in Section 8.

The next theorem provides information about the map $f^*\colon \mathrm{Ph}(X', Y) \to \mathrm{Ph}(X, Y)$ induced by a map $f\colon X \to X'$ which is a little more general than a rational homotopy equivalence. Part (i) is contained in part (ii) but is sufficiently

[6]This condition will be in force throughout this section. However, many of the results in this section are valid for arbitrary finite type domains provided $\mathrm{Ph}(-,-)$ is replaced by $\mathrm{Ph}_{\mathcal{FD}}(-,-)$.

important to deserve highlighting.

THEOREM 6.4 ([**19**, Th. 2 (i)][**20**, Th. 1]). (i) *A map* $f: X \to X'$ *inducing a monomorphism in rational homology induces a surjection*

$$f^*: \mathrm{Ph}(X', Y) \to \mathrm{Ph}(X, Y)$$

for all finite type targets Y;

(ii) *If* $F: \Sigma X \to \Sigma X'$ *is a map inducing a monomorphism in rational homology, then* $\mathrm{Ph}(X', Y) = 0$ *implies* $\mathrm{Ph}(X, Y) = 0$. *If, moreover,* F *is a rational co-*H*-map* (e.g., *part* (i)) *or* Y *is a rational* H*-space, then* F *induces a map*

$$F^*: \mathrm{Ph}(X', Y) \to \mathrm{Ph}(X, Y)$$

(*see the discussion near the end of Section* 3) *which is a surjection.*

Theorems 6.1, 6.2 and 6.4 are linked by their common method of proof, a crucial ingredient being the application of $\varprojlim^1$ methods to inverse systems of finitely generated nilpotent groups [**16**].[7] There is also a direct link between Theorems 6.1 and 6.4; namely, Theorem 6.4 implies the implication (iii) $\Longrightarrow$ (i) in Theorem 6.1. In fact, the adjoint of a rational homotopy equivalence $\Sigma X \to \mathcal{B}$ is a map $X \to \Omega\mathcal{B}$ which is easily seen to induce a monomorphism in rational homology. Theorem 6.4 assures us that $\mathrm{Ph}(\Omega\mathcal{B}, Y)$—which is 0 (see the discussion after the statement of Theorem 6.2)—maps onto $\mathrm{Ph}(X, Y)$, and so the latter must also be 0.

As a corollary of Theorem 6.4(i) and Theorem 4.2, we have:

THEOREM 6.5. *Let* X *be nilpotent with finite fundamental group and rationally elliptic,* $j: X \to L$ *an integral approximation and suppose that* Y *has the homotopy type of a nilpotent, finite CW-complex with finite fundamental group. For any* $k \geq 0$, $\ell \geq 0$, *a map* $f: \Sigma^k X \to \Omega^\ell Y$ *is phantom* $\iff$ f *factors as*

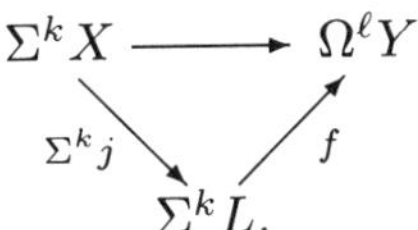

For the proof, we need only remark that, by definition of rational ellipticity, $\pi_n X$ is finite for all but finitely many n, so that L is a finite Postnikov space.

Two especially interesting examples of rationally elliptic spaces—classifying spaces of compact, connected Lie groups and iterated loop spaces of spheres—will be among the examples featured in Sections 8 and 9.

Continuing in the context of Theorem 6.5, we have a surjection

$$f^*: \mathrm{Ph}(\Sigma^k L, \Omega^\ell Y) \to \mathrm{Ph}(\Sigma^k X, \Omega^\ell Y)$$

and $\mathrm{Ph}(\Sigma^k L, \Omega^\ell Y)$ is "completely known". In order to understand $\mathrm{Ph}(\Sigma^k X, \Omega^\ell Y)$, we may therefore try to study the kernel of f^*. In general, this kernel is non-0 and is often very large; the following simple example is typical.

[7] The case of Theorem 5.4(i) where X, X', Y are nilpotent of finite type over $\mathbb{Z}_A$, $\mathbb{Z}_{A'}$, $\mathbb{Z}_B$, $B \neq \phi$ (see footnote 5) and $f: X \to X'$ is a rational homotopy equivalence may also be deduced from Theorem 4.1.

EXAMPLE 6.1 ([**20**, Ex. D]). Let $j\colon S^3 \to K(\mathbb{Z},3)$ represent a generator of $\pi_3K(\mathbb{Z},3)$; j is an integral approximation. The kernel of

$$j^*\colon \operatorname{Ph}\bigl(K(\mathbb{Z},3),S^4\bigr) \to \operatorname{Ph}(S^3,S^4)$$

is clearly all of $\operatorname{Ph}\bigl(K(\mathbb{Z},3),S^4\bigr)$, an uncountable set. (Replacing S^4 by Ω^kS^{4+k}, $k \geq 1$, leads to the same conclusion but now $\operatorname{Ph}\bigl(K(\mathbb{Z},3),\Omega S^{4+k}\bigr)$ has group structure; in fact, $\operatorname{Ph}\bigl(K(\mathbb{Z},3),\Omega^kS^{4+k}\bigr) \cong \mathbb{R}$; cf. Example 4.1.)

Under very special circumstances, one may infer that the kernel of the map

$$f^*\colon \operatorname{Ph}(X',Y) \to \operatorname{Ph}(X,Y)$$

induced by a rational homotopy equivalence $f\colon X \to X'$ is 0.

THEOREM 6.6 ([**20**, Th. 6, Prop. 8]). *Let X, X' and Y be nilpotent with finite fundamental group and let $f\colon X \to X'$ be a rational homotopy equivalence. Under the following sets of conditions, the map*

$$f^*\colon \operatorname{Ph}(X',Y) \to \operatorname{Ph}(X,Y)$$

is a bijection:

(i) *X is a finite Postnikov space, or the iterated suspension of such a space, and Y has the homotopy type of a finite CW-complex, or the iterated loop space of such a space;*

(ii) *X' is a finite Postnikov space, or the iterated suspension of such a space, Y has the homotopy type of a finite CW-complex, or the iterated loop space of such a space, f induces an isomorphism in integral homology in all but finitely many degrees and either f is a co-H-map of cogroups or Y is grouplike.*

We conclude this section with a question suggested by Theorem 6.4 (cf. Question 5.1).

QUESTION 6.1. If $f\colon X \to X'$ induces a monomorphism in rational homology, is the induced map

$$f^*\colon S\operatorname{Ph}(X',Y) \to S\operatorname{Ph}(X,Y)$$

a surjection?

7. Phantom maps and torsion

In this section, we consider certain families of spaces generalizing a space used by Meier in [**24**, Prop. 6(ii)] to show that the divisible Abelian group $\operatorname{Ph}(X,Y)$—for X, Y 1-connected of finite type and Y a loop space—need not be torsion-free, i.e. a $\mathbb{Q}$-vector space. Meier's calculation was based on $\varprojlim{}^1$ considerations, as in Section 3. An alternative approach, based on Theorem 4.2 and yielding a stronger result, is offered in [**34**].

The two basic families, denoted $\{M^{2n-1}\}$, $\{\overline{M}^{2n-1}\}$, $n \geq 2$, are defined by means of homotopy-pullback diagrams

$$\begin{array}{ccc} M^{2n-1} = M^{2n-1}(I,J) & \longrightarrow & \Omega^2 S_J^{2n+1} \\ \downarrow & & \downarrow \\ K(\mathbb{Z}_J, 2n-1) & \longrightarrow & K(\mathbb{Q}, 2n-1), \end{array} \tag{7.1}$$

and

$$\begin{array}{ccc} \overline{M}^{2n-1} = \overline{M}^{2n-1}(I,J) & \longrightarrow & S_J^{2n-1} \\ \downarrow & & \downarrow \\ K(\mathbb{Z}_J, 2n-1) & \longrightarrow & K(\mathbb{Q}, 2n-1), \end{array} \tag{7.2}$$

I and J denoting non-empty complementary sets of primes; miraculously, both (7.1) and (7.2) are also homotopy-pushout diagrams [**34**, Prop. 3]. The spaces in these families are torn between being (iterated loop spaces of) finite CW-complexes and finite Postnikov spaces. For the purpose of creating interesting examples of group structures on $\mathrm{Ph}(X,Y)$, their schizophrenia[8] allows them to serve as either domain or target. However, M^{2n-1} is more suitable as target since it is a (double) loop space and behaves well under looping while $\overline{M}^{2n-1}$ is more suitable as domain since it behaves well under suspension. (Note that $\overline{M}^{2n-1}$ is an H-space provided $2 \in I$ and a loop space $\Leftrightarrow J \subset \{p \mid \mathbb{Z}/n \subset p\text{-adic units}\}$ [**44**, Cor., p. 73]; $\overline{M}^3$, which is a loop space for any choice of I, J, is the original Meier space.)

THEOREM 7.1. (i) *If* $X = K(\mathbb{Z},2)$ *or* $K(\mathbb{Z},2n-2)$ *and* $Y = M^{2n-1}$, *then*

$$\mathrm{Ph}(X,Y) = [X,Y] \cong \mathbb{R} \oplus \bigoplus_{p\in I} \mathbb{Z}_{p^\infty}.$$

Moreover,

$$\mathrm{Ph}(X,\check{Y}) \cong \mathbb{R},$$

so that the torsion subgroup of $\mathrm{Ph}(X,Y)$ *is contained in* $S\,\mathrm{Ph}(X,Y)$ *and*

$$S\,\mathrm{Ph}(X,Y) \cong \mathbb{R} \oplus \bigoplus_{p\in I} \mathbb{Z}_{p^\infty}.$$

(ii) *If* $X = \overline{M}^{2n-1}$ *and* $Y = \Omega^3 S^{2n+3}$, *then*

$$\mathrm{Ph}(X,Y) = [X,Y] \cong \mathbb{R} \oplus \bigoplus_{p\in J} \mathbb{Z}_{p^\infty}.$$

Moreover,

$$\mathrm{Ph}(X,\check{Y}) \cong \mathbb{R},$$

so that the torsion subgroup of $\mathrm{Ph}(X,Y)$ *is contained in* $S\,\mathrm{Ph}(X,Y)$ *and*

$$S\,\mathrm{Ph}(X,Y) \cong \mathbb{R} \oplus \bigoplus_{p\in J} \mathbb{Z}_{p^\infty}.$$

Except for the statements about $\mathrm{Ph}(X,\check{Y})$ and $S\,\mathrm{Ph}(X,Y)$, these results are contained in [**34**, Th.2, Th.3]. The proof in [**34**] applies the homotopical Mayer-Vietoris sequence to (7.1) for (i) and the cohomological Mayer-Vietoris sequence to (7.2) for (ii). Both parts rely essentially on Theorem 4.2 to calculate (as in Example 4.1) various of the terms in these Mayer-Vietoris sequences. However, it is not necessary in these examples (as it will be in later examples) to attempt a direct assault on the crucial term $q^*[\sum X_\tau, Y]$ appearing in Theorem 4.1.

---

[8] We leave it to the reader to decide which of the spaces—M^{2n-1} (or $\overline{M}^{2n-1}$) or the one described in footnote 2—is in greater need of therapy.

As for the calculations of $\operatorname{Ph}(X,\check{Y})$ and $S\operatorname{Ph}(X,Y)$, it suffices to observe that

$$\operatorname{Ph}(K, M^{2n-1}_{(p)}) \cong \begin{cases} \operatorname{Ph}\big(K, K(\mathbb{Z}_{(p)}, 2n-1)\big), & p \in I \\ \operatorname{Ph}(K, \Omega^2 S^{2n+1}_{(p)}), & p \in J, K = K(\mathbb{Z},2) \text{ or } K(\mathbb{Z}, 2n-2) \end{cases}$$
$$\cong \begin{cases} 0, & p \in I \\ \operatorname{Ext}(\mathbb{Q}, \mathbb{Z}_{(p)}) \cong \mathbb{R}, & p \in J; \end{cases}$$

and

$$\operatorname{Ph}\big(\overline{M}^{2n-1}, \Omega^3 S^{2n+3}_{(p)}\big) \cong \operatorname{Ph}\big(\overline{M}^{2n-1}_{(p)}, \Omega^3 S^{2n+3}_{(p)}\big)$$
$$\cong \begin{cases} \operatorname{Ph}\big(K(\mathbb{Z}_{(p)}, 2n-1), \Omega^3 S^{2n+3}_{(p)}\big), & p \in I \\ \operatorname{Ph}\big(S^{2n-1}_{(p)}, \Omega^3 S^{2n+3}_{(p)}\big), & p \in J \end{cases}$$
$$\cong \begin{cases} \operatorname{Ext}(\mathbb{Q}, \mathbb{Z}_{(p)}) \cong \mathbb{R}, & p \in I \\ 0, & p \in J. \end{cases}$$

Taking $X = K(\mathbb{Z}, 2)$ and Y either S^3 or a suitable product of the spaces M^{2n-1} (infinitely many distinct n will, in general, be required) leads to the following result, psychologically important in that it asserts that phantom map theory—restricted to 1-connected finite type domains and targets—is as rich as the algebra allows (see Proposition 3.2).

THEOREM 7.2 ([**34**, Th. 4]). *If G is a direct sum of uncountably many copies of $\mathbb{Q}$ and n_p copies of $\mathbb{Z}_{p^\infty}$ for each prime p, with n_p either finite or uncountable, then there exist 1-connected finite type spaces X and Y, Y a loop space, such that*

$$\operatorname{Ph}(X, Y) \cong G.$$

Moreover,

$$\operatorname{Ph}(X, \check{Y}) \cong \mathbb{R},$$

so $S\operatorname{Ph}(X,Y)$ is a proper subgroup of $\operatorname{Ph}(X,Y)$ and

$$S\operatorname{Ph}(X, Y) \cong G.$$

Only the statement about $\operatorname{Ph}(X,\check{Y})$ and $S\operatorname{Ph}(X,Y)$ is not covered explicitly in [**34**] and that statement follows easily from the corresponding statement in Theorem 7.1(i).

The spaces $\overline{M}^{2n-1}$ are also useful in constructing exotic variations of Example 6.1.

EXAMPLE 7.1 (CF. [**20**, Ex. G]). The rational homotopy equivalence $j\colon S^3 \to K(\mathbb{Z}, 3)$ of Example 6.1 may be factored as

$$S^3 \xrightarrow{f} \overline{M}^3 \xrightarrow{j'} K(\mathbb{Z}, 3),$$

with f and j' rational homotopy equivalences (j' is actually an integral approximation). Then the induced maps

$$f^*\colon \operatorname{Ph}(\overline{M}^3, \Omega^3 S^7) \to \operatorname{Ph}(S^3, \Omega^3 S^7), \quad j'^*\colon \operatorname{Ph}\big(K(\mathbb{Z},3), \Omega^3 S^7\big) \to \operatorname{Ph}(\overline{M}^3, \Omega^3 S^7)$$

satisfy:

$$\ker f^* = \mathrm{Ph}(\overline{M}^3, \Omega^3 S^7) \cong \mathbb{R} \oplus \bigoplus_{p\in J} \mathbb{Z}_{p^\infty},$$

$$\ker j'^* \cong \prod_{p\in J} \widehat{\mathbb{Z}}_p, \widehat{\mathbb{Z}}_p \text{ the } p\text{-adic integers.}$$

The second isomorphism may be seen by identifying $\mathrm{Ph}\big(K(\mathbb{Z},3), \Omega^3 S^7\big)$ with $\mathrm{Ext}(\mathbb{Q},\mathbb{Z})$, $\mathrm{Ph}(\overline{M}^3, \Omega^3 S^7)$ with $\mathrm{Ext}(\mathbb{Z}_J, \mathbb{Z})$ (see [**20**, Ex. G], [**24**, Th. 3]) and j'^* with the epimorphism in the short exact sequence

$$0 \longrightarrow \mathrm{Ext}(\mathbb{Q}/\mathbb{Z}_J, \mathbb{Z}) \longrightarrow \mathrm{Ext}(\mathbb{Q},\mathbb{Z}) \longrightarrow \mathrm{Ext}(\mathbb{Z}_J,\mathbb{Z}) \longrightarrow 0;$$

and by noting the isomorphism $\mathrm{Ext}(\mathbb{Q}/\mathbb{Z}_J, \mathbb{Z}) \cong \prod_{p\in J} \widehat{\mathbb{Z}}_p$ [**5**, Ch. VI, Section 2].

Example 7.2 (cf. [**20**, Ex. E]). Let p^r be a power of a prime p and $X = \Sigma\overline{M}^{2n-1} = X'$, where $J = \{p\}$. Define $f\colon X \to X$ as p^r times the identity map, using the suspension structure on X. Then f is a rational homotopy equivalence and the induced map

$$f^*\colon \mathrm{Ph}(X, \Omega^2 S^{2n+3}) \to \mathrm{Ph}(X, \Omega^2 S^{2n+3})$$

is multiplication by p^r, so that

$$\ker f^* \cong \mathbb{Z}/p^r.$$

More generally, any countable direct sum of cyclic groups of prime power order (for various primes)—in particular, any finite Abelian group—can be realized as $\ker f^*$ where $X = \bigvee_{n\ge 2} \Sigma\overline{M}^{2n-1} = X'$, $f\colon X \to X$ is defined "flowerwise" as above and $Y = \prod_{n\ge 2} \Omega^2 S^{2n+3}$.

For the second part of this example, we need only observe that for any $m \ge 2$,

$$\mathrm{Ph}(\Sigma\overline{M}^{2m-1}, Y) \cong \prod_{n\ge 2} \mathrm{Ph}(\Sigma\overline{M}^{2m-1}, \Omega^2 S^{2n+3}),$$

and that, by Theorem 4.1,

$$\mathrm{Ph}(\Sigma\overline{M}^{2m-1}, \Omega^2 S^{2n+3}) = 0, \quad n \ne m.$$

8. Computing Ph(X, −) for special values of X: I

In this section, we use Theorems 6.1 and 6.4 to establish vanishing and non-vanishing results about the functor $\mathrm{Ph}(X,-)$ for a number of standard spaces arising in homotopy theory. We will find it convenient to say that $\mathrm{Ph}(X,-) \not\equiv 0$ *in the strong sense* if $\mathrm{Ph}(X,Y) \ne 0$ for some finite type target Y.

Example 8.1 (Cf. [**10**, Ex. 3.9]). Fix an infinite set of odd primes $\mathbb{P}$, let $\alpha(p)$ be a generator of the p-component of $\pi_{2p}S^3$ and define C to be the mapping cone of the map $\alpha\colon \bigvee_{p\in\mathbb{P}} S^{2p} \to S^3$ which restricts to $\alpha(p)$ on S^{2p}. Then, for any $k \ge 0$, $\mathrm{Ph}(\Sigma^k C, -) \not\equiv 0$ in the strong sense. However, for any prime p, $\mathrm{Ph}(\Sigma^k C, Y) = 0$ for any nilpotent, p-local target Y; thus, for all nilpotent targets Y,

$$S\,\mathrm{Ph}(\Sigma^k C, Y) = \mathrm{Ph}(\Sigma^k C, Y).$$

To prove the non-vanishing assertion, it suffices, by Theorem 6.1, to rule out the existence of a rational homotopy equivalence,

$$\Sigma^{k+1}C \xrightarrow{f} \mathcal{B} = S^{4+k} \bigvee_{p\in\mathbb{P}} S^{2p+k+2}.$$

If there were such an f, then for all sufficiently large primes p, we would have an isomorphism

$$f^*\colon H^{4+k}(\mathcal{B};\mathbb{Z}/p) \xrightarrow{\cong} H^{4+k}(\Sigma^{k+1}C;\mathbb{Z}/p). \tag{8.1}$$

However, the Steenrod operation $\mathcal{P}^1_p$ acts trivially on $H^{4+k}(\mathcal{B};\mathbb{Z}/p)$ for all p and non-trivially on $H^{4+k}(\Sigma^{k+1}C;\mathbb{Z}/p)$ for all $p\in\mathbb{P}$, and so (8.1) is impossible.

The vanishing assertion results from the fact that, for any prime p, $\Sigma C_{(p)}$ splits as the p-localization of a bouquet of finite CW-complexes; namely

$$\begin{aligned} &C_{\alpha(p)} \bigvee_{q\in\mathbb{P}-\{p\}} S^{2q+1}, && p\in\mathbb{P}\\ &S^3 \bigvee_{q\in\mathbb{P}} S^{2q+1}, && p\notin\mathbb{P}, \end{aligned}$$

where $C_{\alpha(p)}$ denotes the mapping cone of $\alpha(p)$.

EXAMPLE 8.2 (CF. [**10**, Discussion after Th. 3.3, Ex. 3.6]). If BG is the classifying space of a compact, connected Lie group G of rank ≥ 1, then, for any $k\geq 0$, $\mathrm{Ph}(\Sigma^k BG,-)\not\equiv 0$ in the strong sense.

The argument is very similar to the first part of the argument in Example 8.1; however, we use $\mathcal{P}^2_p$ in place of $\mathcal{P}^1_p$.

We draw attention to the fact that Example 8.2 is an improvement over [**10**, Ex. 3.6] in the case G has rank ≥ 1; for [**10**, Ex. 3.6] provides a *countable type* target Y such that $\mathrm{Ph}(BG,Y)\neq 0$. On the other hand, [**10**, Ex. 3.6] applies to arbitrary compact Lie groups, in particular finite groups, and in the latter case Theorem 6.1 implies that $\mathrm{Ph}(BG,Y)=0$ for all finite type targets Y. (The special case $G=\mathbb{Z}/2$ is discussed at the end of Section 4.)

We must mention that a considerably sharper version of Example 8.2 is given in [**7**, Th. 3.3] (see also [**35, 36**], follow-up papers to [**7**]). We also refer the reader to [**48, 8**]; among the (many) papers concerned with maps having domain the classifying space of a Lie group, these two are especially germane to phantom map theory.

EXAMPLE 8.3 ([**10**, Ex. 4.7][**13**, Ex. 2.1]). For any $n\geq 2$, $\mathrm{Ph}(\Omega^2S^{2n+1},-)\not\equiv 0$ in the strong sense. However, for any prime p, $\mathrm{Ph}(\Omega^2S^{2n+1}_{(p)},Y)=0$ for any nilpotent, p-local, finite type target Y; thus, for all nilpotent, finite type targets Y,

$$S\,\mathrm{Ph}(\Omega^2S^{2n+1},Y)=\mathrm{Ph}(\Omega^2S^{2n+1},Y).$$

For the non-vanishing assertion, we proceed as in Example 8.1 and rule out the existence of a rational homotopy equivalence

$$\Sigma\Omega^2S^{2n+1} \xrightarrow{f} \mathcal{B}=S^{2n}.$$

Assuming there is such an f, we take the composition

$$g\colon \Omega^2S^{2n+1} \xrightarrow{\mathrm{adj}(f)} \Omega S^{2n} \xrightarrow{\gamma} S^{2n+1},$$

where adj(f) is the adjoint of f and γ is as in (4.5), and observe that g is again a rational homotopy equivalence. However, the existence of such a g would imply that the double suspension $\pi_r S^{2n-1} \to \pi_{r+2} S^{2n+1}$ is a monomorphism on p-components for all sufficiently large primes p, in violation of classical canon (vanishing theorem for the mod p Hopf invariant).

For the vanishing assertion, we follow [**10**] in exploiting the existence of maps $\Omega^2 S^{2n+1} \to S^{2n-1}_{(p)}$ having degree p on the bottom cell. Such maps, which are clearly rational homotopy equivalences, were constructed by F. Cohen, J. Moore and Neisendorfer for $p > 5$, and subsequently for $p = 2$ (F. Cohen) and $p = 3$ (Neisendorfer); references are given in [**10**]. Applying Theorem 6.4 to these maps, taking into account that

$$\mathrm{Ph}(S^{2n-1}_{(p)}, Y) \cong \mathrm{Ph}(S^{2n-1}, Y) = 0$$

for all nilpotent, p-local, finite type targets Y, yields that Ph $(\Omega^2 S^{2n+1}, Y) = 0$ for all such Y.

We note that, with regard to Examples 8.1, 8.2 and 8.3, Theorem 6.1 actually provides more precise information than we have stated; namely, for the non-vanishing assertions, the target Y may be chosen to be a suitable sphere. In Example 8.3, the only possible sphere which works is S^{2n} since, from Theorem 4.1,

$$\mathrm{Ph}(\Omega^2 S^{2n+1}, S^m) = 0, \quad m \neq 2n.$$

There is a striking contrast between Example 8.3 and the other two, made plain in the following:

EXAMPLE 8.4. For any $n \geq 2$, $\mathrm{Ph}(\Sigma\Omega^2 S^{2n+1}, Y) = 0$ for all finite type targets Y.

In fact, thanks to Theorem 6.1, all we need observe is that there is a rational homotopy equivalence

$$\Sigma(\Sigma\Omega^2 S^{2n+1}) \longrightarrow \mathcal{B} = S^{2n+1},$$

namely the double adjoint of the identity map of $\Omega^2 S^{2n+1}$.

One consequence of Example 8.4 is that the suspension map

$$\mathrm{Ph}(\Omega^2 S^{2n+1}, S^{2n}) \to \mathrm{Ph}(\Sigma\Omega^2 S^{2n+1}, S^{2n+1})$$

is the 0 map. (For other non-trivial examples where $\mathrm{Ph}(X, Y) \to \mathrm{Ph}(\Sigma X, \Sigma Y)$ is 0, see [**32**, Ex. 4.2, Ex. 4.3].)

Another consequence of Example 8.4 is that $\mathrm{Ph}(\Omega^2 S^{2n+1}, Y) = 0$ whenever Y is a loop space ΩZ; indeed,

$$\mathrm{Ph}(\Omega^2 S^{2n+1}, \Omega Z) \cong \mathrm{Ph}(\Sigma\Omega^2 S^{2n+1}, Z) = 0.$$

It would thus appear to be difficult to find a target Y yielding an interesting group structure on $\mathrm{Ph}(\Omega^2 S^{2n+1}, Y)$. Nevertheless, at least for $n = 2$, we may find such a Y, namely (the rational H-space) BS^3; a description of the group $\mathrm{Ph}(\Omega^2 S^5, BS^3)$ will follow in Section 9.

A more general version of Example 8.3 may be obtained with the aid of Theorem 6.4.

EXAMPLE 8.5 ([**19**, Ex. 3]). For any $n \geq 2$, $k \geq 1$, $\mathrm{Ph}(\Omega^{k+1}S^{2n+k}, S^{2n}) \neq 0$. The proof is based on the existence of rational homotopy equivalences

$$\begin{aligned} \Omega^2 S^{2n+1} &\to \Omega^{k+1}S^{2n+k}, & k \text{ odd},\\ \Omega^k S^{2n+k-1} \times \Omega^{k+1}S^{4n+2k-1} &\to \Omega^{k+1}S^{2n+k}, & k \text{ even};\end{aligned}$$

details are omitted.

9. Computing Ph(X, $-$) for special values of X: II

We now wish to refine the conclusions in Examples 8.1 and 8.3 and obtain computations of the groups $\mathrm{Ph}(\Sigma C, S^5)$ and $\mathrm{Ph}(\Omega^2 S^5, BS^3)$. The computations of these groups are implicit, though not explicit, in [**10**] and I am grateful to McGibbon for outlining a $\varprojlim^1$ computation of $\mathrm{Ph}(\Omega^2 S^5, BS^3)$ to me. The route we follow here, based on Theorem 4.1, is different, and it is reassuring that the end results agree!

EXAMPLE 9.1 (EXAMPLE 8.1 ELABORATED). $\mathrm{Ph}(\Sigma C, S^5) \cong \mathbb{R} \oplus \bigoplus_{p \in \mathbb{P}} \mathbb{Z}_{p^\infty}$.

EXAMPLE 9.2 (EXAMPLE 8.3 ELABORATED). $\mathrm{Ph}(\Omega^2 S^5, BS^3) \cong \mathbb{R} \oplus \bigoplus_p \mathbb{Z}_{p^\infty}$, the latter sum extending over all primes.

The arguments for Examples 9.1 and 9.2 are very similar. We sketch the computation for Example 9.1 and then indicate the modification necessary to handle Example 9.2.

Let $j\colon S^5 \to K(\mathbb{Z}, 5)$ represent a generator of $\pi_5 K(\mathbb{Z}, 5)$; j is an integral approximation. We abbreviate $S^5 = S$ and write $L = K(\mathbb{Z}, 5)$ to conform to the notation of (4.4). Recall that Theorem 4.1 asserts that there is an isomorphism

$$\mathrm{Ph}(\Sigma C, S) \cong [\Sigma C_{(0)}, S]/q^*[\Sigma^2 C_\tau, S].$$

The middle square of (4.4) becomes

$$\begin{array}{ccc} [\Sigma C_{(0)}, S] & \xleftarrow{q^*} & [\Sigma^2 C_\tau, S] \\ \downarrow{\cong} & & \downarrow{j_*} \\ [\Sigma C_{(0)}, L] & \xleftarrow{q^*} & [\Sigma^2 C_\tau, L], \end{array} \tag{9.1}$$

and the bottom arrow may be identified with the canonical epimorphism

$$\widehat{\mathbb{Z}}/\mathbb{Z} \cong \mathrm{Ext}(\mathbb{Q}, \mathbb{Z}) \longleftarrow \mathrm{Ext}(\mathbb{Q}/\mathbb{Z}, \mathbb{Z}) \cong \widehat{\mathbb{Z}}.$$

To understand the top arrow in (9.1), it thus suffices to understand the right arrow in (9.1). Now we claim that

$$[\Sigma^2 C_\tau, Y] \underset{\cong}{\xrightarrow{\check{e}_*}} [\Sigma^2 C_\tau, \check{Y}] \cong \prod_p [\Sigma^2 C_\tau, Y_{(p)}], \tag{9.2}$$

where $Y = S$ or L. This follows from the fact that the homotopy-fiber of $\check{e}\colon Y \to \check{Y}$ is a rational space together with the fact that $\Sigma^2 C_\tau$ is a torsion space. Therefore, the right arrow in (9.1) may be investigated one prime at a time.

Since $\mathrm{Ph}(\Sigma C, S_{(p)}) = 0$ and since

$$[\Sigma^2 C_{(0)}, S_{(p)}] = 0 = [\Sigma^2 C_{(0)}, L_{(p)}],$$

(4.4) gives us a commutative diagram with short exact rows

$$
\begin{array}{ccccccccc}
0 & \leftarrow & [\Sigma C_{(0)}, S_{(p)}] & \xleftarrow{q^*} & [\Sigma^2 C_\tau, S_{(p)}] & \leftarrow & [\Sigma^2 C, S_{(p)}] & \leftarrow & 0 \\
 & & \downarrow \cong & & \downarrow j_* & & \downarrow j_* & & \\
0 & \leftarrow & [\Sigma C_{(0)}, L_{(p)}] & \xleftarrow{q^*} & [\Sigma^2 C_\tau, L_{(p)}] & \leftarrow & [\Sigma^2 C, L_{(p)}] & \leftarrow & 0,
\end{array}
\tag{9.3}
$$

whose bottom row may be identified with

$$
\begin{array}{ccccccccc}
0 & \longleftarrow & \operatorname{Ext}(\mathbb{Q}, \mathbb{Z}_{(p)}) & \longleftarrow & \operatorname{Ext}(\mathbb{Q}/\mathbb{Z}, \mathbb{Z}_{(p)}) & \longleftarrow & \operatorname{Hom}(\mathbb{Z}, \mathbb{Z}_{(p)}) & \longleftarrow & 0 \\
 & & \cong \downarrow & & \cong \downarrow & & \cong \downarrow & & \\
 & & \widehat{\mathbb{Z}}_p/\mathbb{Z}_{(p)} & & \widehat{\mathbb{Z}}_p & & \mathbb{Z}_{(p)}. & &
\end{array}
$$

The splitting of $\Sigma^2 C_{(p)}$ described in Example 8.1 readily implies that

$$
j_*[\Sigma^2 C, S_{(p)}] = \begin{cases} p \cdot [\Sigma^2 C, L_{(p)}] \cong p\mathbb{Z}_{(p)}, & p \in \mathbb{P} \\ [\Sigma^2 C, L_{(p)}] \cong \mathbb{Z}_{(p)}, & p \notin \mathbb{P}, \end{cases}
\tag{9.4}
$$

and a diagram chase of (9.3) then shows that

$$
j_*[\Sigma^2 C_\tau, S_{(p)}] = \begin{cases} p \cdot [\Sigma^2 C_\tau, L_{(p)}] \cong p\widehat{\mathbb{Z}}_{(p)}, & p \in \mathbb{P} \\ [\Sigma^2 C_\tau, L_{(p)}] \cong \widehat{\mathbb{Z}}_{(p)}, & p \notin \mathbb{P}. \end{cases}
\tag{9.5}
$$

Reverting to (9.2), we infer from (9.5) that

$$
j_*[\Sigma^2 C_\tau, S] \cong \prod_p p^{\epsilon(p)} \widehat{\mathbb{Z}}_p, \epsilon(p) = \begin{cases} 1, & p \in \mathbb{P} \\ 0, & p \notin \mathbb{P}, \end{cases}
$$

and then

$$
\begin{aligned}
[\Sigma C_{(0)}, S]/q^*[\Sigma^2 C_\tau, S] &\cong [\Sigma C_{(0)}, L]/q^* j_*[\Sigma^2 C_\tau, S] \quad (\text{eq. } (9.1)) \\
&\cong \frac{\prod_p \widehat{\mathbb{Z}}_p / \prod_p p^{\epsilon(p)} \widehat{\mathbb{Z}}_p}{\mathbb{Z}} \\
&\cong \frac{\prod_{p \in \mathbb{P}} \mathbb{Z}/p}{\mathbb{Z}} \\
&\cong \mathbb{R} \oplus \bigoplus_{p \in J} \mathbb{Z}_{p^\infty}.
\end{aligned}
$$

The computation of $\operatorname{Ph}(\Omega^2 S^5, BS^3)$ is virtually a verbatim repetition of the one just given. The only modification necessary is to find the appropriate analogue of (9.4). That analogue states that the image of

$$
j_* \colon [\Sigma \Omega^2 S^5, BS^3_{(p)}] \to [\Sigma \Omega^2 S^5, K(\mathbb{Z}_{(p)}, 4)] \cong \mathbb{Z}_{(p)}
$$

is precisely $p \cdot [\Sigma \Omega^2 S^5, K(\mathbb{Z}_{(p)}, 4)] \cong p\mathbb{Z}_{(p)}$, for all primes p. In fact, this is precisely the information received from the existence of the Cohen-Moore-Neisendorfer maps $\Omega^2 S^5 \to S^3_{(p)}$ alluded to in Example 8.3.

The above method for computing $\mathrm{Ph}(\Sigma C, S^5)$ and $\mathrm{Ph}(\Omega^2 S^5, BS^3)$ may also be applied to extend the examples of Section 7. For instance, if $\overline{M}^3 = \overline{M}^3(I, J)$ is the Meier space, then

$$\mathrm{Ph}(\overline{M}^3, BS^3) \cong \mathbb{R} \oplus \bigoplus_{p \in J} \mathbb{Z}_{p^\infty};$$

in particular, if $J = \{p\}$, a single prime, then

$$\mathrm{Ph}(\overline{M}^3, BS^3) \cong \mathbb{R} \oplus \mathbb{Z}_{p^\infty}.$$

Now the Cohen-Moore-Neisendorfer map $\Omega^2 S^5 \to S^3_{(p)}$ allows us to construct a rational homotopy equivalence

$$g \colon \Omega^2 S^5 \to \overline{M}^3, J = \{p\}.$$

We also have the rational homotopy equivalence $j' \colon \overline{M}^3 \to K(\mathbb{Z}, 3)$; see Example 7.1. Composing these two maps, we obtain a rational homotopy equivalence

$$g' \colon \Omega^2 S^5 \xrightarrow{g} \overline{M}^3 \xrightarrow{j'} K(\mathbb{Z}, 3).$$

Because the three groups $\mathrm{Ph}(\Omega^2 S^5, BS^3)$, $\mathrm{Ph}(\overline{M}^3, BS^3)$ and $\mathrm{Ph}\big(K(\mathbb{Z}, 3), BS^3\big)$ are mutually non-isomorphic, we immediately infer that the surjections (see Theorem 6.4)

$$g^* \colon \mathrm{Ph}(\overline{M}^3, BS^3) \to \mathrm{Ph}(\Omega^2 S^5, BS^3), \quad g'^* \colon \mathrm{Ph}\big(K(\mathbb{Z}, 3), BS^3\big) \to \mathrm{Ph}(\Omega^2 S^5, BS^3)$$

have non-0 kernels.

10. Eckmann-Hilton duality

Several of the results in this paper may be dualized à la Eckmann-Hilton. And some of these dual results are even true.

Examples of the latter kind are Theorems 6.1 and 6.4. Here are the statements of their duals:

THEOREM 10.1 ([**19**, Th. 1′]). *If Y is a finite type target, then the following statements are equivalent:*

- (i) $\mathrm{Ph}(X, Y) = 0$ *for all X having the homotopy type of a skeleton-finite CW-complex (see footnote 6):*
- (ii) $\mathrm{Ph}\big(K(\mathbb{Z}, n), Y\big) = 0$ *for all n;*
- (iii) *There exists a rational homotopy equivalence from a product of $K(\mathbb{Z}, m)$'s to the basepoint component of ΩY.*

THEOREM 10.2 ([**19**, Th. 2(ii)]). *A map $g \colon Y \to Y'$ of finite type targets, inducing an epimorphism on the higher rational homotopy groups, induces a surjection*

$$g_* \colon \mathrm{Ph}(X, Y) \to \mathrm{Ph}(X, Y')$$

for all X having the homotopy type of a skeleton-finite CW-complex.

It is tempting to consider the dual of the vanishing result,

$$\mathrm{Ph}(\Omega S^n, -) \equiv 0, \quad n \geq 2,$$

discussed immediately after the statement of Theorem 6.2. The dual result, which (presumably) would assert that

$$\mathrm{Ph}\big(-, \Sigma K(\mathbb{Z}, n)\big) \equiv 0, \quad n \geq 1,$$

is, however, false, at least when $n = 1$. For $\Sigma K(\mathbb{Z}, 1) = S^2$ and, by Theorem 4.1,

$$\mathrm{Ph}\big(K(\mathbb{Z}, 2), S^2\big) \neq 0.$$

Of course, the case $n = 1$ is exceptional in that $K(\mathbb{Z}, 1)$, and hence also $\Sigma K(\mathbb{Z}, 1)$, is a finite CW-complex. We ask:

QUESTION 10.1. Is $\mathrm{Ph}\big(-, \Sigma K(\mathbb{Z}, n)\big) \not\equiv 0$ when $n \geq 2$?

It is a straightforward task to formulate and prove valid duals of Examples 7.1 and 7.2; we leave the details to the interested reader. As for the examples in Section 8 and Section 9, it strikes us as plausible that Examples 8.1 and 8.3 are dualizable but that Example 8.2 is not. However, we make no attempt to pursue these questions in this paper.

We should mention that Theorem 6.3 has a correct dualization. We refer to [**14**] for that statement and for a discussion of the success/failure to obtain valid duals of various of the results in [**10**].

11. Spaces of the same n-type for all n

Following Wilkerson [**46**], we write $\mathrm{SNT}(Z)$ for the set of all homotopy types of CW-spaces having the same n-type as the CW-space Z for all n. According to [**46**, Th. I], there is a bijection

$$\mathrm{SNT}(Z) \cong \varprojlim{}^1 \mathrm{Aut}(Z^{(n)}), \tag{11.1}$$

where $\mathrm{Aut}(Z^{(n)})$ refers to the group of homotopy classes of self-homotopy equivalences of $Z^{(n)}$. The identification (11.1) is the starting point for a systematic study of $\mathrm{SNT}(Z)$ by McGibbon and Moller [**16, 17, 18, 15**]. In this section, we content ourselves with briefly surveying some points of contact of SNT-theory with phantom map theory.

One way of establishing a contact between the two theories is to relate the $\varprojlim{}^1$ description of $\mathrm{Ph}(X, Y)$ (Corollary 3.1) with the $\varprojlim{}^1$ description of $\mathrm{SNT}(Z)$ ((11.1)), for suitable Z. We take $Z = X \times \Omega Y$, where X has the homotopy type of a skeleton-finite CW-complex, and define a homomorphism of inverse systems

$$\big\{[X, \Omega Y^{(n)}]\big\} \to \big\{\mathrm{Aut}(Z^{(n)})\big\} \tag{11.2}$$

by associating to the map $f\colon X \to \Omega Y^{(n)}$ the homotopy equivalence $h\colon X^{(n)} \times \Omega Y^{(n)} \to X^{(n)} \times \Omega Y^{(n)}$ given by

$$(x, \lambda) \to \big(x, \lambda * f^{(n)}(x)\big);$$

here, $f^{(n)}\colon X^{(n)} \to \Omega Y^{(n)}$ is induced by f and $\lambda * f^{(n)}(x)$ denotes loop multiplication. Of course, (11.2) gives rise to a map

$$\begin{aligned}(11.3) \quad \mathrm{Ph}(X, Y) &\cong^{9} \varprojlim{}^1 [X, \Omega Y^{(n)}] \\ &\to \varprojlim{}^1 \mathrm{Aut}\big((X \times \Omega Y)^{(n)}\big) \cong \mathrm{SNT}(X \times \Omega Y).\end{aligned}$$

[9]See footnote 6.

There is another, more direct, way of obtaining a map from $\mathrm{Ph}(X,Y)$ to $\mathrm{SNT}(X\times\Omega Y)$; namely, to a phantom map $\phi\colon X\to Y$, associate the homotopy-fiber F_ϕ[10] of ϕ. The resulting map

$$(11.4) \qquad F\colon \mathrm{Ph}(X,Y)\to \mathrm{SNT}(X\times\Omega Y),$$

modeled on (the dual of) an idea in [**9**], was introduced and utilized in [**11**] to obtain results about $\mathrm{SNT}(X\times\Omega Y)$ augmenting the work of [**16**].

We do not know whether the maps (11.3) and (11.4) are equivalent and we have not attempted to make calculations based on (11.3). However, calculations based on (11.4) have proved to be successful and we now describe the main result on (11.4) obtained in [**11**].

THEOREM 11.1 ([**11**, Th. 3.4]). *If the grouplike target Y has the homotopy type of (an iterated loop space of) a finite CW-complex and if* $\mathrm{Ph}\big(K(\mathbb{Z},n),Y\big)\neq 0$, *then the image of F is an uncountable subset of* $\mathrm{SNT}\big(K(\mathbb{Z},n)\times\Omega Y\big)$. *If, in addition,* $\mathrm{Ph}\big(K(\mathbb{Z},n)\wedge K(\mathbb{Z},n),Y\big)=0$, *then the image of F consists of H-spaces.*

We remark that the proof of the uncountability of the image of F in [**11**] depends on the fact that $\mathrm{Ph}\big(K(\mathbb{Z},n),\check{Y}\big)$ is uncountable; see Theorem 5.2. Under the conditions of Theorem 11.1, it is also true that $S\,\mathrm{Ph}\big(K(\mathbb{Z},n),Y\big)\neq 0$; see Theorems 5.1 and 5.2. For ϕ in $S\,\mathrm{Ph}\big(K(\mathbb{Z},n),Y\big)$, the homotopy-fiber F_ϕ is, in the terminology of [**18, 15**], a "clone" of $K(\mathbb{Z},n)\times\Omega Y$; that is, F_ϕ and $K(\mathbb{Z},n)\times\Omega Y$ have the same m-type for all m and are p-equivalent for all primes p.

A number of specific examples illustrating Theorem 11.1 are described in [**11**, Section 4]. We mention here an especially simple special case of [**11**, Ex. 4.3] which follows from the proof of the first part of Theorem 11.1 (though not from the actual statement of Theorem 11.1).

EXAMPLE 11.1. $\mathrm{SNT}\big(K(\mathbb{Z},3)\times S^3\big)$ is an uncountable set.

Less is claimed in [**11**, Ex. 4.3], namely that $\mathrm{SNT}\big(K(\mathbb{Z},3)\times S^3\big)\neq 0$. However, the stronger assertion holds thanks to the fact that Theorem 5.2 extends to show that

$$\mathrm{Ph}\big(K(\mathbb{Z},3),BS^3\big)\to \mathrm{Ph}\big(K(\mathbb{Z},3),\check{B}S^3\big)$$

is an epimorphism of $\mathbb{Q}$-vector spaces of uncountable dimension. It is interesting to note that McGibbon [**15**, Ex. 1] constructs uncountably many clones of $K(\mathbb{Z},3)\times S^3$.[11] On the other hand, the uncountable subset of $\mathrm{SNT}\big(K(\mathbb{Z},3)\times S^3\big)$ obtained above consists entirely of non-clones of $K(\mathbb{Z},3)\times S^3$!

Not surprisingly, the foregoing discussion may be dualized to obtain maps

$$\mathrm{Ph}(X,Y)\cong \varprojlim{}^1[\Sigma X, Y^{(n)}]\to \varprojlim{}^1\,\mathrm{Aut}\big((Y\vee\Sigma X)^{(n)}\big)\cong \mathrm{SNT}(Y\vee\Sigma X),$$

and

$$C\colon \mathrm{Ph}(X,Y)\to \mathrm{SNT}(Y\vee\Sigma X),$$

with $C(\phi)$ being the homotopy type of the mapping cone C_ϕ of ϕ. The use of C_ϕ in connection with SNT-theory goes back to [**9**]; see also the recent papers of Shitanda [**37, 38, 39**]. The dual of the first assertion of Theorem 11.1 states:

---

[10] To ensure that F_ϕ is path-connected, we should and do require Y to be 1-connected for the remainder of this section.

[11] It is not clear from McGibbon's construction whether his clones of $K(\mathbb{Z},3)\times S^3$ are of the form F_ϕ with $\phi\colon K(\mathbb{Z},3)\to BS^3$ a special phantom map.

THEOREM 11.1′ ([**11**, Th. 3.4′]). *If the domain X is of the form $\Sigma^k W$, where W is a 1-connected finite type, finite Postnikov space, $k \geq 1$, and* $\mathrm{Ph}(X, S^n) \neq 0$, *then the image of C is an uncountable subset of* $\mathrm{SNT}(S^n \vee \Sigma X)$.

Theorem 11.1′ bears the same relationship to [**17**] as Theorem 11.1 does to [**16**]. Once again, we refer to [**11**, Section 4] for specific illustrations of Theorem 11.1′.

12. Weak identities

In this final section, we discuss a natural generalization of the notion of S-phantom map which has proved to be of utility in homotopy theory.

DEFINITION 12.1 (CF. DEFINITION 1.1). Two maps $u, v \colon X \to Y$ are said to be weakly S-homotopic if, for any $W \in S$ and any $j \colon W \to X$, the compositions

$$W \xrightarrow{j} X \xrightarrow{u} Y, \quad W \xrightarrow{j} X \xrightarrow{v} Y$$

are homotopic.

For a given $u \colon X \to Y$, we write $\mathrm{Ph}_S(X, Y; u)$ for the subset of $[X, Y]$ consisting of homotopy classes of maps $v \colon X \to Y$ such that u and v are weakly S-homotopic (cf. [**31**, Section 1]). If u is the constant map to the basepoint, then obviously $\mathrm{Ph}_S(X, Y; u)$ is none other than $\mathrm{Ph}_S(X, Y)$. When either X is a cogroup or Y is grouplike, so that $[X, Y]$ has a natural group structure, it is readily seen that $\mathrm{Ph}_S(X, Y)$ is a normal subgroup of $[X, Y]$ and that $\mathrm{Ph}_S(X, Y; u)$ is the coset of u with respect to $\mathrm{Ph}_S(X, Y)$; in particular, there is a bijection

$$\mathrm{Ph}_S(X, Y; u) \cong \mathrm{Ph}_S(X, Y). \tag{12.1}$$

In case $\mathcal{S} = \mathcal{F}$, we abandon the terminology "weakly F-homotopic" in favor of the classical terminology "weakly homotopic"[12] and adopt the abbreviated notation $\mathrm{Ph}(X, Y; u)$ for $\mathrm{Ph}_{\mathcal{F}}(X, Y; u)$. We are particularly interested in the situation where $X = Y$ and $u = 1$, the identity map. Following [**30, 31**], we write

$$WI(X) = \mathrm{Ph}(X, X; 1)$$

and refer to $WI(X)$ as the set of *weak identities* of X. Composition of self-maps of X induces a group structure on $WI(X)$, whether or not X is a cogroup or a grouplike space; in fact, $WI(X)$ is a normal subgroup of $\mathrm{Aut}(X)$. By analogy with Corollary 3.1, McGibbon and Moller [**18**] note:

PROPOSITION 12.1. *There is an isomorphism*

$$WI(X) \cong \varprojlim{}^1 \pi_1\big(\mathrm{aut}(X^{(n)})\big), \tag{12.2}$$

$\mathrm{aut}(X^{(n)})$ *denoting the H-space of self-homotopy equivalences of $X^{(n)}$; thus $WI(X)$ is Abelian.*

Moreover, if X is nilpotent of finite type, then (12.2) displays $WI(X)$ as a divisible, Abelian group of the type described in Proposition 3.2.

Proposition 12.1 enables us to answer the question raised after [**33**, Th. 3.2] affirmatively and thus sharpen that theorem.

[12] The notions of "weakly homotopic maps", "weak H-spaces", etc. figure in Adams' work on E.H. Brown's representability theorem [**2**] and also appear in the 1959–60 Séminaire Cartan-Moore, in connection with a homological proof of the Bott periodicity theorem.

THEOREM 12.1. *If X is nilpotent of finite type and N is a normal subgroup of* $\mathrm{Aut}(X)$ *such that the quotient group* $\mathrm{Aut}(X)/N$ *is residually finite, then N contains* $WI(X)$.

(According to [**30**, Th. 4.4], $\mathrm{Aut}(X)/WI(X)$ is actually residually finite under rather more general finiteness conditions on X, which we do not spell out here. Thus $WI(X)$ is minimal among normal subgroups of $\mathrm{Aut}(X)$ yielding a residually finite quotient group $\mathrm{Aut}(X)/N$).

In [**31**, Th. 3.2], Theorem 12.1 was proved under the additional assumption that X is a grouplike space (actually a homotopy-associative H-space) with $\pi_1 X$ finite. The proof was based on part (i) of the following theorem.

THEOREM 12.2 ([**31**, Th. 3.1],[**45**]). (i) *If X is a homotopy-associative H-space with $\pi_1 X$ finite, then the bijection*

$$\mathrm{Ph}(X,X) \to WI(X)$$

defined by

$$\phi \to \phi + 1 \quad (\text{Cf. } (12.1)),$$

is an isomorphism, where $\phi+1$ denotes addition in the (non-Abelian) group $[X,X]$;

(ii) *If X is a 1-connected cogroup with $\pi_2 X$ finite, then the bijection*

$$\mathrm{Ph}(X,X) \to WI(X)$$

defined by

$$\phi \to 1 + \phi$$

is an isomorphism.

A version of Theorem 12.2(i) appropriate to the case where X is a rational H-space is given in [**40**, Th. 1.7].

Theorem 12.2 opens up the possibility of computing $WI(X)$ and perhaps learning some of the darker secrets of $\mathrm{Aut}(X)$. Here is a sample computation.

EXAMPLE 12.1 ([**33**, Th. 2.1]). If $X = K(\mathbb{Z},2) \times S^3$, $\mathrm{Aut}(X)$ is the semidirect product of the normal subgroup $WI(X) \cong \mathbb{R}$ and a subgroup $C \cong \mathbb{Z}/2 \times \mathbb{Z}/2$. Generators ξ, η may be chosen for C so that ξ and η act on $WI(X)$ by

$$\xi * w = w^{-1} = \eta * w, \quad w \in WI(X).$$

Related examples may be found in [**40, 41**].

An interesting application of $WI(X)$ to SNT-theory was made by McGibbon and Moller.

EXAMPLE 12.2 ([**18**, Ex. B]). If $Z = K(\mathbb{Z},2) \times \Omega S^3 \times S^3$, there exists Z' in $\mathrm{SNT}(Z)$ such that

$$WI(Z) \cong \mathbb{R}, \quad WI(Z') \cong \mathbb{R} \oplus \bigoplus_p \mathbb{Z}_{p^\infty}.$$

For further results about $WI(X)$, we refer to [**10**, Cor. 4.6] and [**15**, Th. 5].

ACKNOWLEDGEMENTS. I want to thank Ross Geoghegan for his comments in connection with the possibility of extending phantom map theory beyond the homotopy category of CW-spaces, to Joe Neisendorfer for his comments regarding Example 8.3 and to Chuck McGibbon for showing me his computations regarding Example 9.2 and, more generally, for sharing his ideas on phantom map theory with me for the past three years. Much of what is reported on here is drawn from the work of McGibbon and his collaborators; I justify this shameless borrowing on the grounds that I am one of those collaborators.

Finally, I want to thank Peter Hilton for twenty-five years of collaboration and friendship. I find it especially appropriate that the subject matter of my talk at a conference honoring him should rely so much on ideas we and Guido Mislin worked out together some twenty years ago, and I offer this paper to him as a birthday gift.

References

1. J. F. Adams, *An example in homotopy theory*, Proc. Camb. Phil. Soc. **53** (1957), 922–923.
2. ——, *A variant of E.H. Brown's representability theorem*, Topology **10** (1971), 185–198.
3. J. F. Adams and G. Walker, *An example in homotopy theory*, Proc. Camb. Phil. Soc. **60** (1964), 699–700.
4. D. W. Anderson and L. Hodgkin, *The K-theory of Eilenberg-MacLane complexes*, Topology **7** (1968), 317–329.
5. A. K. Bousfield and D. M. Kan, *Homotopy limits, completions and localizations*, Lecture Notes in Math., vol. 304, Springer-Verlag, Berlin-Heidelberg-New York, 1972.
6. F. R. Cohen, J. C. Moore and J. A. Neisendorfer, *Torsion in homotopy groups*, Ann. Math. **109** (1979), 121–168.
7. E. Friedlander and G. Mislin, *Locally finite approximation of Lie groups*, I, Invent. math. **83** (1986), 425–436.
8. ——, *Locally finite approximation of Lie groups*, II, Math. Proc. Camb. Phil. Soc. **100** (1986), 505–517.
9. B. Gray, *Spaces of the same n-type, for all n*, Topology **5** (1966), 241–243.
10. B. Gray and C. A. McGibbon, *Universal phantom maps*, Topology **32** (1993), 371–394.
11. J. R. Harper and J. Roitberg, *Phantom maps and spaces of the same n-type for all n*, J. Pure and Applied Algebra **80** (1992), 123–137.
12. P. Hilton, G. Mislin and J. Roitberg, *Localization of nilpotent groups and spaces*, Notas de Matemática, North Holland Math. Studies, vol. 15, Amsterdam, 1975.
13. C. U. Jensen, *Les foncteurs dérivés de $\varprojlim$ et leurs applications en théorie des modules*, Lecture Notes in Math., vol. 254, Springer-Verlag, Berlin-Heidelberg-New York, 1972.
14. C. A. McGibbon, *Loop spaces and phantom maps*, Contemporary Math. **146** (1993), 297–308.
15. ——, *Clones of spaces and maps in homotopy theory*, Comment. Math. Helvetici **68** (1993), 263–277.
16. C. A. McGibbon and J. M. Moller, *On spaces with the same n-type for all n*, Topology **31** (1992), 177–201.
17. ——, *On infinite dimensional spaces that are rationally equivalent to a bouquet of spheres*, Lecture Notes in Math. **1509** (1992), Springer-Verlag, Berlin-Heidelberg-New York, 285–293.
18. ——, *How can you tell two spaces apart when they have the same n-type for all n?*, London Math. Soc. Lecture Note Series **176** (1992), Cambridge Univ. Press, 131–143.
19. C. A. McGibbon and J. Roitberg, *Phantom maps and rational equivalences*, Amer. J. Math. (to appear).
20. ——, *Phantom maps and rational equivalences*: II, Bol. Soc. Mat. Mex. (to appear).
21. C. A. McGibbon and R.J. Steiner, *Some questions about the first derived functor of the inverse limit*, Preprint.
22. W. Meier, *Localisation, complétion, et applications fantômes*, C.R. Acad. Sci. Paris **281** (1975), 787–789.
23. ——, *Détermination de certains groupes d'applications fantômes*, C.R. Acad. Sci. Paris **283** (1976), 971–974.
24. ——, *Pullback theorems and phantom maps*, Quart. J. Math. **29** (1978), 469–481.
25. H. Miller, *The Sullivan fixed point conjecture on maps from classifying spaces*, Ann. Math. **120** (1984), 39–87.
26. J. Milnor, *On characteristic classes for spherical fibre spaces*, Comment. Math. Helvetici **43** (1968), 51–77.

27. N. Oda and Y. Shitanda, *On the unstable homotopy spectral sequences*, Manuscripta Math. **56** (1986), 19–35.
28. ——, *Equivariant phantom maps*, Publ. RIMS, Kyoto Univ. **24** (1988), 811–820.
29. ——, *Localization, completion and detecting equivariant maps on skeletons*, Manuscripta Math. **65** (1989), 1–18.
30. J. Roitberg, *Residually finite, Hopfian and co-Hopfian spaces*, Contemporary Math. **37** (1985), 131–144.
31. ——, *Weak identities, phantom maps and H-spaces*, Israel J. Math. **66** (1989), 319–329.
32. ——, *Phantom maps, cogroups and the suspension map*, Quaestiones Math. **13** (1990), 335–347.
33. ——, *Note on phantom phenomena and groups of self-homotopy equivalences*, Comment. Math. Helvetici **66** (1991), 448–457.
34. ——, *Phantom maps and torsion*, Topology and its Applications (to appear).
35. Y. Shitanda, *On the homotopy groups of $Map_*(BH, X)$*, Memoirs of the Faculty of Science, Kyushu Univ. Series A, Math. **43** (1989), 1–5.
36. ——, *Fibrations over classifying spaces*, Yokohama Math. J. **38** (1991), 121–127.
37. ——, *Uncountably many loop spaces of the same n-type for all n*: I, Yokohama Math. J. (to appear).
38. ——, *Uncountably many loop spaces of the same n-type for all n*: II, Yokohama Math. J. (to appear).
39. ——, *Uncountably many infinite loop spaces of the same n-type for all n*, Math. J. of Okayama Univ. (to appear).
40. ——, *Phantom maps and monoid structure of endomorphisms of $K(\mathbb{Z}, m) \times S^n$*, Publ. RIMS, Kyoto Univ. (to appear).
41. ——, *Monoid structure of endomorphisms of $HP^\infty \times S^n$*, Yokohama Math. J. (to appear).
42. E. H. Spanier, *Algebraic Topology*, McGraw-Hill, 1966.
43. R. J. Steiner, *Localization, completion and infinite complexes*, Mathematika **24** (1977), 1–15.
44. D. Sullivan, *Genetics of homotopy theory and the Adams conjecture*, Ann. Math. **100** (1974), 1–79.
45. P. Touhey, CUNY Doctoral Dissertation.
46. C. Wilkerson, *Classification of spaces of the same n-type for all n*, Proc. Amer. Math. Soc. **60** (1976), 279–285.
47. A. Zabrodsky, *On phantom maps and a theorem of H. Miller*, Israel J. Math. **58** (1987), 129–143.
48. ——, *Maps between classifying spaces*, Algebraic Topology and Algebraic K-theory, Ann. Math. Studies **113** (1987), 228–246.

Department of Mathematics and Statistics, Hunter College, CUNY, 695 Park Avenue, NYC 10021 and Department of Mathematics, Graduate School and University Center, CUNY, 33 West 42 Street, NYC 10036

E-mail address: jorhc@cunyvm.cuny.edu

Centre de Recherches Mathématiques
CRM Proceedings and Lecture Notes
Volume **6**, 1994

On Groups Generated by Three Involutions, Two of which Commute

Denis Sjerve and Michael Cherkassoff

ABSTRACT. We completely determine those alternating groups A_n, symmetric groups S_n, and projective groups $PSL_2(q)$ and $PGL_2(q)$, which can be generated by three involutions, two of which commute.

1. Introduction

In this paper we investigate the possibility of generating certain finite groups G by three involutions, two of which commute. That is, when does G have a presentation of the form

$$G \cong \langle R_1, R_2, R_3 \mid R_1^2 = R_2^2 = R_3^2 = 1, R_1R_2 = R_2R_1, \mathrm{ETC}\rangle? \tag{1}$$

Here ETC denotes the extra relations needed to present the finite group. To avoid trivial cases we assume throughout this paper that the R_j, as elements of G, are not the identity and are mutually distinct.

Remarks. (a) In the trivial cases the groups we obtain are trivial, $\mathbb{Z}_2 \oplus \mathbb{Z}_2$ or dihedral D_{2k}, where k is order of the product of two non-commuting involutions. Note that in particular, when $k = 3$ we get the symmetric group S_3.

(b) Without any extra relations the group presented in (1) is the free product $(\mathbb{Z}_2 \oplus \mathbb{Z}_2) * \mathbb{Z}_2$, hence infinite.

We will answer this question for the alternating groups A_n, the symmetric groups S_n, the projective special linear groups $PSL_2(q)$ over finite fields $GF(q)$, and the projective general linear groups $PGL_2(q)$ over $GF(q)$, where $q = p^n$ and p is any prime. Our results are given in the following four theorems.

THEOREM 1.1. *The alternating group A_n has a presentation as in (1) if, and only if, $n = 5$ or $n \geq 9$.*

THEOREM 1.2. *The symmetric group S_n has a presentation as in (1) if, and only if, $n \geq 4$.*

THEOREM 1.3. *The projective special linear group $PSL_2(q)$ has a presentation as in (1) if, and only if, $q \neq 2, 3, 7, 9$.*

THEOREM 1.4. *The projective general linear group $PGL_2(q)$ has a presentation as in (1) if, and only if, $q \neq 2$.*

1991 *Mathematics Subject Classification.* Primary: 20G40; Secondary: 05C45.
This is the final form of the paper.

The motivation for this paper arose in combinatorics, more specifically, in the area of hamiltonian cycles. A conjecture in graph theory, arising from a question of Lovász [**7**], is that every vertex transitive graph, with just 4 exceptions, has a hamiltonian path. In particular we can consider this conjecture for Cayley graphs associated to finite presentations of finite groups. It turns out that the above presentation on three involutions is very amenable to this conjecture. As far as Cayley graphs are concerned a reasonable conjecture is

CONJECTURE. For any finite presentation of any finite group G (with the trivial exception of $\mathbb{Z}_2$) there is a Hamiltonian cycle in the associated Cayley graph.

The paper [**10**] by Witte and Gallian contains an excellent survey on this subject.

In Section 2 of this paper we collect the preliminaries and background material we need. Section 3 contains the proofs of Theorems 1.1 and 1.2 and Section 4 contains the proofs of Theorems 1.3 and 1.4. And finally in Section 5 we prove results on the existence of Hamiltonian cycles.

2. Preliminaries

First we give some definitions.

DEFINITION 2.1. The *projective general linear group* over the finite field of $q = p^n$ elements, $PGL_2(q)$, is the quotient group of the group of invertible 2×2 matrices, with entries in the finite field of q elements, by its center.

DEFINITION 2.2. The *projective special linear group* over the finite field of $q = p^n$ elements, $PSL_2(q)$, is the quotient group of the group of 2×2 matrices of determinant 1, with entries in the finite field of q elements, by its center.

DEFINITION 2.3. Let the group G be generated by distinct elements g_1, g_2, $\dots$, g_n all different from the identity. Then *the Cayley graph* for G with this set of generators is the graph with vertices corresponding to the group elements and with vertices s and t connected by an edge if and only if there is i, such that $s = tg_i$.

In the case when g_i is an involution it is customary to consider the graph as having only one edge between s and t.

DEFINITION 2.4. A *Hamiltonian path* in a graph is an edge-path which visits each vertex exactly once.

DEFINITION 2.5. A *Hamiltonian cycle* is a Hamiltonian path having the same vertex as its beginning and end.

DEFINITION 2.6. The *triangle group* $T(r, s, t)$ is the group with the presentation

$$\langle x_1, x_2, x_3 \mid x_1^r = x_2^s = x_3^t = x_1x_2x_3 = 1\rangle.$$

It is known that a triangle group is finite if and only if $1/r + 1/s + 1/t > 1$.

The triangle groups are closely related to the geometry of Riemann surfaces. To see this consider the triangle, whose angles are $\pi/r, \pi/s$ and π/t. This triangle is on the sphere if $1/r + 1/s + 1/t > 1$, the Euclidean plane if $1/r + 1/s + 1/t = 1$, and the upper half plane in all other cases.

Letting x_1, x_2, x_3 be rotations by $2\pi/r$, $2\pi/s$, $2\pi/t$ about the respective vertices, we see that the group generated by x_1, x_2, x_3 is isomorphic to $T(r, s, t)$. It is a group of symmetries of the tesselation produced by repeated reflections of the triangle in free sides.

Secondly, we need some group-theoretic results. The most important result for us is the Dickson classification of subgroups of $PSL_2(q)$ [**4, 6, 9**].

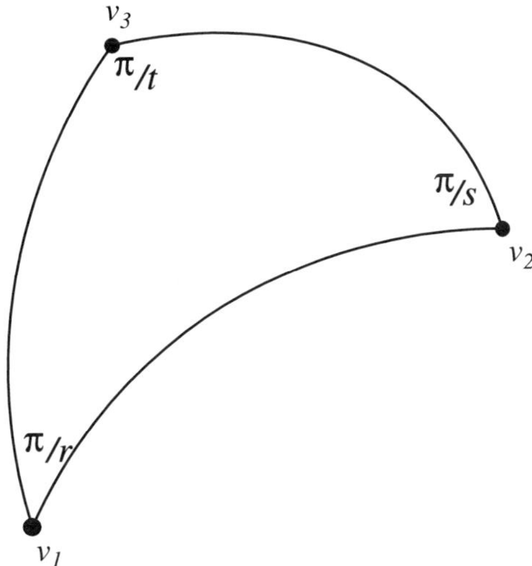

FIGURE 1. Fundamental triangle.

THEOREM 2.7. *The subgroups of $PSL_2(q)$ fall into three types:*

The projective subgroups: *If $m \mid n$ then $GF(p^m)$ is a subfield of $GF(q)$ and therefore $PSL_2(p^m)$ is a subgroup of $PSL_2(q)$. If also $2m \mid n$ and $p > 2$ then $PGL_2(p^m)$ is a subgroup of $PSL_2(q)$. Any subgroup isomorphic to either $PSL_2(p^m)$ or $PGL_2(p^m)$ is called a projective subgroup.*

The affine subgroups: *Consider the subgroups of $PSL_2(p^{2n})$ which have one of the following forms: either*

$$\left\{\begin{bmatrix} a & b \\ 0 & a^{-1} \end{bmatrix} : a, b \in GF(q), a \neq 0\right\},$$

or

$$\left\{\begin{bmatrix} \lambda & 0 \\ 0 & \bar{\lambda} \end{bmatrix} : \lambda \in GF(p^{2n})^*, \lambda^{q+1} = 1\right\},$$

where $\lambda \mapsto \bar{\lambda} = \lambda^q$ is the involution in the field $GF(p^{2n})$. Then any subgroup of $PSL_2(q)$ isomorphic to a subgroup of one of the above is called an affine subgroup.

The exceptional subgroups: *The finite non-cyclic triangle groups are:*

$T(2,2,t)$: *the dihedral group of order $2t$, $t \geq 2$;*
$T(2,3,3)$: *the tetrahedral group $\cong A_4$;*
$T(2,3,4)$: *the octahedral group $\cong S_4$;*
$T(2,3,5)$: *the icosahedral group $\cong A_5$.*

Subgroups of $PSL_2(q)$ isomorphic to one of these are called exceptional groups.

We also need a result, due to Dyck, on representations of $PSL_2(q)$ as permutation groups [**1, 4**]

THEOREM 2.8. *$PSL_2(q)$ may be represented as a transitive permutation group on $q+1$ symbols but on no fewer, except when $q = 5, 7, 9, 11$, for which the minimum number of symbols is $5, 7, 6, 11$ respectively.*

The next two lemmas are obvious and useful in investigating the generation of groups:

LEMMA 2.9. *Suppose some permutations* $t_1, \ldots, t_l$, *on the set of symbols* $\{i_1, \ldots, i_k\}$, *generate a subgroup isomorphic to* A_k. *Let* $t = (i_p, i_q, j)$ *be a 3-cycle, where* i_p, i_q *are from this set and* j *is not. Then the permutations* $t_1, \ldots, t_l, t$, *on the set of symbols* $\{i_1, \ldots, i_k, j\}$, *generate a subgroup isomorphic to* A_{k+1}.

LEMMA 2.10. *Suppose we have three involutions, two of which commute, and the third does not commute with them or their product. Then the group these involutions generate is not dihedral.*

We also need to point out some isomorphisms:

PROPOSITION 2.11. $PSL_2(2) = PGL_2(2) \cong S_3$, $PSL_2(3) \cong A_4$, $PGL_2(3) \cong S_4$, $PSL_2(5) \cong A_5$, $PGL_2(5) \cong S_5$ *and* $PSL_2(9) \cong A_6$.

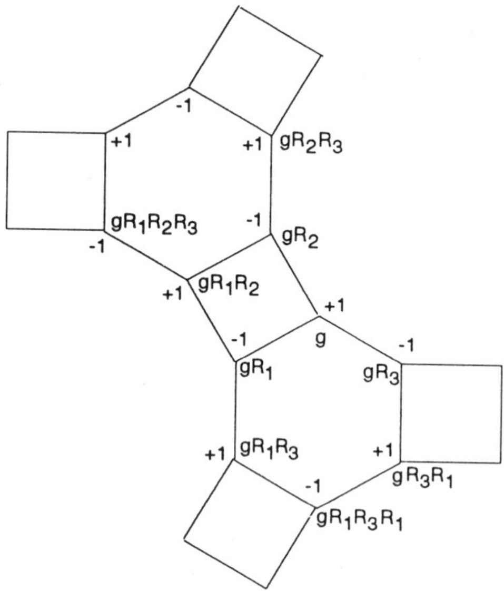

FIGURE 2. The Cayley graph with local orientation.

The Cayley graph is a 1-dimensional simplicial complex constructed from the vertex set G by right multiplication by the generators $g_1, \ldots, g_n$. The group G acts on itself by left multiplication, and we can extend this action to the Cayley graph by making it linear on the edges. Thus the Cayley graph has a natural left G-action. We now want to add 2-cells to the Cayley graph to obtain a surface, and then extend the G-action to the surface. We limit our attention to trivalent Cayley graphs, that is graphs such that each vertex has exactly 3 edges coming into it.

A Cayley graph that is trivalent must come from a presentation in one of two ways: either the group is presented by 3 involutions, or it is generated by an involution and one other element of order bigger than two (see fig. 3); therefore having a presentation of one of the following forms:

(a) $$G = \langle R_1, R_2, R_3 \mid R_1^2 = R_2^2 = R_3^2 = 1, \text{ETC} \rangle$$

(b) $$G = \langle R, S | R^2 = S^n = 1, \text{ETC} \rangle, \quad \text{where } n > 2.$$

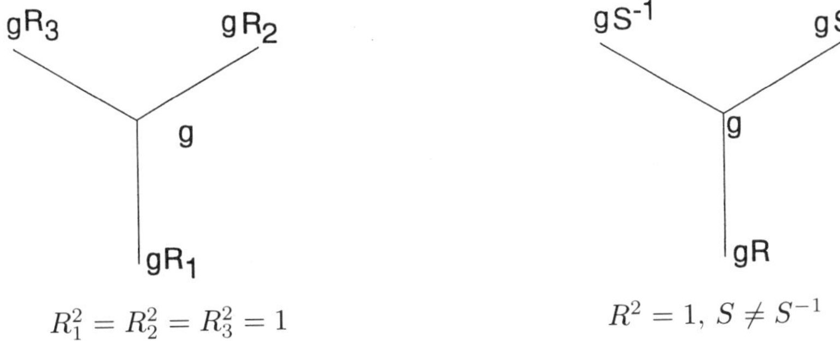

$R_1^2 = R_2^2 = R_3^2 = 1$ $\qquad$ $R^2 = 1,\ S \neq S^{-1}$

FIGURE 3.

As far as the first case is concerned let the orders of R_1R_2, R_2R_3 and R_3R_1 be r, s, t respectively. Then we attach three 2-cells to the cycles

$$g, gR_1, gR_1R_2, \ldots, g(R_1R_2)^{r-1}R_1,$$
$$g, gR_2, gR_2R_3, \ldots, g(R_2R_3)^{s-1}R_2,$$
$$g, gR_3, gR_3R_1, \ldots, g(R_3R_1)^{t-1}R_3.$$

Thus the 2-cells attached are $2r$-gons, $2s$-gons and $2t$-gons respectively.

In the second case the relevant cycles are:

$$g, gR, gRS, \ldots,$$
$$g, gR, gRS^{-1}, \ldots,$$
$$g, gS, gS^2, \ldots,$$

and the 2-cells are $2r$-gons, $2r$-gons and s-gons, where $r =$ the order of $RS =$ the order of RS^{-1} and $s =$ the order of S. Thus trivalent Cayley graphs can be completed to surfaces, and the left G-action on the Cayley graph can be extended linearly to the cells. In these cases we refer to the surface as the Cayley surface.

In this paper we are only dealing with case (a). For the presentation of $PSL_2(q)$ as in (b) see [**5, 6**]. The existence of Hamiltonian cycles in case (b) is still open in most of the cases; however, Tzu-Yi Yang[**11**] has proved that the Cayley graph of the presentation of $PSL_2(p)$ on the generators

$$R = \begin{pmatrix} 0 & 1 \\ -1 & 0 \end{pmatrix}, \quad S = \begin{pmatrix} 1 & 1 \\ 0 & 1 \end{pmatrix}$$

does have Hamiltonian cycles.

The Cayley surface of the group G, presented as in (1), is composed of $|G|/4$ 4-gons, $|G|/(2s)$ $2s$-gons and $|G|/(2t)$ $2t$-gons where $s =$ the order of R_2R_3 and $t =$ the order of R_3R_1.

It is worth noting which of the Cayley surfaces are non-orientable.

LEMMA 2.12. *If a finite group is generated by three involutions, two of which commute, then the Cayley surface of this group is orientable if and only if there is no product of an odd number of generators equal to* 1.

PROOF. First we note that the graph is locally planar, so locally we can select an orientation around each vertex. We choose the orientation that crosses the edges

R_1, R_2, R_3 in that order. Next note that adjacent edges have opposite orientation (see fig. 2).

A word W in the R_j can be represented by a path along the edges of the Cayley graph. This path is a loop, if and only if, $W = 1$ in G. If this word W has odd length in the R_j, then going once around a loop representing W reverses the local orientation, and so the Cayley surface is non-orientable.

Conversely, if the Cayley surface is non-orientable, then there are loops which reverse the orientation. Any such loop can be homotoped to an edge path, which must therefore correspond to a word W in the R_j which has odd length and represents 1 in the group □

THEOREM 2.13. *All Cayley surfaces arising from presentation* (1) *of simple groups are non-orientable.*

PROOF. According to lemma 2.12 we have to find a product of an odd number of generators which is equal to 1. Consider the diagram

$$\begin{array}{ccccccccc} 1 & \longrightarrow & T(2,s,t) & \longrightarrow & \Gamma & \longrightarrow & \mathbb{Z}_2 & \longrightarrow & 1 \\ & & \downarrow & & \downarrow & & \downarrow & & \\ 1 & \longrightarrow & G' & \longrightarrow & G & \longrightarrow & G/G' & \longrightarrow & 1 \end{array}$$

where $T(2, s, t)$ is the triangle group, Γ is the group generated by reflections through the sides of the triangle with the angles $\pi/2$, π/s and π/t, G is the group generated by three involutions, two of which commute and G' is the subgroup generated by their pairwise products. Then the downarrows are epimorphisms and since $T(2, s, t)$ has index 2 in Γ G' has index 1 or 2 in G. But if G is simple, index 2 can not happen, so $G' = G$. Therefore pairwise products generate the whole group and in particular R_1 can be written as a product of R_1R_2's, R_1R_3's and R_2R_3's. Multiplying both sides by R_1 we obtain a product of an odd number of generators equal to 1. Hence the surface is non-orientable □

3. Permutation groups

In this section we shall prove Theorems 1.1 and 1.2. The proof of Theorem 1.1 has two components. First we show that A_n has a presentation as in (1) if either $n = 5$ or $n \geq 9$, and then we show that the remaining A_n do not have such a presentation (the negative half).

First we want to develop some notation. The symbols R_1, R_2, R_3 will always denote involutions. They will always be chosen so that the R_j are distinct, all are different from the identity, and so that $R_1R_2{=}R_2R_1$. The symbols c_i will always denote 3-cycles. All permutations are written in cycle notation.

For technical reasons we deal with the case $n \doteq 5$ separately. In this case it is easy to see, that by choosing:

$$R_1 = (1,2)(3,4), \quad R_2 = (1,3)(2,4), \quad R_3 = (1,2)(4,5).$$

we generate all of A_5. Indeed let

$$c_1 = R_1R_3 = (3,5,4), \quad c_3 = R_2c_1R_2 = (1,5,2), \quad c_2 = c_3^2c_1c_3 = (2,4,3),$$

then c_1, c_2, c_3 are clearly generators of A_5. To prove the rest of the positive half of Theorem 1.1 we show that it holds for $n = 9$, 10, 11, 12, 13, 14, 15, 16 and then we show that it holds for A_{n+8} if it is true for A_n.

Before giving the proof of the inductive step we establish the base case of the induction. In all cases we define three involutions R_1, R_2, R_3 and exhibit an appropriate power $(R_1R_3)^m$ which is a 3-cycle c_1. Then we find another 3-cycle $c_2 = (i_p, i_q, j)$, where i_p, i_q are involved in c_1 but j is not. The 3-cycle c_1 generates the group A_3 and adjoining c_2 then gives A_4, according to Lemma 2.9. The idea is to then find a succession of 3-cycles c_3, c_4, ... , adjoin them to the previous generators to produce a succession of groups isomorphic to A_5, A_6, ... These extra 3-cycles are obtained from the previous ones by conjugation. Table 1 contains some of the details.

TABLE 1. Initial case for induction.

n	Involutions	Cycles	A_k on
9	$R_1 = (1,2)(3,4)(5,6)(7,8)$	$c_1 = (R_1R_3)^4 = (7,9,8)$	$\{7,8,9\}$
	$R_2 = (1,3)(2,4)(5,7)(6,8)$	$c_2 = (R_2c_1^2R_2)c_1(R_2c_1R_2)$	$\{6,7,8,9\}$
	$R_3 = (4,5)(8,9)$	$c_3 = R_2c_1R_2$	$\{5,\dots,9\}$
		$c_4 = R_3(5,6,7)R_3$	$\{4,\dots,9\}$
		$c_5 = R_1(4,5,6)R_1$	$\{3,\dots,9\}$
		$c_6 = R_2(4,5,6)R_2$	$\{2,\dots,9\}$
		$c_7 = R_1(2,3,4)R_1$	$G = A_9$
10	$R_1 = (1,2)(3,4)(5,6)(9,10)$	$c_1 = (R_1R_3)^5 = (8,10,9)$	$\{8,9,10\}$
	$R_2 = (1,3)(2,4)(5,6)(7,8)$	$c_2 = R_2c_1R_2$	$\{7,8,9,10\}$
	$R_3 = (1,2)(4,5)(6,7)(8,9)$	$c_3 = R_3(7,8,9)R_3$	$\{6,\dots,10\}$
		$\vdots$	
11	$R_1 = (1,2)(3,4)(5,6)(7,8)$	$c_1 = (R_1R_3)^7 = (1,2,11)$	$\{1,2,11\}$
	$R_2 = (1,3)(2,4)(5,6)(9,10)$	$c_2 = (R_2c_1^2R_2)c_1(R_2c_1R_2)$	$\{1,2,3,11\}$
	$R_3 = (1,11)(4,5)(6,7)(8,9)$	$c_3 = R_2c_1R_2$	$\{1,\dots,4,11\}$
		$\vdots$	
12	$R_1 = (1,2)(3,4)(5,6)(7,8)$	$c_1 = (R_1R_3)^{10} = (7,9,8)$	$\{7,8,9\}$
	$R_2 = (1,3)(2,4)(9,10)(11,12)$	$c_2 = R_2c_1R_2$	$\{7,8,9,10\}$
	$R_3 = (4,5)(6,12)(8,9)(10,11)$	$c_3 = R_3c_2R_3$	$\{7,\dots,11\}$
		$\vdots$	
13	$R_1 = (1,2)(3,4)(5,6)(7,8)(9,10)(11,12)$	$c_1 = (R_1R_3)^4 = (11,13,12)$	$\{11,12,13\}$
	$R_2 = (1,3)(2,4)(5,7)(6,8)(9,11)(10,12)$	$c_2 = (R_2c_1^2R_2)c_1(R_2c_1R_2)$	$\{10,\dots,13\}$
	$R_3 = (1,2)(4,5)(8,9)(12,13)$	$c_3 = R_2c_1R_2$	$\{9,\dots,13\}$
		$\vdots$	
14	$R_1 = (1,2)(3,4)(5,6)(7,8)$	$c_1 = (R_1R_3)^{14} = (1,14,2)$	$\{1,2,14\}$
	$R_2 = (1,3)(2,4)(9,10)(11,12)$	$c_2 = (R_2c_1R_2)c_1(R_2c_1^2R_2)$	$\{1,2,4,14\}$
	$R_3 = (1,14)(4,5)(6,7)(8,9)(10,11)(12,13)$	$c_3 = R_2c_1R_2$	$\{1,\dots,4,14\}$
		$\vdots$	
15	$R_1 = (1,2)(3,4)(5,6)(7,8)(9,10)(13,14)$	$c_1 = (R_1R_3)^{35} = (12,14,13)$	$\{12,13,14\}$
	$R_2 = (1,3)(2,4)(5,7)(6,8)(9,10)(11,12)$	$c_2 = R_2c_1R_2$	$\{11,\dots,14\}$
	$R_3 = (1,15)(2,3)(4,5)(8,9)(10,11)(12,13)$	$c_3 = R_3c_2R_3$	$\{10,\dots,14\}$
		$\vdots$	
16	$R_1 = (1,2)(3,4)(5,6)(7,8)(9,10)(11,12)$	$c_1 = (R_1R_3)^{28} = (1,2,16)$	$\{1,2,16\}$
	$R_2 = (1,3)(2,4)(5,7)(6,8)(9,10)(13,14)$	$c_2 = (R_2c_1^2R_2)c_1(R_2c_1R_2)$	$\{1,2,3,16\}$
	$R_3 = (1,16)(4,5)(8,9)(10,11)(12,13)(14,15)$	$c_3 = R_1(3,2,1)R_31$	$\{1,\dots,4,16\}$
		$\vdots$	

This establishes the base of the induction. To prove the inductive step we make following hypotheses:

There are involutions R_1, R_2, R_3 in A_n, a partition of $\{1, 2, \dots, n\}$ into disjoint non-empty subsets S_1, S_2, and elements $i \in S_1, j \in S_2$ so that

(a) $R_1R_2 = R_2R_1$.
(b) $R_1(S_1) = S_1$, $R_1(S_2) = S_2$, $R_2(S_1) = S_1$, $R_2(S_2) = S_2$.
(c) $(i, j) \in R_3$ and $R_3(S_1 \setminus i) = S_1 \setminus i$, $R_3(S_2 \setminus j) = S_2 \setminus j$.
(d) $(k, i) \in R_1$ for some $k \neq i$ and $R_3(k) = k$.
(e) $(R_1R_3)^m$ is a 3-cycle c, where m is some integer.

Condition (e) is equivalent to the cycle decomposition of R_1R_3 having one 3-cycle and all other cycles having lengths relatively prime to three. Next we extend this data to A_{n+8} as follows:

$$\begin{aligned}
R_1' &= R_1(n+1, n+2)(n+3, n+4)(n+5, n+6)(n+7, n+8),\\
R_2' &= R_2(n+1, n+3)(n+2, n+4)(n+5, n+7)(n+6, n+8),\\
R_3' &= R_3(i, j)(i, n+1)(n+4, n+5)(n+8, j),\\
S_1' &= S_1 \cup \{n+1, n+2, n+3, n+4\},\\
S_2' &= S_2 \cup \{n+5, n+6, n+7, n+8\},\\
i' &= n+4, j' = n+5, k' = n+3.
\end{aligned}$$

PROPOSITION 3.1. *The elements R_1', R_2', R_3' are involutions in A_{n+8} satisfying conditions* (a)–(e) *above.*

PROOF. It is clear that R_1' and R_2' are involutions in A_{n+8}. Note that $(i, j) \in R_3$ and so $R_3(i, j)$ is a product of disjoint transpositions involving only elements from the set $\{1, 2, \dots, n\} \setminus \{i, j\}$. Therefore, R_3 is also an involution in A_{n+8}.

Properties (a),(b),(c) and (d) are easy to check. To prove (e) we compute $R_1'R_3'$:

$$R_1'R_3' = (n+1, n+2, i, k)(n+3, n+5, n+6, n+4)(n+8, n+7, j, \dots)(\text{other cycles}).$$

The "other cycles" in $R_1'R_3'$ are identical to some of those in R_1R_3, specifically those not including i or j.

Now the cycle of R_1R_3 containing j has the form $(i, k, j, j_2, \dots, j_t)$, where $\{j_2, \dots, j_t\}$ is a subset (possibly empty) of $S_2 \setminus \{j\}$ and $R_1(j_t) = j$ (or $R_1(j) = j$ in which case the above cycle is just the 3-cycle (i, k, j)). The form of $R_1'R_3'$ now reveals that the cycle of $R_1'R_3'$ containing j is $(n+8, n+7, j, j_2, \dots, j_t)$ and so has the same length as the cycle of R_1R_3 containing j. It now follows that $(R_1'R_3')^{m'}$ is a 3-cycle, where $m' = \operatorname{lcm}(4, m)$ □

Remark. The involutions R_1, R_2, R_3 defined in the Table 1 for the initial cases $n = 9$, 10, 11, 12, 13, 14, 15, 16 satisfy (a), … , (e) for the choices of S_1, S_2, i, j, k, m in Table 2 (S_2 is always the complement of S_1).

Thus it follows that (a), … , (e) hold for A_n if $n \geq 9$. We will show that these involutions generate A_n. The method of proof will be by "extending 3-cycles", as illustrated in the Table 2. That is, we start with the 3-cycle $c_1 = (R_1R_3)^m$, which generates A_3, and then inductively construct additional 3-cycles $c_2, c_3, \dots$ so that each c_r involves only one letter different from those present in $c_1, \dots, c_{r-1}$, each c_r is of the form WcW^{-1} where c is a 3-cycle in the group generated by $c_1, \dots, c_{r-1}$, and W is a word in R_1, R_2, R_3.

PROPOSITION 3.2. *The involutions R_1, R_2, R_3 generate A_n for $n \geq 9$.*

TABLE 2.

n	S_1	i	j	k	m
9	$\{1,2,3,4\}$	4	5	3	4
10	$\{1,2,3,4\}$	4	5	3	5
11	$\{1,2,3,4,11\}$	4	5	3	7
12	$\{1,2,3,4\}$	4	5	3	10
13	$\{1,2,3,4\}$	4	5	3	4
14	$\{1,2,3,4,14\}$	4	5	3	14
15	$\{5,\ldots,14\}$	5	4	6	35
16	$\{1,2,3,4,16\}$	4	5	3	28

PROOF. The proof is by induction on n. The initial step is provided by the Table 2, that is for $n =$ 9,10, 11, 12, 13, 14, 15, 16. So assume that A_n is generated by R_1, R_2, R_3 for some value of n. Then we must show that A_{n+8} is generated by R_1', R_2', R_3'.

First note that the 3-cycle $c_1 = (R_1R_3)^m$ does not involve i or j. Therefore $c_1' = (R_1'R_3')^{m'}$ is identical to c_1. Moreover, c_1 involves only letters from either S_1 or S_2, say S_1, the case S_2 being similar.

Now consider the sequence of 3-cycles in A_n up to the point where j is adjoined, say $c_1, c_2, \ldots, c_{r+1}$, where $c_1 = (R_1R_3)^m$ and $c_t = W_t\gamma_tW_t^{-1}$, $2 \le t \le r+1$, for some word W_t in R_1, R_2, R_3 and some $\gamma_t \in \langle c_1, c_2, \ldots, c_{t-1}\rangle$. Necessarily $\gamma_r = (i_1, i_2, i)$ for some i_1, $i_2 \in S_1 \setminus \{i\}$ and $c_{r+1} = (i_3, i_4, j)$ for some i_3, $i_4 \in S_1$.

Examination of the table (the base case of the induction) reveals that we have $W_{r+1} = R_3$ and that all subsequent conjugations are also by generators. We make the inductive assumption that this also occurs for A_n.

Let W_t' denote the word in R_1', R_2', R_3' obtained from W_t by replacing each occurence of R_j by R_j'. Then the 3-cycle $c_t' = W_t'\gamma_tW_t'^{-1}$ is identical to c_t for $2 \le t \le r$, and $c_{r+1}' = R_3'\gamma_{r+1}R_3' = (i_3, i_4, n+1)$.

Thus we have added the new letter $n+1$. Then conjugating by R_1', R_2', R_1', R_3', R_1', R_2', R_1', R_3' in turn yields the new letters $n+2$, $n+4$, $n+3$, $n+5$, $n+6$, $n+8$, $n+7$, j respectively. Notice that all new conjugations are also by generators.

All that remains now is to add the letters in $S_2 \setminus \{j\}$. To do this we merely follow the corresponding sequence in A_n, replacing each occurence of R_j in every conjugation by R_j'. □

Now the "negative" part.

PROPOSITION 3.3 *A_6, A_7 and A_8 can not be generated by such involutions.*

PROOF. The easiest case is A_7. Up to conjugation, the first involution can be chosen only one way—namely $(1,2)(3,4)$. The second is either $(1,3)(2,4)$, or $(1,2)(5,6)$. In both cases we have four subsets stable under both permutations, namely $\{1,2,3,4\}$, $\{5\}$, $\{6\}$, $\{7\}$ and $\{1,2\}$, $\{3,4\}$, $\{5,6\}$, $\{7\}$ respectively. The third involution consists of only two transpositions, so it can not connect all these four subsets into one, therefore we do not have a transitive action on the set $\{1,2,3,4,5,6,7\}$. Hence the subgroup generated by R_1, R_2 and R_3 is not A_7.

In the case of A_6 for the first two involutions we have again only two essentially different choices: $R_1 = (1,2)(3,4)$, $R_2 = (1,2)(5,6)$ or $R_2 = (1,3)(2,4)$. The transitivity argument forces the choice of R_3: $R_3 = (2,3)(4,5)$ in the first case and $R_3 = (1,5)(2,6)$ in the second. In both cases we compute the order of $\langle R_1, R_2, R_3\rangle$.

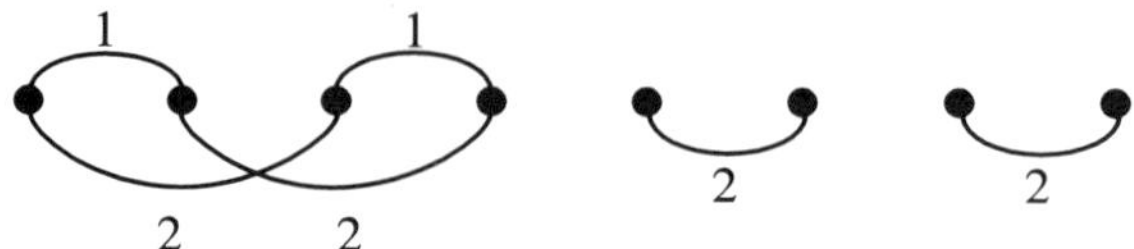

FIGURE 4. R_1, R_2 acting on 8 symbols. Case (IV).

It is 60 in the first case and 24 in the second. Hence the subgroup is not A_6. This was done by computer.

The most extensive case is A_8. Up to conjugacy there are only two choices for R_1, namely $R_1 = (1,2)(3,4)$ and $R_1 = (1,2)(3,4)(5,6)(7,8)$. Now we must choose an involution R_2 so that $R_1R_2 = R_2R_1$. When choosing R_2 we do not distinguish amongst choices that are conjugate by any conjugation that leaves R_1 invariant. Nor do we distinguish amongst choices that give the same subgroup $\langle R_1, R_2\rangle \cong \mathbb{Z}_2 \oplus \mathbb{Z}_2$. There are then only five essentially different choices for R_1 and R_2:

(I)	$R_1 = (1,2)(3,4)$,	$R_2 = (1,3)(2,4)$;
(II)	$R_1 = (1,2)(3,4)$,	$R_2 = (1,2)(5,6)$;
(III)	$R_1 = (1,2)(3,4)$,	$R_2 = (5,6)(7,8)$;
(IV)	$R_1 = (1,2)(3,4)$,	$R_2 = (1,3)(2,4)(5,6)(7,8)$;
(V)	$R_1 = (1,2)(3,4)(5,6)(7,8)$,	$R_2 = (1,3)(2,4)(5,7)(6,8)$.

In case (V) there is only one choice for R_2 since any other choice is either equivalent to this one or reduces to case (IV).

We must now choose the third involution R_3. Again we do not distinguish amongst choices that differ only by a conjugation not altering the subgroup $\langle R_1, R_2\rangle = \mathbb{Z}_2 \oplus \mathbb{Z}_2$. These conjugations can be by elements of S_8. Our choices for R_3 are restricted by the fact that the group $\langle R_1, R_2, R_3\rangle$ is supposed to be transitive on $\{1,2,3,4,5,6,7,8\}$. In each choice of R_3 we used a computer to determine the order of the subgroup $\langle R_1, R_2, R_3\rangle$. The maximum order turns out to be 576, and therefore A_8 can not be generated by three involutions as in (1).

	Choice of R_3	order of $\langle R_1, R_2, R_3\rangle$
(I)	$R_3 = (1,5)(2,6)(3,7)(4,8)$	32
(II)	$R_3 = (1,7)(2,3)(4,5)(6,8)$	192
(III)	$R_3 = (1,7)(2,3)(4,5)(6,8)$	48

In case (IV) there are many choices for R_3 leading to groups of different orders. Figure 4 indicates how R_1 and R_2 are acting on the eight elements 1, 2, ..., 8.

R_3 must necessarily involve a transposition containing one letter from 1, 2, 3, 4 and one from 5, 6, 7, 8. Up to conjugacy we may assume $(4,5)$ is in R_3. Since R_3 is an even permutation it must involve either one more transposition or three

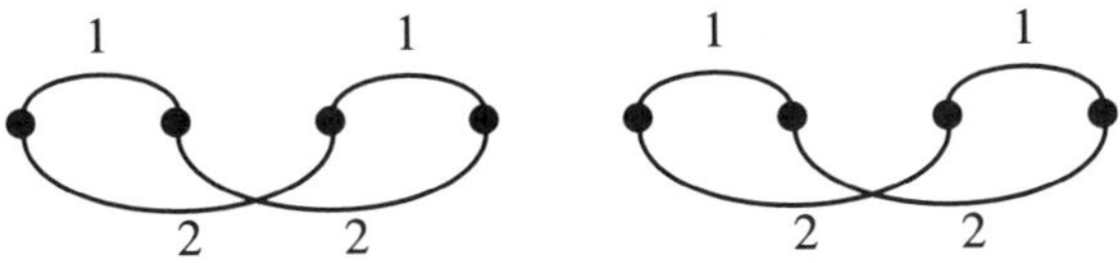

FIGURE 5. R_1, R_2 acting on 8 symbols. Case (V).

more. Thus the cases are:

	Choice of R_3	order of $\langle R_1, R_2, R_3\rangle$
(IVa)	$R_3 = (4,5)(1,7)$	48
(IVb)	$R_3 = (4,5)(3,7)$	64
(IVc)	$R_3 = (4,5)(6,7)$	576

We should point out that either 7 or 8 must be involved in the other transposition and that the choice $R_3 = (4,5)(2,7)$ is equivalent to $R_3 = (4,5)(1,7)$. To see this note that switching 1 and 2 leaves $\langle R_1, R_2\rangle$ invariant. The same remarks apply when the choices for R_3 involve 4 disjoint transpositions.

	Choice of R_3	order of $\langle R_1, R_2, R_3\rangle$
(IVd)	$R_3 = (4,5)(1,7)(2,3)(6,8)$	16
(IVe)	$R_3 = (4,5)(1,7)(2,6)(3,8)$	16
(IVf)	$R_3 = (4,5)(1,7)(2,8)(3,6)$	32
(IVg)	$R_3 = (4,5)(3,7)(1,2)(6,8)$	64
(IVh)	$R_3 = (4,5)(3,7)(1,6)(2,8)$	16
(IVi)	$R_3 = (4,5)(6,7)(1,2)(3,8)$	64
(IVj)	$R_3 = (4,5)(6,7)(1,8)(2,3)$	48

In case (V) the action of R_1, R_2 is given by Figure 5.
All choices for R_3 must involve $(4,5)$, at least up to conjugacy. The possibilities

are:

	Choice of R_3	order of $\langle R_1, R_2, R_3\rangle$
(Va)	$R_3 = (4,5)(1,2)$	192
(Vb)	$R_3 = (4,5)(1,6)$	48
(Vc)	$R_3 = (4,5)(1,8)$	16
(Vd)	$R_3 = (4,5)(1,2)(3,6)(7,8)$	16
(Ve)	$R_3 = (4,5)(1,2)(3,7)(6,8)$	24
(Vf)	$R_3 = (4,5)(1,2)(3,8)(6,7)$	24
(Vg)	$R_3 = (4,5)(1,6)(2,3)(7,8)$	24
(Vh)	$R_3 = (4,5)(1,6)(2,7)(3,8)$	8
(Vi)	$R_3 = (4,5)(1,6)(2,8)(3,7)$	32
(Vj)	$R_3 = (4,5)(1,8)(2,3)(6,7)$	16
(Vk)	$R_3 = (4,5)(1,8)(2,6)(3,7)$	8
(Vl)	$R_3 = (4,5)(1,8)(2,7)(3,6)$	8 □

The symmetric groups.

THEOREM 3.4. *Any symmetric group S_n with $n \geq 4$ can be generated by three involutions, two of which commute.*

PROOF. If $n = 2k+1$ we choose

$$R_1 = (3,4)(5,6)\dots(2k-1,2k), \quad R_2 = (1,2), \quad R_3 = (2,3)(4,5)\dots(2k,2k+1).$$

Now, $(1,3) = R_3R_2R_3$, $(1,4) = R_1(1,3)R_1$, $(1,5) = R_3(1,4)R_3$, etc. So we can produce all $(1,i)$, for $2 \leq i \leq n$, which obviously generate S_n. The case where n is even is similar to this. The cases $n < 4$ are trivial (there are not enough involutions). □

4. The projective linear groups

In this section we shall prove Theorems 1.3 and 1.4. Again we prove the "positive" part first. We start with the case $p = 2$. Thus we must prove the following

PROPOSITION 4.1. *$SL_2(2^n)$ can be generated by three involutions, two of which commute, if $n \geq 2$.*

PROOF. Consider the matrices

$$R_1 = \begin{pmatrix} 0 & 1 \\ 1 & 0 \end{pmatrix}, \quad R_2 = \begin{pmatrix} x & 1+x \\ 1+x & x \end{pmatrix}, \quad R_3 = \begin{pmatrix} 1 & y \\ 0 & 1 \end{pmatrix}.$$

Here x and y are elements of the finite field $GF(2^n)$. We want these matrices to be different from the identity and from each other, and this happens if, and only if, $x \neq 0$, 1 and $y \neq 0$. Thus we need $n \geq 2$ so that such elements exist.

It is easy to check that $R_1R_2 = R_2R_1$ and that R_3 does not commute with any of R_1, R_2, R_1R_2. In fact

$$R_1R_3 = \begin{pmatrix} 0 & 1 \\ 1 & y \end{pmatrix}, \quad R_2R_3 = \begin{pmatrix} x & xy+1+x \\ 1+x & y+xy+x \end{pmatrix},$$
$$R_1R_2R_3 = \begin{pmatrix} 1+x & y+xy+x \\ x & xy+1+x \end{pmatrix}$$

and the respective traces are y, $y+xy$, and xy. Since these traces are not zero it follows that R_3 does not commute with any of R_1, R_2, R_1R_2.

Now we suppose that $y = z+z^{-1}$ for some $z \in GF(2^n)$, where $z \neq 0$ and $z \neq 1$. Again this assumes that $n \geq 2$. Then R_1R_3 is similar to the diagonal matrix

$$\begin{pmatrix} z & 0 \\ 0 & z^{-1} \end{pmatrix}$$

and so its order is the least integer k so that $z^k = 1$. Choosing z to be a primitive root of GF(2^n) we see that the order of R_1R_3 is $2^n - 1$. According to Dickson's theorem (Theorem 2.7) the possible subgroups of $SL_2(2^n)$ are: A_4 if n is even, A_5 if n is even, the dihedral groups $T(2,2,t)$ of order $2t$ for $t = 2$ or $t \mid (2^n - 1)$, the affine groups, and the projective groups $SL_2(2^m)$ for $m \mid n$. Let G be the subgroup generated by R_1, R_2, R_3.

G can not be dihedral since R_3 does not commute with any of R_1, R_2, R_1R_2 (see Lemma 2.10). Also G can not be cyclic since it contains the subgroup $\mathbb{Z}_2 \oplus \mathbb{Z}_2 \cong \langle R_1, R_2 \rangle$. Thus G must be A_4, A_5, affine or projective. In fact G can not be A_4 according to Theorem 1.1.

Suppose G is affine, that is G is isomorphic to a subgroup of the group

$$A = \left\{ \begin{pmatrix} a & b \\ 0 & a^{-1} \end{pmatrix} \;\middle|\; a, b \in GF(2^n), a \neq 0 \right\}$$

There exists a split short exact sequence $1 \to \mathbb{Z}_{2^n} \to A \to \mathbb{Z}_{2^n-1} \to 1$, where the normal subgroup $\mathbb{Z}_{2^n}$ consists of all matrices of the form

$$\begin{pmatrix} 1 & b \\ 0 & 1 \end{pmatrix}.$$

Any subgroup of A generated by elements of order 2 would have to be a subgroup of $\mathbb{Z}_2^n$ and therefore be Abelian. G is not Abelian and so can not be affine. Thus G is A_5 or projective.

For the moment assume that $n > 2$. Then $2^n - 1 \geq 7$ and so G must be projective. The proper projective subgroups $SL_2(2^m)$, where m is a proper divisor of $2^n - 1$, do not have elements of order $2^n - 1$, and therefore G must be $SL_2(2^n)$. This proves the proposition for $n > 2$.

To prove the proposition for $q = 4$ we can use the isomorphism $SL_2(4) \cong A_5$ and then apply Theorem 1.1. Or we can consider the three matrices

$$R_1 = \begin{pmatrix} 0 & 1 \\ 1 & 0 \end{pmatrix}, \quad R_2 = \begin{pmatrix} x & 1+x \\ 1+x & x \end{pmatrix}, \quad R_3 = \begin{pmatrix} 1 & x \\ 0 & 1 \end{pmatrix}.$$

Here x is any element of $GF(4)$ satisfying $x^2 + x + 1 = 0$. Then it is easy to check that the orders of R_1R_3 and R_2R_3 are 5 and 3 respectively. Since the subgroup generated by R_1 and R_2 has order 4 it follows that the order of G is at least 60. On the other hand $SL_2(4)$ has order exactly equal to 60 and so G is all of $SL_2(4)$. □

Now we will prove Theorem 1.3 for odd primes p. The proof is broken down into two cases, the case where -1 is a square in $GF(q)$ and the case where it is not. First we consider the case where -1 is a square. This happens if, and only if, either $p \equiv 1 \pmod 4$ and n is arbitrary, or $p \equiv -1 \pmod 4$ and n is even.

PROPOSITION 4.2. *Suppose p is an odd prime and -1 is a square in $GF(q)$. Then $PSl_2(q)$ can be generated by three involutions, two of which commute, if $q \neq 9$.*

PROOF. Let a be any element of $GF(q)$ such that $a^2 = -1$. Then consider the three matrices

$$R_1 = \begin{pmatrix} 0 & 1 \\ -1 & 0 \end{pmatrix}, \quad R_2 = \begin{pmatrix} a & 0 \\ 0 & -a \end{pmatrix}, \quad R_3 = \begin{pmatrix} a & x \\ 0 & -a \end{pmatrix}.$$

Here x is any non-zero element of $GF(q)$. Then we can check that these three matrices are different from the identity and each other. Moreover $R_1R_2 = R_2R_1$ and R_3 does not commute with any of R_1, R_2, R_1R_2. We want to show that G, the subgroup of $PSL_2(q)$ generated by R_1, R_2, R_3, is the entire group, provided x is chosen appropriately.

The idea of the proof is to use Dickson's theorem (Theorem 2.7) to rule out all proper subgroups. First of all note that G can not be dihedral since if it were R_3 would have to commute with at least one of R_1, R_2, R_1R_2, and it does not. Next we note that G can not be affine since $\mathbb{Z}_2 \oplus \mathbb{Z}_2$ is never a subgroup of an affine group when p is odd. Therefore G must be either projective or exceptional, that is G is isomorphic to one of the following:

(a) $G \cong PSL_2(p^m)$ for some $m \mid n$, or $G \cong PGL_2(p^m)$ for some $2m \mid n$.

(b) $G \cong A_4$, S_4, or A_5.

Now consider the matrix

$$R_1R_3 = \begin{pmatrix} 0 & -a \\ -a & -x \end{pmatrix}.$$

We choose x equal to $z + z^{-1}$, for some $z \in GF(q), z \neq 0, \pm 1$. Then the matrix R_1R_3 is similar to

$$\begin{pmatrix} z & 0 \\ 0 & z^{-1} \end{pmatrix}$$

and therefore its order is the least positive integer k so that $z^k = \pm 1$. In particular if we choose z to be a primitive root of $GF(q)$ then the order of R_1R_3 will be $(q-1)/2$. For the moment assume that $q > 11$ so that $(q-1)/2 > 5$. Thus G can not be exceptional and so must be projective. In fact G must be all of $PSL_2(q)$ since any proper projective subgroup can not have an element of order $(q-1)/2$. This proves the proposition when -1 is a square, except for $q = 5$.

If $q = 5$ we choose matrices

$$R_1 = \begin{pmatrix} 0 & 1 \\ -1 & 0 \end{pmatrix}, \quad R_2 = \begin{pmatrix} 2 & 0 \\ 0 & -2 \end{pmatrix}, \quad R_3 = \begin{pmatrix} 2 & x \\ 0 & -2 \end{pmatrix},$$

where x is any non-zero element of $GF(5)$. The order of R_2R_3 is 5, the order of the subgroup $\langle R_1, R_2 \rangle$ is 4, and so the order of G is at least 20. This is sufficient to guarantee that G is all of $PSL_2(5)$. We could also argue as follows:

$$R_1R_3 = \begin{pmatrix} 0 & -2 \\ -2 & -x \end{pmatrix}, \quad R_1R_2R_3 = \begin{pmatrix} 0 & -1 \\ 1 & -2x \end{pmatrix}.$$

The respective traces are $-x$ and $-2x$ and so one of these matrices will have order three. Therefore G will have order at least 60, and so must be all of $PSL_2(5)$ since $PSL_2(5)$ has order 60.

Another approach would be to use Theorem 1.1 together with Proposition 2.11. This completes the proof of the proposition. □

PROPOSITION 4.3. *Suppose $p \equiv -1 \pmod 4$ and n is odd. Then $PSl_2(q)$ can be generated by three involutions, two of which commute, if $q \geq 11$.*

PROOF. We choose elements a, $b \in GF(q)$ satisfying $a^2 + b^2 = -1$. This is always possible. Then let R_1, R_2, R_3 be the matrices

$$R_1 = \begin{pmatrix} 0 & 1 \\ -1 & 0 \end{pmatrix}, \quad R_2 = \begin{pmatrix} a & b \\ b & -a \end{pmatrix}, \quad R_3 = \begin{pmatrix} 0 & y \\ -y^{-1} & 0 \end{pmatrix},$$

where y is any element of $GF(q)$ such that $y \neq 0, \pm 1$. For any such y it is easy to check that the R_j are all distinct and different from the identity, $R_1R_2 = R_2R_1$, and R_3 does not commute with any of R_1, R_2, R_1R_2.

Let G denote the group generated by R_1, R_2, R_3. G can not be affine since it contains the subgroup $\langle R_1, R_2\rangle \cong \mathbb{Z}_2 \oplus \mathbb{Z}_2$. Moreover G can not be dihedral since R_3 does not commute with any of R_1, R_2, R_1R_2 (see Lemma 2.10). Therefore G must be projective or one of A_4, S_4, A_5. The actual type of G will depend on how we choose y.

$$R_1R_3 = \begin{pmatrix} -y^{-1} & 0 \\ 0 & -y \end{pmatrix}$$

has order equal to the smallest positive integer k such that $y^k = \pm 1$.

Choosing y to be a primitive root of $GF(q)$ this order will be $(q-1)/2$. If $(q-1)/2 > 5$ Then G will be projective, and in fact all of $PSL_2(q)$. This proves the proposition, except for $q = 11$.

Now we consider the case $q = 11$. Consider the matrices

$$R_1 = \begin{pmatrix} 0 & 1 \\ -1 & 0 \end{pmatrix}, \quad R_2 = \begin{pmatrix} 1 & 3 \\ 3 & -1 \end{pmatrix}, \quad R_3 = \begin{pmatrix} 1 & -1 \\ 2 & -1 \end{pmatrix}.$$

Then we see that the products

$$R_1R_3 = \begin{pmatrix} 2 & -1 \\ -1 & 1 \end{pmatrix}, \quad R_2R_3 = \begin{pmatrix} 7 & -4 \\ 1 & -2 \end{pmatrix}$$

have respective orders 5, 6. Since R_3 does not commute with any of R_1, R_2, R_1R_2 the only possibilities for G are $PSL_2(11)$ and one of A_4, S_4, A_5. But the groups A_4, S_4, A_5 do not have elements of order 6 and so G must be $PSL_2(11)$. □

The "negative" part of the Theorem 1.3 is just a combination of previous results.

PROPOSITION 4.4. *$PSL_2(q)$ can not be generated by three involutions two of which commute if $q = 2$, 3, 7 or 9.*

PROOF. For $q = 2$ or 3 there are not enough involutions (or we can use Proposition 2.11 in conjunction with Theorem 1.1). For $q = 9$ we refer to Propositions 2.11 and 3.3. For $q = 7$ we refer to Theorem 2.8 and the proof of Proposition 3.3 noting the fact that any transitive permutation representation $\rho\colon PSL_2(7) \to S_7$ must have its image in A_7 since $PSL_2(7)$ can be generated by two elements, one of order 3, and the other of order 7. Now, if $PSL_2(7)$ could be generated by three involutions, two of which commute, then it would follow that these involutions in A_7 would act transitively; but the proof of Proposition 3.3 shows that this is impossible. □

Now we turn our attention to the projective general linear groups $PGL_2(q)$.

PROPOSITION 4.5. *$PGL_2(q)$ can be generated by three involutions, two of which commute, if and only if, $q > 2$.*

PROOF. First notice that for even q's the result is Theorem 1.3 since $PGL_2(q) \cong PSL_2(q)$. For odd q's we again heavily rely on Dickson's Classfication Theorem (Theorem 2.7). We choose involutions to be images in $PGL_2(q)$ of

$$R_1 = \begin{pmatrix} 1 & 0 \\ 0 & -1 \end{pmatrix}, \quad R_2 = \begin{pmatrix} 0 & 1 \\ x & 0 \end{pmatrix}, \quad R_3 = \begin{pmatrix} 1 & y \\ 0 & -1 \end{pmatrix},$$

with x and y to be defined later. As in the case of the projective special linear groups, the cyclic, affine and dihedral subgroups are easily eliminated, so long as $y \neq 0$.

We are going to produce an element of order $q-1$ in the subgroup generated by R_1, R_2 and R_3, thus eliminating all proper projective subgroups and all exceptional subgroups for $q \geq 7$. Note that the result for $q = 3$ and $q = 5$ follows from Proposition 2.11 and Theorem 1.2.

Consider the matrix $A = R_2R_3 = \left(\begin{smallmatrix} 0 & -1 \\ x & xy \end{smallmatrix}\right)$. It has trace xy and determinant x. We choose x to be a primitive root in $GF(q)$ and $y = 1 + x^{-1}$. Then A is conjugate to $\left(\begin{smallmatrix} 1 & 0 \\ 0 & x \end{smallmatrix}\right)$ and therefore A has order $q-1$. Thus R_1, R_2 and R_3 generate all of $PGL_2(q)$. □

5. Hamiltonian cycles

In this section we shall show that many graphs admit Hamiltonian cycles, thus corroborating the Lovász Conjecture. The graphs we consider are the Cayley graphs of finite groups G having a presentation as in (1). Any such graph is trivalent and so, in some sense, represents the hardest possible case for the existence of Hamiltonian cycles.

Requiring two of the involutions to commute allows us to give a very simple inductive proof for the existence of Hamiltonian cycles. The idea was first mentioned to us by George Maxwell [**8**], and was also used by Conway, Sloane and Wilks [**3**] to prove that every finite Coxeter group has Hamiltonian cycles.

THEOREM 5.1. *Suppose G has a presentation as in* (1). *Then the Cayley graph of this presentation has a Hamiltonian cycle.*

PROOF. The Cayley graph is trivalent. We can extend the graph to a surface by adding 4-gons, $2s$-gons and $2t$-gons, where s and t are the orders of R_1R_3 and R_2R_3 respectively, as was described in Section 2. We will obtain the Hamiltonian cycle as the boundary of a union of faces of the Cayley surface.

The first step is to color all the $2t$-gons, with the same color, and then take the boundary of all colored faces. This will involve all vertices, but in $|G|/(2t)$ disjoint cycles.

The next step is to join the colored $2t$-gons by coloring certain 4-gons. Pick any 4-gon and color it. Now the boundary of the union of all colored faces will have one less component, but still involve all vertices. The idea is to repeat this last step until there is only one colored component whose boundary involves all vertices. Suppose at some stage of the construction we still have several colored components. Then there will be a pair of adjacent $2t$-gons in different components, that is there will be an uncolored 4-gon joining these components. Coloring this 4-gon then reduces the number of components. This construction will end in a contractible union of faces whose boundary is a Hamiltonian cycle. □

Note added in proof. After this paper was written we came upon another paper containing results overlapping with ours. In **[2]** Conder proves that every A_n, S_n for n sufficiently large, can be generated by three involutions, two of which commute. His proof is different from ours, and the bound on n is not explicit, but much bigger than necessary (see Theorems 1.1 and 1.2).

References

1. W. Burnside, *Theory of groups of finite order*, Dover Publications Inc., 1955.
2. Marston Conder, *More on generators for alternating and symmetric groups*, Quart. J. Math. Oxford Ser. (2) **32** (1981), 137–163.
3. J. H. Conway, N. J. A. Sloane, and Allan R. Wilks, *Gray codes for reflection groups*, Graphs Combin. **5** (1989), 315–325.
4. Leonard Dickson, *Linear groups, with an exposition of the Galois field theory*, Dover Publications Inc., New York, 1958.
5. Henry Glover and Denis Sjerve, *Representing $PSl_2(p)$ on a Riemann surface of least genus*, Enseign. Math. (2) **31** (1985), 305–325.
6. ——, *The genus of $PSl_2(q)$*, J. Reine Angew. Math. **380** (1987), 59–86.
7. Laslo Lovász, *Problem* 11, Combinatorial structures and their applications (New York) (Richard Guy, Haim Hanani, Norbert Sauer, and Johanan Schonheim, eds.), Gordon and Breach, 1970.
8. George Maxwell, Private communication.
9. Michio Suzuki, *Group theory* I, Springer-Verlag, Berlin-Heidelberg-New York, 1982.
10. D. Witte and J. A. Gallian, *A survey — Hamiltonian cycles in Cayley graphs*, Discrete Math. **51** (1984), 293–304.
11. Tzu-Yi Yang, *On hamilton cycles in cayley graphs*, Ph.D. thesis, Ohio State University, 1993.

Mathematics Department, University of British Columbia, Vancouver, BC, Canada, V6T 1Z2

E-mail address, D. Sjerve: sjer@math.ubc.ca

Mathematics Department, University of British Columbia, Vancouver, BC, Canada, V6T 1Z2

E-mail address, Michael Cherkassoff: mikec@math.ubc.ca

Centre de Recherches Mathématiques
CRM Proceedings and Lecture Notes
Volume 6, 1994

Types of Projective Resolutions for Finite Groups

Urs Stammbach

In order to define the cohomology of a group G all the books on homological algebra describe at some place an explicit free resolution of $\mathbb{Z}$. In this paper I shall first make some historical remarks about these standard resolutions. Then I shall turn to various types of projective resolutions of k, where k is a field of characteristic p and p divides the order of the (finite) group G. In particular, the growth rate of these resolutions will be discussed, and the question whether there are resolutions that can be written as multiple complexes.

1. Heinz Hopf introduced the integral homology of a group G in his 1944 paper [**H**].[1] Hopf's starting point was a result of Hurewicz who had shown in the thirties that the homology of an aspherical space X only depends on its fundamental group G. Hopf's aim was to analyse in detail how the group G determines the homology groups of X. Guided by the way the chain complex of the universal cover of an aspherical space looks like, he was led to construct, in a purely algebraic fashion, a complex of free modules over G. In today's terminology of homological algebra this complex is precisely a *free resolution* of $\mathbb{Z}$ over $\mathbb{Z}G$. To define the integral homology of G, which corresponds to the homology of the aspherical space with fundamental group G, he then divided out the action of G and took the homology of the resulting complex of quotient groups. To make this definition work, Hopf had to prove the algebraic equivalent of the result of Hurewicz, namely, that his homology groups are independent of the choice of the free resolution of $\mathbb{Z}$ over $\mathbb{Z}G$. His result thus contained the result of Hurewicz, at least for suitable spaces. As we know today, resolutions and the corresponding independence result lie at the heart of homological algebra.

Independently of Hopf's paper, Samuel Eilenberg and Saunders Mac Lane [**EML**] defined the cohomology of a group G by considering a specific such free resolution, the *standard resolution in inhomogeneous form*:

$$\mathbf{B}\colon \cdots \to B_i \to B_{i-1} \to \cdots \to B_1 \to B_0 \to \mathbb{Z} \to 0\ .$$

Here B_i is defined to be the free $\mathbb{Z}G$-module with basis consisting of all the symbols

1991 *Mathematics Subject Classification.* Primary: 20J05; Secondary: 20C20.
This is the final form of the paper.

[1]I am greatly indebted to Beno Eckmann for crucial information about the early stages of the development of homological algebra.

$[x_1|x_2|\dots|x_i]$, $x_j \in G$. The differential is given by the formula

$$d[x_1|x_2|\dots|x_i] = x_1[x_2|x_3|\cdots|x_i] + \\ + \sum_{j=1}^{i-1}(-1)^j[x_1|\dots|x_j x_{j+1}|\dots|x_i] + (-1)^i[x_1|x_2|\cdots|x_{i-1}].$$

In their paper Eilenberg and Mac Lane also mentioned that the resolution can equivalently be described in "homogeneous form". This homogeneous standard resolution also formed the basis of a paper by Beno Eckmann [**E**] who independently had arrived at results similar to those of Eilenberg and Mac Lane. He too had noticed the two equivalent descriptions; in his paper he favored the homogeneous form because it made the description of the product structure in the cohomology simpler.[2] In the *standard resolution in homogeneous form*

$$\mathbf{B}': \cdots \to B'_i \to B'_{i-1} \to \cdots \to B'_1 \to B'_0 \to \mathbb{Z} \to 0,$$

B'_i is the free Abelian group on all symbols $(x_0, x_1, \dots, x_i)$, $x_j \in G$ with G-action defined by

$$x(x_0, x_1, \dots, x_i) = (xx_0, xx_1, \dots, xx_i).$$

This action makes B'_i into a free $\mathbb{Z}G$-module. The differential corresponds to the boundary operator in simplicial homology:

$$d(x_0, x_1, \dots, x_i) = \sum_{j=0}^{i}(-1)^j(x_0, x_1, \dots, \hat{x}_j, \dots, x_i)\ .$$

These standard (or bar) resolutions can be found in any text book on homological algebra or on cohomology of groups (see for example [**HSt**]). There exist a number of variants, but it is rather disappointing that all of them appear to be of limited use. The main reason for this seems to be that they are extremely big. The size of the free modules in the resolution, measured for example by their rank, grows exponentially with the dimension i.

When I was a student I was somewhat surprised to see that the (co)homology theory of groups had not developed faster in the ten or fifteen years after its first definition. I compared of course with the rapid development in other fields, for example with the development of algebraic topology in the twenties and thirties with its wealth of applications in many different areas. *Some* applications of the (co)homology of groups outside group theory (class field theory, topology, etc) have been discussed already in the fifties, and *some* homological notions, like projectivity and relative projectivity, have been studied early on. But in the first ten or even fifteen years the (co)homology theory of groups has very rarely been used as a tool to study group theoretical properties. The papers of Eilenberg and Mac Lane on extension theory remained rather isolated (although very successful) attempts in this direction. May it be that the emphasis put on the big standard resolutions[3] for the definition of cohomology groups contributed to this fact? Not only computer people have problems with exponential growth!

[2]Note that a description of the product structure without making use of the standard resolution did not seem possible at the time.

[3]Compare for example the 1949 survey article of Eilenberg [**Ei**].

2. Topologists knew better than algebraists, at least in certain instances: If the topological space X has fundamental group G and if the universal cover $\widetilde{X}$ is contractible, then the chain complex of $\widetilde{X}$ gives rise to a free resolution of $\mathbb{Z}$ which sometimes is good, *i.e.* small.

Let's look at a trivial example. The fundamental group of a bouquet X of n circles is a free group F on n generators. Since the universal cover $\widetilde{X}$ of X is contractible by trivial reasons, its chain complex is a nice and small free resolution of $\mathbb{Z}$ for the free group F; it has the form

$$0 \to C_1 \to C_0 \to \mathbb{Z} \to 0.$$

This clearly implies that F has cohomological dimension 1, and also it allows one to compute the cohomology of F with any coefficient module.

Another explicit example is this. If the group G acts freely and simplicially on an odd dimensional sphere S^{2k-1} then it is easy to see, for example by an application of the Lefschetz fixed point formula, that the action on the homology is trivial. Therefore the chain complex of S^{2k-1} yields an exact sequence

$$0 \to \mathbb{Z} \to C_{2k-1} \to \cdots \to C_1 \to C_0 \to \mathbb{Z} \to 0$$

where the C_i's are free $\mathbb{Z}G$-modules. Splicing the finite exact sequence together one obtains a periodic free resolution of $\mathbb{Z}$ over G. It is small in a sense different from the one in the previous example: since the resolution is periodic the size of the modules is globally bounded.

These two examples strongly suggest that small projective resolutions of $\mathbb{Z}$ over $\mathbb{Z}G$ provide a lot of information about the cohomology of G. There is thus considerable interest in such resolutions, the smaller the better!

Our two examples show two different ways in which a resolution may be small. In this paper we shall almost exclusively consider finite groups, where smallness will always have to be understood in a sense related to the second example. The reason for this is that a finite group has infinite cohomological dimension. Smallness in the sense of our first example can only occur for infinite groups. The corresponding theory which is intimately connected with algebraic topology and algebraic group theory has seen a rapid development in recent years, initiated by Serre's 1971 monograph [**S**].

3. For the rest of the paper I shall restrict attention to projective resolutions of a field k over kG instead of free resolutions of the integers $\mathbb{Z}$ over the integral group ring $\mathbb{Z}G$. Of course the field should be of finite characteristic p, where p divides the group order, for otherwise the group algebra kG is semisimple and its homological algebra is trivial. Let

$$\mathbf{P}\colon \cdots \to P_i \to P_{i-1} \to \cdots \to P_1 \to P_0 \to k \to 0$$

be such a projective resolution. Small is beautiful! We thus try to choose the modules P_i such that the k-dimension is minimal at each stage. This then is a resolution which is minimal in an obvious sense. Not surprisingly it turns out that such resolutions enjoy rather nice properties. For example it turns out that over kG the term P_{i+1} is nothing else than the projective cover of $\ker(P_i \to P_{i-1})$. Also, it is not too hard to see that the minimal projective resolution of k over kG is uniquely determined.

Clearly for any projective resolution $\mathbf{P}$ of k over kG we have the inequality $\dim_k P_i \geq \dim_k H^i(G,k)$. Already Swan, in his 1964 paper [**Sw2**], gave sharper estimates in some cases. He showed for example that for a p-group S the minimal projective resolution of k satisfies $\dim_k P_i = |S| \cdot \dim_k H^i(S,k)$. We shall come back to this result below.

The concept of a minimal resolution can be generalized. In particular we can ask for minimal resolutions over the integral group ring $\mathbb{Z}G$. In fact the 1964 paper by Swan [**Sw2**] is concerned with this case; in this paper minimal resolution seem to have been considered for the first time. It must be mentioned here that the integral theory is much more complicated than the theory over a field, since, for example, projective covers do not exist over $\mathbb{Z}G$, and minimal $\mathbb{Z}G$-resolutions are not unique. We refer to [**G1**], [**GL**], [**G2**] for recent results in this area.

4. Next we introduce the notion of growth. Let $\mathbf{m}\colon m_0, m_1, m_2, \ldots$ be a sequence of natural numbers. We say that the growth rate of $\mathbf{m}$ is less than or equal to n, $\operatorname{gr}(\mathbf{m}) \leq n$, if there exists $C > 0$ such that $m_i \leq Ci^{n-1}$. The least integer n satisfying the inequality is called the growth rate of $\mathbf{m}$, and denoted by $\operatorname{gr}(\mathbf{m})$.

We define the growth rate $\operatorname{gr}(\mathbf{V})$ of a graded k-vector space $\mathbf{V} = \{V_i\}$, $i \geq 0$ to be the growth rate of the sequence $\mathbf{m}$ with $m_i = \dim_k V_i$. If G is a group and if

$$\mathbf{P}\colon \cdots \to P_i \to P_{i-1} \to \cdots \to P_0 \to k \to 0,$$

is a projective resolution then the growth rate $\operatorname{gr}(\mathbf{P})$ is by definition the growth rate of the sequence $\mathbf{m}$ with $m_i = \dim_k P_i$. Finally, we denote the minimum of the growth rates over all projective kG-resolutions $\mathbf{P}$ of k by $\operatorname{gr}(G)$. Alternatively, $\operatorname{gr}(G)$ may be defined as the growth rate of the minimal projective resolution of k.

Clearly, if G, $G \neq 1$, admits a periodic resolution of k over kG, then we have $\operatorname{gr}(G) = 1$. By the result of Swan mentioned above we have $\operatorname{gr}(S) = \operatorname{gr}\bigl(\mathbf{H}(S,k)\bigr)$ if S is a p-group. Here we consider the cohomology algebra $\mathbf{H}(S,k) = \bigl\{H^i(S,k)\bigr\}$, $i \geq 0$, as graded vector space over k. For the group $C_p \times C_p$ there is a resolution which is the tensor product of the periodic resolutions of the two factors. This immediately yields $\operatorname{gr}(C_p \times C_p) \leq 2$. Since $\mathbf{H}(C_p \times C_p, k)$ contains a polynomial subalgebra in two indeterminates we have $\operatorname{gr}(S) \geq \operatorname{gr}\bigl(\mathbf{H}(C_p \times C_p, k)\bigr) \geq 2$. Thus $\operatorname{gr}(C_p \times C_p) = 2$.

What kind of group invariant is $\operatorname{gr}(G)$? The answer to this question has been known for a long time in the special case where the group G is such that k admits a periodic resolution. If for the trivial module k for G admits a periodic resolution, then the same is true for every subgroup U of G. Now, we have seen that $\operatorname{gr}(C_p \times C_p) = \operatorname{gr}\bigl(\mathbf{H}(C_p \times C_p, k)\bigr)$ is 2. It thus follows that our group G cannot contain any elementary Abelian p-subgroup of rank bigger than one. From this, together with some group theory it is then easy to conclude that k admits a periodic projective resolution if and only if the Sylow p-subgoup of G is cyclic, or possibly, if $p = 2$, generalized quaternion.

5. A result of Quillen handles the general case. Quillen considered the even dimensional (and hence commutative) part of the cohomology algebra $\mathbf{H}^*(G,k) = \bigoplus_{i\geq 0} H^{2i}(G,k)$. Proving a conjecture of Atyiah and Swan, he obtained the following theorem (see [**Q1**], [**Q2**]):

THEOREM 1. *The Krull dimension of* $\mathbf{H}^*(G,k)$ *equal the* p*-rank of* G.

The p-rank of a group G is the maximal rank of elementary Abelian p-subgroups E of G. In our context it is best to interpret the Krull dimension as the maximal number of indeterminates of a polynomial subalgebra. It is not hard to see that for a polynomial algebra, with grading induced by assigning arbitrarily an even degree to each indeterminate, the growth rate coincides with the number of indeterminates. We can now use a little commutative algebra, and the fact that the cohomology algebra is finitely generated (see [**E**]) to show from Quillen's result that the growth rate of $\mathbf{H}^*(G,k)$, and hence also of $\mathbf{H}(G,k)$, equals the p-rank of G.

We finally claim that Quillen's result also determines $\operatorname{gr}(G)$.

COROLLARY 1. $\operatorname{gr}(G) = p$*-rank of* G.

To see this we first recall the fact mentioned above that $\dim_k H^i(G,k)$ is bounded above by the dimension of P_i in a minimal resolution of k. On the other hand we note that if S is a Sylow p-subgroup and $\mathbf{Q}$ is the minimal projective resolution of k over S then $kG \otimes_{kS} \mathbf{Q}$ contains the minimal projective resolution of k over G as a direct summand. We thus obtain

$$\begin{aligned} p\text{-rank of } G = \operatorname{gr}\mathbf{H}(G,k)) &\le \operatorname{gr}(G) \le \operatorname{gr}(kG \otimes_{kS} \mathbf{Q}) \\ &= \operatorname{gr}(\mathbf{Q}) = p\text{-rank of } S = p\text{-rank of } G. \end{aligned}$$

We thus have equality at each step. This proves the Corollary.

6. We next use some elementary modular representation theory to obtain further results. If P is a projective module over kG it is a direct sum of so called *principal indecomposable modules* $P^{(0)}, P^{(1)}, \ldots, P^{(l)}$. These are by definition the direct summands of kG, one for each isomorphism class. It is a simple, but fundamental result of modular representation theory that the radical quotient of a principal indecomposable module is simple. Taking the radical quotient thus establishes a one-to-one correspondence between the principal indecomposable modules and the (isomorphism classes of) simple modules $M^{(0)}, M^{(1)}, \ldots, M^{(l)}$. Of course, we choose the notation in such a way that $M^{(0)} = k$.

From this it is almost immediate how to construct the projective cover P of an arbitrary kG-module M. If M has radical quotient $M/\operatorname{rad} M = \bigoplus_j m_j M^{(j)}$ then we take $P = \bigoplus_j m_j P^{(j)}$. Since P is projective, the canonical projection $P \to M/\operatorname{rad} M$ factors through M. Since the image of the map $P \to M$ so constructed generates M modulo its radical, the map must be surjective, and hence P is the projective cover of M.

From this remark we may want to deduce the following fact. Let

$$\mathbf{P}\colon \cdots \to P_i \to P_{i-1} \to P_{i-2} \to \cdots \to P_0 \to k \to 0,$$

be the minimal projective resolution of k. Setting $\Omega_i = \ker(P_{i-1} \to P_{i-2})$ we have the short exact sequence

$$0 \to \Omega_i \to P_{i-1} \to \Omega_{i-1} \to 0,$$

where, by minimality, P_{i-1} is the projective cover of the radical quotient of Ω_{i-1}. Applying $\operatorname{Hom}_{kG}(-,k)$ we obtain

$$0 \to \operatorname{Hom}_{kG}(\Omega_{i-1},k) \to \operatorname{Hom}_{kG}(P_{i-1},k) \to \operatorname{Hom}_{kG}(\Omega_i,k) \to \mathrm{H}^i(G,k) \to 0\ .$$

Since every homomorphism in $\operatorname{Hom}_{kG}(P_{i-1},k)$ factors through the radical quotient of P_{i-1}, which by construction coincides with the radical quotient of Ω_{i-1}, the exact sequence reduces to two isomorphisms. We thus obtain

$$\mathrm{H}^i(G,k) \cong \mathrm{Hom}_{kG}(\Omega_i,k) \cong \mathrm{Hom}_{kG}(P_i,k).$$

We next look at the special case where G is a p-group, $G = S$. In that case k is the only simple kS-module, and kS is the corresponding principal indecomposable module. Every projective module over kS is thus free. Taking as above the minimal projective resolution, we have

$$\dim_k \mathrm{H}^i(G,k) = \dim_k \mathrm{Hom}_{kG}(P_i,k).$$

Hence we obtain

$$\dim_k P_i = |S| \cdot \dim_k \mathrm{H}^i(S,k).$$

This is the result due to Swan [**Sw2**] mentioned in Section 4.

If G is arbitrary with Sylow p-subgroup S, and if P is a projective kG-module, we have that $\dim_k P$ is a multiple of $|S|$, since the restriction of P to S is a projective, and hence free, kS-module. In that case we obtain for the minimal projective resolution of k over kG the inequality

$$\dim_k P_i \geq |S| \cdot \dim_k \mathrm{H}^i(G,k).$$

More precisely we may write

$$P_i = \bigoplus_{j=1}^{l} m_{ij} P^{(j)}$$

where the natural number m_{ij} is the multiplicity of $P^{(j)}$ in P_i. Then

$$\dim_k \mathrm{H}^i(G,k) = m_{i0},$$

since the radical quotient of $P^{(0)}$ is k. Quillen's theorem therefore determines the growth rate of the sequence of natural numbers $m_{00}, m_{10}, m_{20} \ldots$ It tells us something about the frequency by which $P^{(0)}$ occurs in the minimal projective resolution of k.

There is a natural question here. What can be said about the occurrence of the principal indecomposable modules $P^{(j)}$, $j \neq 0$, in the minimal projective resolution of k? About this question much less is known. It is clear, again from representation theory, that only principal indecomposable modules can occur that belong to the so called principal block of kG, but which of these really occur, and how often, is unclear. In some special cases some partial results are known. If G is p-constrained (note that solvable, and p-solvable groups are p-constrained), then it is known [**LS1**] that each of the principal indecomposable modules in the principal block occurs infinitely often. Nothing, apart from trivial bounds, seems to be known about the growth rate of the corresponding sequences $m_{0j}, m_{1j}, m_{2j}, \ldots$, $j \neq 0$. For other groups, for example for simple groups, it is – apart from specific small examples – not even known whether each principal indecomposable module in the principal block occurs in the minimal projective resolution of k.

7. Given two graded vector spaces $\mathbf{A}$ and $\mathbf{B}$. The two following results are almost obvious:

$$\operatorname{gr}(\mathbf{A}\oplus\mathbf{B}) = \max\bigl(\operatorname{gr}(\mathbf{A}),\operatorname{gr}(\mathbf{B})\bigr),$$
$$\operatorname{gr}(\mathbf{A}\otimes\mathbf{B}) = \operatorname{gr}(\mathbf{A})+\operatorname{gr}(\mathbf{B}).$$

Using the second of these statements it is possible to realize the growth rate of the minimal resolution of k in some special cases in an easy way. This can be explained in case G is the direct product of cyclic p-groups, $G = C^{(1)}\times C^{(2)}\times\cdots\times C^{(l)}$. If $\mathbf{Q}^{(j)}$ is a periodic projective resolution of k over the cyclic factor $C^{(j)}$, then the (total complex of the) tensor product of the l projective resolutions of k over the factors $C^{(j)}$, *i.e.* $\mathbf{Q}=\bigotimes\mathbf{Q}^{(j)}$, is a projective resolution of k over G, and one has

$$\operatorname{gr}(\mathbf{Q}) = \operatorname{gr}(\mathbf{Q}^{(1)})+\operatorname{gr}(\mathbf{Q}^{(2)})+\cdots+\operatorname{gr}(\mathbf{Q}^{(l)}) = l.$$

On the other hand, the rank of G is also l. Our construction thus realizes $\operatorname{gr}(G)$.

This idea can be made to work in more general situations. For example, if G is a group with normal Abelian Sylow p-subgroup S and if in addition the G/S module S splits as a direct sum of G/S-submodules whose underlying Abelian group is cyclic, the above procedure applied to S yields a projective resolution in the form of a tensor product of periodic exact kG-sequences. This resolution thus clearly realizes $\operatorname{gr}(G)$. Actually the hypotheses can be weakened even further: it is enough if, for a field extension k, the $k(G/S)$-module $k\otimes_{\mathbb{F}_p} S$ splits into a direct sum of 1-dimensional submodules.

A simple specific example of this more general case is the alternating group A_4 of degree 4 for the prime 2. The group A_4 is the split extension of C_3 by the Klein four-group V. The normal Abelian subgroup V considered as module over $\mathbb{F}_2(C_3)$ is simple. But its tensor product with an extension field k, which contains at least 4 elements, splits into a direct sum of two 1-dimensional kC_3-modules.

The explicit construction of the projective resolution of k over A_4 is then as follows (see [**Sch**], Teil I[4]). It is not hard to see that kV carries a $k\,\mathrm{A}_4$-action which is obtained by extending the regular action of V by letting the quotient C_3 act by conjugation on the basis elements of kV. Since

$$kV = k(C_2\times C_2)\cong k[x,y]//(x^2,y^2)$$

and since the kC_3-module $k\otimes V$ splits into a direct sum 1-dimensional modules we may perform a base change, and obtain

$$kV\cong k[x',y']/(x'^2,y'^2) = k[x']/(x'^2)\otimes_k k[y']/(y'^2),$$

where both tensor product factors are $k\,\mathrm{A}_4$-modules.

It is then easy to write down periodic exact sequences $\mathbf{P}'$ and $\mathbf{Q}'$ for the factors $k[x']/(x'^2)$ and $k[y']/(y'^2)$ respectively, such that $\mathbf{P}'\otimes_k\mathbf{Q}'$ is a double complex whose associated total complex is a projective resolution of k over $k\mathrm{A}_4$.

Apart from the trivial module k we have two 1-dimensional simple modules M and N. We denote their projective covers by P_k, P_M, P_N, respectively. Then our

[4]The second part of [**Sch**] unfortunately contains a mistake which invalidates many of the general results claimed there.

double complex looks as follows:

$$\begin{array}{ccccccccccc}
\downarrow & & \downarrow & & \downarrow & & \downarrow & & \downarrow & & \\
P_k & \leftarrow & P_M & \leftarrow & P_N & \leftarrow & P_k & \leftarrow & P_M & \leftarrow & \\
\downarrow & & \downarrow & & \downarrow & & \downarrow & & \downarrow & & \\
P_M & \leftarrow & P_N & \leftarrow & P_k & \leftarrow & P_M & \leftarrow & P_N & \leftarrow & \\
\downarrow & & \downarrow & & \downarrow & & \downarrow & & \downarrow & & \\
P_N & \leftarrow & P_k & \leftarrow & P_M & \leftarrow & P_N & \leftarrow & P_k & \leftarrow & \\
\downarrow & & \downarrow & & \downarrow & & \downarrow & & \downarrow & & \\
P_k & \leftarrow & P_M & \leftarrow & P_N & \leftarrow & P_k & \leftarrow & P_M & \leftarrow &
\end{array}$$

This double complex resolution is not minimal since it can be shown to differ from the (easily constructed) minimal resolution.

8. In the cases mentioned above, the resolutions can explicitly be constructed by the procedure which we have described. But abstractly, and for an algebraically closed field k, a much more general result is true.

THEOREM 2 (D. BENSON, J. CARLSON [**BC1**]). *Let G be a group with p-rank l. Then there are periodic exact sequences, $j = 1, 2, \dots, l$,*

$$\mathbf{Q}^{(j)}\colon \cdots \to Q_i^{(j)} \to Q_{i-1}^{(j)} \to \cdots \to Q_1^{(j)} \to Q_0^{(j)} \to k \to 0$$

such that $\mathbf{Q} = \bigotimes \mathbf{Q}^{(j)}$ is a projective resolution of k over kG.

The proof of this remarkable result requires the theory of varieties, due to Alperin, Benson, Carlson, Evens, *et al.*, in which the methods of commutative algebra and algebraic geometry are applied to the cohomology algebra of G. In the proof of the above result the construction of the periodic sequences $\mathbf{Q}^{(j)}$, $j = 1, 2, \dots, l$, requires a detailed knowledge of the cohomology algebra of G: one needs to know a set of l homogeneous elements $\zeta_1, \zeta_2, \dots, \zeta_l$ of $\mathbf{H}(G, k)$ that freely generate a polynomial subalgebra $k[\zeta_1, \zeta_2, \dots, \zeta_l]$ over which $\mathbf{H}(G, k)$ is finitely generated as a module.

The periodic sequence $\mathbf{Q}^{(j)}$ is then obtained as follows. Let $\mathbf{P}$ be the minimal resolution of k over kG. If ζ_j is an element in $H^m(G, k)$ (m must be even) then it can be realized as a surjective map $\zeta_j\colon \Omega_m \to k$. Denoting its kernel by L_{ζ_j} we may construct B_{ζ_j} as pushout and obtain the following commutative diagram with exact rows:

$$\begin{array}{ccccccccccccccc}
 & & 0 & & 0 & & & & & & & & & & \\
 & & \downarrow & & \downarrow & & & & & & & & & & \\
 & & L_{\zeta_j} & = & L_{\zeta_j} & & & & & & & & & & \\
 & & \downarrow & & \downarrow & & & & & & & & & & \\
0 & \to & \Omega_m & \to & P_{m-1} & \to & \cdots & \to & P_1 & \to & P_0 & \to & k & \to & 0 \\
 & & \downarrow & & \downarrow & & & & \| & & \| & & \| & & \\
0 & \to & k & \to & B_{\zeta_j} & \to & \cdots & \to & P_1 & \to & P_0 & \to & k & \to & 0 \\
 & & \downarrow & & \downarrow & & & & & & & & & & \\
 & & 0 & & 0 & & & & & & & & & &
\end{array}$$

The bottom exact sequence starts and ends with k. Splicing it together produces a periodic exact sequence which by definition is $\mathbf{Q}^{(j)}$.

The complex $\bigotimes \mathbf{Q}^{(j)}$ is obviously exact, and the occuring tensor products are projective. The latter is clear except for the tensor product $B_{\zeta_1} \otimes B_{\zeta_2} \otimes \cdots \otimes B_{\zeta_l}$. It is this point which crucially requires the theory of varieties.

It should be added that it cannot be expected in general that the resolution **Q** is minimal, although it realizes the growth rate of the minimal resolution. As a matter of fact it is rarely the case that it is minimal. It has even been shown (see [**B**]), that the minimal resolution cannot in general be written as a tensor product of periodic sequences.

I should also mention here that the role of the kG-module k is not special. Instead of the trivial module k, any kG-module M could be taken. The Krull dimension of the cohomology algebra $\mathbf{H}(G,k)$ is then replaced by the so-called *complexity of the module* M, which can be defined as the growth rate of the minimal projective resolution of M. Finally it should be mentioned that there is also an integral version of the result, in which a projective tensor product resolution of $\mathbb{Z}$ over $\mathbb{Z}G$ is constructed (see [**BC1**]).

9. Benson and Carlson [**BC1**] have shown that their procedure to obtain the integral version of the result mentioned in Section 8 also yields a new proof of a result of Carlsson [**C**] about group actions on products of spheres. Basing ourself on [**BC1**] we briefly explain this, we thus take up one of the examples in Section 2.

THEOREM 3 (G. CARLSSON[**C**]). *Let X be a finite CW-complex with the homotopy type of a product $(S^r)^n$ of n spheres of the same dimension r, and let G be a group acting freely on X such that the induced action on the homology of X is trivial. Then for any prime p the p-rank of G is at most n.*

Our example in Section 2 is concerned with the case $n = 1$. By a result of Swan [**Sw1**] the converse of the statement is known to be true too in this case. Note that it is essential for the converse that X may be homotopy equivalent (and not homeomorphic) to a sphere. As for $n \geq 2$ one may ask whether the statement of the theorem remains true if products of spheres of different dimensions are allowed. This does not seem to be known. That group actions of that kind exist has been shown by R. Oliver: the alternating group A_4 admits a free action on $S^2 \times S^3$ such that the induced action on homology is trivial, but no such free action on any product of spheres of the same dimension. For a possible converse to the statement of the theorem one thus has certainly to allow for products of spheres of *different* dimensions. Whether such a converse indeed exists still seems to be open.

In the proof Benson and Carlson first show that the complex of cellular chains on X can be replaced by a homology isomorphic complex which is an n-fold tensor product of complexes with the homology of S^r. This gives rise to n elements $\xi_1, \xi_2, \ldots, \xi_n$ of $H^{r+1}(G,\mathbb{Z})$. In a similar way as in the proof of the Theorem of Benson-Carlson (see Section 8) one then realizes the cohomology class ξ_i as map in the cochain complex, and considers the corresponding kernel $\widetilde{L}_{\xi_i}$. In order to be able to apply the theory of varieties one then looks at $L_{\zeta_i} = k \otimes_{\mathbb{Z}} \widetilde{L}_{\xi_i}$ where k is an algebraically closed field of characteristic p. Using the theory of varieties one then may conclude from the construction of these kG-modules L_{ζ_i} that their tensor product is kG-projective. This is enough to prove that the p-rank of G is at most n.

10. The theorem of Benson and Carlson on the construction of multiple complex resolutions is clearly not the end of the story. In many explicit examples Alperin, Benson and Carlson have discovered ways to write the minimal resolution as the total complex of a multiple, more precisely, as an l-fold complex where $l = \text{rank}(G)$. I would like to record here some of these examples. They all concern

p-rank 2, *i.e.* double complexes. In fact, due to the complications introduced by higher dimensions, I do not know similar pictures of triple or even larger complexes.

The first example is the minimal resolution for A_5 at the prime 2 (see [**A**]). Apart from the trivial module k there are two 2-dimensional simple modules N and M in the principal block (if k contains at least 4 elements). The corresponding principal indecomposable modules are denoted by P_k, P_M and P_N. There are maps between these modules, which are obvious from the well-known structure of P_k, P_M and P_N. The following is then a double complex whose associated total complex is the minimal projective resolution of k. Note that the 2-rank of A_5 is 2.

$$\begin{array}{cccccccccccc}
 & & & & & & \downarrow & & \downarrow & & \downarrow & \\
 & & & & P_N & \leftarrow & P_k & \leftarrow & P_k & \leftarrow & P_k & \leftarrow \\
 & & & & \downarrow & & \downarrow & & \downarrow & & \downarrow & \\
 & & & & P_k & \leftarrow & P_k & \leftarrow & P_k & \leftarrow & P_k & \leftarrow \\
 & & & & \downarrow & & \downarrow & & \downarrow & & \downarrow & \\
 & & P_N & \leftarrow & P_k & \leftarrow & P_k & \leftarrow & P_k & \leftarrow & P_k & \leftarrow \\
 & & \downarrow & & \downarrow & & \downarrow & & \downarrow & & \downarrow & \\
 & & P_k & \leftarrow & P_k & \leftarrow & P_k & \leftarrow & P_k & \leftarrow & P_M & \\
 & & \downarrow & & \downarrow & & \downarrow & & & & & \\
P_N & \leftarrow & P_k & \leftarrow & P_k & \leftarrow & P_M & & & & & \\
\downarrow & & \downarrow & & & & & & & & & \\
P_k & \leftarrow & P_M & & & & & & & & &
\end{array}$$

We note that one gets exactly the same picture for the group A_6 at the prime 2. In this case the symbols M and N denote the two 4-dimensional simple A_6-modules. This is rather surprising since the Sylow 2-subgroups of A_5 and A_6 are different, although both have, of course, 2-rank 2.

In the second example we consider the group $L_3(2)$, the simple group of order 168, again at the prime 2 (see [**A**], [**B**]). The simple modules in the principal block are k, A and B, where A is the natural 3-dimensional module and B is its dual. The associated principal indecomposable modules are denoted by P_k, P_A, P_B, respectively. The obvious maps between these fit together in the following double complex whose total complex is the minimal projective resolution of k. Note that the 2-rank of $L_3(2)$ is 2.

$$\begin{array}{cccccccccc}
\downarrow & & \downarrow & & \downarrow & & \downarrow & & \downarrow & \\
P_k & \leftarrow & P_B & \leftarrow & P_A & \leftarrow & P_k & \leftarrow & P_B & \leftarrow \\
\downarrow & & \downarrow & & \downarrow & & \downarrow & & \downarrow & \\
P_B & \leftarrow & P_A & \leftarrow & P_k & \leftarrow & P_B & \leftarrow & P_A & \leftarrow \\
\downarrow & & \downarrow & & \downarrow & & \downarrow & & \downarrow & \\
P_A & \leftarrow & P_k & \leftarrow & P_B & \leftarrow & P_A & \leftarrow & P_k & \leftarrow \\
\downarrow & & \downarrow & & \downarrow & & \downarrow & & \downarrow & \\
P_k & \leftarrow & P_B & \leftarrow & P_A & \leftarrow & P_k & \leftarrow & P_B & \leftarrow
\end{array}$$

The third example concerns the Mathieu group M_{11} and $p = 2$ (see [**BC2**]). Apart from the trivial module k there is a 10-dimensional and a 44-dimensional simple module. We denote the associated principal indecomposable modules by P_1, P_{10}, P_{44}, respectively. Since the simple modules are selfdual, the principal indecomposable modules are selfdual, also. The structure of these modules show that there are obvious non-trivial maps $P_{10} \to P_1$, $P_{44} \to P_1$, $P_{10} \to P_{10}$, $P_{44} \to P_{44}$, $P_{10} \to P_{44}$, and the maps dual to these. The following then is a description of the minimal resolution of k in terms of a double complex. Note that the 2-rank of M_{11} is 2.

$$\begin{array}{cccccccccccccccc}
 & & & & & & & & & & & & P_{10} & \leftarrow & P_{10} & \leftarrow \\
 & & & & & & & & & & & & \downarrow & & \downarrow & \\
 & & & & & & & & & & & & P_{1} & \leftarrow & P_{44} & \leftarrow \\
 & & & & & & & & & & & & \downarrow & & \downarrow & \\
 & & & & & & & & P_{10} & \leftarrow & P_{10} & \leftarrow & P_{44} & \leftarrow & P_{44} & \leftarrow \\
 & & & & & & & & \downarrow & & \downarrow & & \downarrow & & \downarrow & \\
 & & & & & & & & P_{1} & \leftarrow & P_{44} & \leftarrow & P_{44} & \leftarrow & P_{1} & \leftarrow \\
 & & & & & & & & \downarrow & & \downarrow & & \downarrow & & \downarrow & \\
 & & & & P_{10} & \leftarrow & P_{10} & \leftarrow & P_{44} & \leftarrow & P_{44} & \leftarrow & P_{1} & \leftarrow & P_{44} & \leftarrow \\
 & & & & \downarrow & & \downarrow & & \downarrow & & \downarrow & & \downarrow & & \downarrow & \\
 & & & & P_{1} & \leftarrow & P_{44} & \leftarrow & P_{44} & \leftarrow & P_{1} & \leftarrow & P_{44} & \leftarrow & P_{44} & \leftarrow \\
 & & & & \downarrow & & \downarrow & & \downarrow & & \downarrow & & \downarrow & & \downarrow & \\
P_{10} & \leftarrow & P_{10} & \leftarrow & P_{44} & \leftarrow & P_{44} & \leftarrow & P_{1} & \leftarrow & P_{44} & \leftarrow & P_{44} & \leftarrow & P_{1} & \leftarrow \\
\downarrow & & \downarrow & & \downarrow & & \downarrow & & \downarrow & & \downarrow & & \downarrow & & \downarrow & \\
P_{1} & \leftarrow & P_{44} & \leftarrow & P_{44} & \leftarrow & P_{1} & \leftarrow & P_{44} & \leftarrow & P_{44} & \leftarrow & P_{1} & \leftarrow & P_{44} & \leftarrow
\end{array}$$

In the examples treated here the minimal resolution of k over kG can be written as the total complex of an l-fold complex where l is the p-rank of G and where the dimension of the individual entries is globally bounded by a constant.[5] The question of course is whether this can always be achieved. To the best of my knowledge this question is still open. The pictures shown here have a high degree of symmetry and a rather pleasing combinatorial structure. It is absolutely clear to me that this cannot be an accident. And it is likely that many nice things in this area still await discovery.

References

[A] J.L. Alperin, *Resolutions for finite groups*, Proceedings of the Conference on Finite Groups University of Utah (Park City, Utah 1975), Academic Press, New York, 1976, p. 341–356.

[B] D. J. Benson, *Representations and cohomology* II: *Cohomology of groups and module*, Cambridge University Press, Cambridge, 1991.

[BC1] D.J. Benson and J.F. Carlson, *Complexity and multiple complexes*. Math. Zeit. **195** (1987), 221–238.

[BC2] ——, *Diagrammatic methods for modular representations and cohomology*, Comm. Algebra **15** (1987), 53–121.

[C] G. Carlsson, *On the rank of Abelian groups acting freely on* $(S^n)^k$, Invent. Math. **69** (1982) 393–400.

[Eck] B. Eckmann, *Der Cohomologie-Ring einer beliebigen Gruppe*, Comment. Math. Helv. **18** (1945-46), 232–282.

[Ei] S. Eilenberg, *Topological methods in abstract algebra*, Cohomology Theory of Groups. Bull. Amer. Math. Soc. **55** (1949), 3–37.

[EML] S. Eilenberg, S. Mac Lane, *Relations between homology and homotopy groups*, Proc. Nat. Acad. Sci. USA **29** (1943), 155–195.

[E] L. Evens, *The cohomology ring of a finite group*, Trans. Amer. Math. Soc. **101** (1961), 224–239.

[HSt] P.J. Hilton and U. Stammbach, *A course in homological algebra*, Graduate Texts in Mathematics **4**, Springer-Verlag, Berlin-New York, 1971.

[H] H. Hopf, *Über die Bettischen Gruppen, die zu einer beliebigen Gruppe gehören*, Comment. Math. Helv. **17** (1944-45), 39–79.

[5] It may be worth remarking that we cannot expect the entries to be indecomposable, as can been seen from the minimal (periodic) resolution for the quaternion group.

[G1] K.W. Gruenberg, *Partial Euler characteristics of finite groups and the decomposition of lattices*, Proc. London Math. Soc. (3) **48** (1984), 91–107.
[G2] ——, *Stably free resolutions of lattices over finite groups*, J. Austral. Math. Soc. **49** (1990), 364–385.
[GL] K.W. Gruenberg and P.A. Linnell, *Minimal free resolutions of lattices over finite groups* III, J. Math. **32** (1988), 361–374.
[LS1] P.A. Linnell, U. Stammbach, *On the cohomology of p-constrained groups*, Proc. Symp. Pure Math. **47** (1987), 467–469.
[LS2] ——, *The cohomology of p-constrained groups*, J. Pure Appl. Algebra **49** (1987), 273–279.
[O] R. Oliver, *Free compact group actions on products of spheres*, Algebraic Topology (Aarhus 1978), Lecture Notes in Math. **763**, Springer, New York, 1979.
[Q1] D. Quillen, *The spectrum of an equivariant cohomology ring.* I, Ann. of Math. **94** (1971), 549–572.
[Q2] —— *The spectrum of an equivariant cohomology ring* II, Ann. of Math. **94** (1971), 573–602.
[Sch] R.J. Schmid, *Projektive Auflösungen als mehrfache Komplexe. Teil* I, Diss. ETH Zürich, 1987.
[S] J-P. Serre, *Cohomologie des groupes discrets*, Ann. of Math. Studies **70** (1971), 77–169.
[Sw1] R.G. Swan, *Periodic resolutions for finite groups*, Anal. Math. **72** (1960), 267–291.
[Sw2] ——, *Minimal resolutions for finite groups*, Topology **4** (1964), 193–208.

MATHEMATIK, ETH-ZENTRUM, CH–8092 ZURICH, SWITZERLAND
E-mail address: stammb@math.ethz.ch